Cellular Senescence and Brain Aging

Gaurav Gupta
Editor

Cellular Senescence and Brain Aging

Mechanisms, Therapeutic Strategies, and Implications for Neurodegenerative Diseases

Springer

Editor
Gaurav Gupta
Centre for Research Impact & Outcome,
Chitkara College of Pharmacy
Chitkara University
Rajpura, Punjab, India

ISBN 978-981-96-8872-2 ISBN 978-981-96-8873-9 (eBook)
https://doi.org/10.1007/978-981-96-8873-9

This Springer imprint is published by the registered company Springer Nature Singapore Pte Ltd.
The registered company address is: 152 Beach Road, #21-01/04 Gateway East, Singapore 189721, Singapore

Preface

The aging process invariably affects the structure and function of the human brain. Increasingly, scientists recognize cellular senescence as a key driver behind the progressive deterioration of brain regions and the accompanying decline in cognition, memory, and other neurological functions. While cellular senescence was understood as a cell's irreversible cessation of division, it is now appreciated as a dynamic and influential process with profound implications for brain health.

In *Cellular Senescence and Brain Aging*, leading scientists and experts examine how brain aging relates to senescent cells in the body. This book combines cellular biology and neuroscience, combining molecular, experimental, and translational information. All these chapters highlight the connection between aging brain cells and the brain's environment, disclosing the reasons behind neuroinflammation, dysfunction in brain connections, and the death of brain cells. One main idea explored in the book is that cellular senescence has a complex nature. Even though being senescent prevents cancer cell proliferation, too many senescent cells in the brain can eventually cause harm. When cells enter senescence, they secrete various inflammatory molecules and enzymes collectively known as the senescence-associated secretory phenotype (SASP). The SASP disrupts normal cellular function, promotes further damage to brain tissue, and impairs the brain's capacity for repair and regeneration.

It looks at all brain cell types, such as neurons, astrocytes, microglia, and oligodendrocyte progenitors, and their role in brain aging. Aged astrocytes no longer provide the same metabolic support, and aged microglia keep triggering inflammation. Identifying the particular features in each cell allows us to target the most appropriate treatments. This field relies on experimental techniques, and several chapters discuss animal systems that imitate the aging of the brain and cells. Because of these models, scientists have studied senescence's progression over time and examined whether senescent cell removal can benefit the body. So far, these investigations point out that by working with senescent cells, it may be possible to address the early signs of aging in the brain and prevent diseases such as Alzheimer's and Parkinson's.

The book also explores how aging affects the brain and discusses the neurovascular unit's role in brain well-being. The impact of senescent cells in circulation and the breakdown of the blood–brain barrier demonstrate that brain aging is tied to the whole body's aging. The contributors describe various current problems, such as finding accurate markers for brain aging, exploring whether senescence in the brain can be tuned back, and ensuring the safety and success of senolytic drugs over the years. There are many ethical matters relating to anti-aging therapies, which highlight why it is not straightforward to implement these laboratory findings clinically.

This book was produced to help anyone interested in aging biology, brain aging, neurological disease, and treatment with regenerative medicine. Since scientists are recognizing cellular senescence in brain aging, new treatment approaches are appearing that might help improve the mental well-being and memory of older adults.

We thank the authors and the readers for their involvement with this fast-changing area. We hope that the information in this book will encourage further partnership and progress, helping battle brain aging by exploring cellular senescence.

Chandigarh, Punjab, India Gaurav Gupta

Contents

Aging and Neurodegeneration

1

Gaurav Gupta, Raihan Siddique, Riya Thapa, Asif Ahmad Bhat, Md Sadique Hussain, Thakur Gurjeet Singh, Sachin Kumar Singh, Saurabh Gupta, Pran Kishore Deb, and Moayad Al Shahwan

Introduction

The natural process of biological development results in molecular cellular and systemic functional decline, which progressively escalates neurodegenerative diseases to become the primary aging risk factors. The analysis of oxidative stress needs to be combined with studies on mitochondrial dysfunction telomere shortening and protein aggregation because these mechanisms create an imbalance in cells while causing neuronal death [1, 2]. Cells undernourished in reactive oxygen species activity indicator view result in essential biomolecule damage, yet the medical

G. Gupta (✉)
Centre for Research Impact & Outcome, Chitkara College of Pharmacy, Chitkara University, Rajpura, Punjab, India
e-mail: Gaurav.gupta@chitkara.edu.in

R. Siddique · R. Thapa
School of Pharmacy, Suresh Gyan Vihar University, Jaipur, India

A. A. Bhat
BM College of Pharmacy, Farrukhnagar, Gurugram, Haryana, India

M. S. Hussain
Uttaranchal Institute of Pharmaceutical Sciences, Uttaranchal University, Dehradun, India

T. G. Singh
Chitkara College of Pharmacy, Chitkara University, Rajpura, Punjab, India
e-mail: gurjeet.singh@chitkara.edu.in

S. K. Singh
School of Pharmaceutical Sciences, Lovely Professional University, Phagwara, Punjab, India
e-mail: sachin.16030@lpu.co.in

G. Gupta (ed.), *Cellular Senescence and Brain Aging*,
https://doi.org/10.1007/978-981-96-8873-9_1

community names this condition oxidative stress. Mitochondrial dysfunction impairs ATP production and elevates oxidative stress, accelerating neurodegeneration. The breakdown of the autophagy and proteostasis pathways results in misfolding and aggregation of proteins as the basic symptoms in both Alzheimer's and Parkinson's disease. The process of neural damage continues until it affects brain functionality as well as the fundamental structure of the neuron cells [3–5]. Neurodegenerative disorders have emerged as a global medical crisis because population changes lead to higher numbers of dementia and Parkinson's disease patients, as well as various additional brain diseases. Your healthcare expenses rise caregivers face burnout and working people face disruption as a result of these disorders, creating significant financial troubles. Although neuroscience research has advanced, the field current disease management lacks effective treatment for stopping disease progression indicating the urgent need for improved therapies [6, 7]. The core issues of neurodegenerative conditions establish major obstacles in medical evaluation and therapeutic procedures for drug development of these diseases. Research about brain disorders that occur during aging occupies a vital standing because of its vital significance. Progressive improvements in patient results depend on studies focused on early disease indicator detection, specific disease mechanism exams, and treatment-specific drug creation. Research shows that cognitive decline can be slowed down through primary prevention strategies before preclinical stage starting points if patients make lifestyle adjustments and participate in mental activities with proper medication [8, 9]. Multiple combinations of modern high-tech methods that analyze multi-omics and utilize systems and imaging systems enable a better visual understanding of neurodegenerative processes. A team comprising molecular biologists, pharmacologists, neurologists, and computational scientists must work together to treat these intricate conditions. The combination of aging-related neurodegeneration with increased life expectancy requires immediate scientific effort to develop modern preventive holistic solutions against this condition [10, 11].

S. Gupta
Department of Pharmacology, Chameli Devi Institute of Pharmacy,
Indore, Madhya Pradesh, India

P. K. Deb
Department of Pharmaceutical Sciences and Technology, Birla Institute of Technology,
Mesra, Ranchi, India

M. Al Shahwan
Centre of Medical and Bio-allied Health Sciences Research, Ajman University,
Ajman, United Arab Emirates

Department of Clinical Sciences, College of Pharmacy and Health Sciences,
Ajman University, Ajman, United Arab Emirates
e-mail: m.shahwan@ajman.ac.ae

The Biology of Aging

Cellular and Molecular Mechanisms of Aging

Molecular Drivers of Aging: Oxidative Damage, Senescence, and Protein Aggregation

A series of biological processes during aging leads to functional decline and makes organisms more susceptible to diseases. Faulty mitochondria functioning along with oxidative stress generate ROS excess through their dysfunctional mitochondria, thereby causing DNA protein and lipid oxidation damage. The process accelerates both cell age-related processes as well as brain tissue decline [12]. Telomere shortening creates a sequence of biochemical reactions that reduce cell multiplication potential until it causes ongoing inflammation accompanied by impaired tissue healing ability. During aging, the accumulating senescent cells produce pro-inflammatory cytokines within tissues that cause deterioration of metabolic conditions together with worsening neurodegenerative diseases. Neurodegenerative diseases affecting elderly individuals like Alzheimer's disease and Parkinson's disease present misfolded and aggregating proteins as their main feature because deficient protein elimination pathways result in destructive protein clusters damaging neural cells [13, 14]. Both telomere reduction and protein aggregation experience direct connections due to oxidative stress while senescent cells establish conditions that activate oxidative stress and release inflammatory responses. Medical interventions for aging-related pathologies include antioxidant treatments that target specific mitochondrial proteins as well as telomerase activation schemes with drugs that support proteostatic health. Research must concentrate on designing integrated approaches that deliver both aging delay and prevention of neurodegenerative conditions development [15].

Genetic and Epigenetic Factors in Aging

Genetic and Epigenetic Influences on Brain Aging and Neurodegeneration

Brain performance together with the aging process is controlled through genetic and epigenetic factors that affect cellular life expectancy and mental operations. The aging-related genes FOXO and SIRTUINS together with mTOR control base cellular operations from stress defense to DNA maintenance to protein equilibrium. Inadequate operation of age control systems drives quick brain aging thus leading to Alzheimer's and Parkinson's disease through neuron damage and synaptic disruption [16]. The balance of DNA methylation histone modifications and noncoding RNA mechanisms collaborate to modify gene expression to affect neural plasticity and manage both inflammatory and oxidizing stress responses. The age-related modifications of epigenetic chemical markers called epigenetic drift result in dysfunctional gene patterns which intensify neurodegeneration. The evaluation of current investigation findings suggests DNA regulatory research could help slow down natural brain aging processes while potentially stopping neurodegenerative processes [17]. There are two types of treatment for neurodegeneration in development: caloric restriction mimetics that work with sirtuin activators alongside histone deacetylase inhibitors and DNA methylation modulators. The development of novel antiaging medical interventions and neuroprotective solutions depends on a total understanding of genetic inheritances' effects on epigenetic changes because this knowledge enables the creation of treatments that enhance cognitive health durations through neurodegenerative disease reduction [18].

Aging and Its Impact on the Brain

Structural and Functional Changes in the Aging Brain

Neurodegeneration at the Cellular Level: Synaptic Loss, Neuronal Death, and Barrier Breakdown

Elderly brain structure and function strongly influence the development of cognitive decline, which eventually causes neurodegenerative diseases. Synaptic degradation functions as the first significant change because it disrupts neural communication that results in decreased neural networks which are responsible for memory learning and executive function control. Synaptic breakdown occurs within the brain before doctors can measure neuronal damage while linking the brain changes directly to cognitive dysfunction [19]. The combination of brain deterioration and neuronal cell death results in decreased gray and white matter volumes, thereby reducing cognitive operations while affecting emotional control aspects. The hippocampus along with the prefrontal cortex demonstrates considerable decreases in the processes of memory and decision-making. Old age weakens the blood–brain barrier, so its protective features function less effectively. The dysfunction of the brain leads destructive substances to enter brain tissue through new pathways that generate neuroinflammation and disrupt waste systems before they increase neuron susceptibility to damage [20, 21]. The progressive deterioration of neurological structures heightens disability as well as accelerates further structural worsening. The rapid deterioration of Alzheimer's and Parkinson's diseases accelerates when health professionals understand better how synaptic degeneration relates to blood–brain barrier degradation and neurodegeneration with neuronal loss. The development of specialized treatments depends on complete brain modification knowledge to safeguard brain structure and enhance resistance to aging as well as delay age-related disease development [22].

Neuroinflammation and Immune Response

Neuroinflammation in Aging: The Dual Role of Microglia and Astrocytes in Neurodegeneration

Neurodegenerative illnesses develop pathophysiological characteristics through immune-mediated inflammatory responses where microglia and astrocytes take control roles in regulating brain immune environments. These glial cells function to protect neurons through their actions including cellular waste clearance and synaptic protection and positive influence on neuronal activation in typical brain environments [23]. Microglia and astrocytes reach ongoing pro-inflammatory states in aging and disease that lead to lasting production of TNF-α, IL-1β, and IL-6 cytokines. A persistent inflammation causes faster neural damage, adds to oxidative pressure, and weakens blood–brain barrier performance. The death of dopaminergic substantia nigra neurons in Parkinson's disease results from neuroinflammation, while Alzheimer's disease relates to neurotoxic chemical release from activated microglia that promotes tau pathology development around amyloid plaques [24, 25]. Glial cells exhibit a complex nature because they provide protection initially but become detrimental after the initial response period, which makes the treatment of neuroinflammation an extremely challenging prospect. The newest treatment method includes immunotherapies and microglial modulators alongside anti-inflammatory drugs that change glial cell phenotypes into protective states. New research focuses on immunological response refinement by studying nanocarrier medications and the NF-κB and NLRP3 inflammasome signaling pathways to improve brain function preservation. The long-term progression of neurodegenerative conditions could be prevented through an understanding of glial cell behavior, which leads to behavioral modifications [26, 27] (Fig. 1.1).

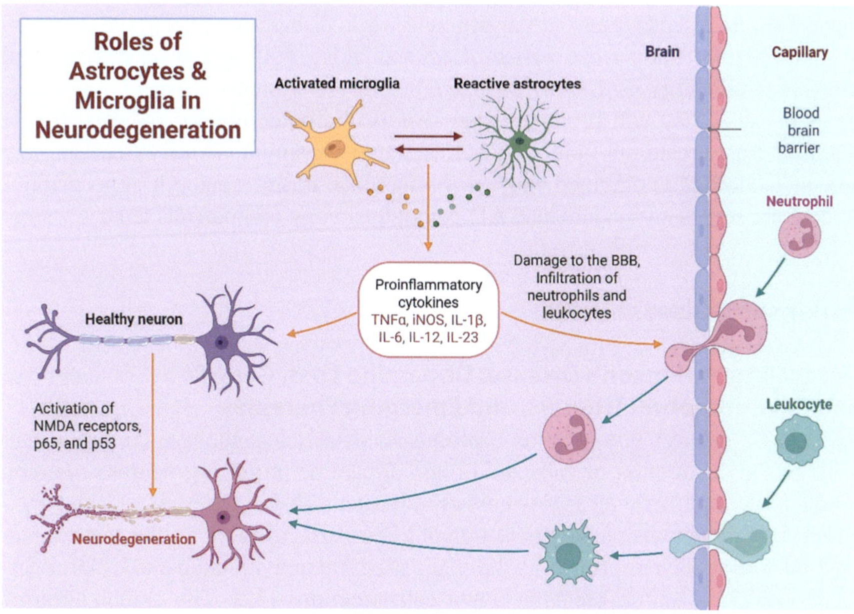

Fig. 1.1 Glial cell activation and neuroinflammation. The figure shows the functions conducted by astrocytes and microglia in neurodegeneration. Activation of glial cells triggers their release of proinflammatory cytokines, which harm the blood–brain barrier, allowing neutrophil and leukocyte cells to pass through. Neuronal cellular damage occurs as the inflammatory cascade activates NMDA receptors and triggers both p65 and p53 pathways, which result in neurodegeneration and brain dysfunction in aging-related diseases

Major Age-Related Neurodegenerative Diseases

Alzheimer's Disease (AD)

Unraveling Alzheimer's Disease: Pathophysiology, Risk Factors, and Therapeutic Challenges

The pathophysiological consequences of Alzheimer's disease (AD) stem from amyloid-beta plaques, in addition to neurofibrillary tangles, which form from tau protein hyperphosphorylation, until brain damage becomes extensive. Major atrophy of the brain occurs because pathological elements destroy neurons through interrupted synaptic communication and produce neuroinflammatory conditions [28]. APICE4 allele carriers develop a higher risk for Alzheimer's disease although early familial AD occurs in people carrying gene defects of APP and PSEN1 or PSEN2. Advanced age and heart health status physical inactivity and unhealthful eating habits serve as essential contributors to both Alzheimer's disease initiation and its disease progression. Current AD treatment includes cholinesterase inhibitors (donepezil and rivastigmine) as well as memantine as an NMDA receptor antagonist although these medicines alleviate symptoms without medicine progression. Medical authorities approved aducanumab and lecanemab monoclonal antibodies for treating amyloid-beta proteins, yet their true efficacy remains unclear because of their hefty price tags [29, 30]. Diagnosis of early disease evolution remains challenging because treatments for late disease stages are insufficient while disease-modifying prescriptions are unavailable. Early-stage patient diagnosis along with improved neurological imaging technology and individualized healthcare systems demand immediate focus because they will produce better therapeutic outcomes for patients. Future research must create multiple approach strategies to simultaneously treat amyloid and tau pathologies combined with neuroinflammation and metabolic abnormalities since these represent fundamental AD pathophysiology components [31].

Parkinson's Disease (PD)

Unraveling Parkinson's Disease: Dopamine Loss, Genetic and Environmental Triggers, and Emerging Therapies

Parkinson's disease functions as a progressive neurodegenerative condition that leads to the destruction of substantia nigra dopamine-producing neurons, thereby producing symptoms such as slowness of movement, shaking movements, and muscle stiffness. The main pathological sign of PD consists of Lewy bodies that contain misfolded α-synuclein protein as the main structure and simultaneously harm cellular operations before breaking down cell structures [32]. The combination of inherited defects in SNCA, LRRK2, PARK7, PINK1, and PRKN, together with environmental exposures to pesticides and heavy metals, determines the clinical presentation of Parkinson's disease. The current treatment of Parkinson's disease combines levodopa and dopamine agonists, which help restore dopamine function as main symptom management drugs. Patients with PD receive medication consisting of MAO-B inhibitors combined with COMT inhibitors to extend dopamine effectiveness along with deep brain stimulation (DBS) when more advanced treatment for motor circuit control is needed. The available treatments do not stop the progression of Parkinson's disease symptoms [33]. Research investigators work to create novel therapies for interacting with α-synuclein aggregation as well as enhancing mitochondrial functionality and controlling neuroinflammatory responses. Both gene therapy and neuroprotective agents serve as research developments that protect dopaminergic neurons from destruction. Treatment solutions for Parkinson's disease need to be personalized because this condition shows multiple clinical expressions thus genetic and environmental assessments are vital for successful results. Research on PD molecular processes will remain important to create treatments that protect affected neurons and reverse the course of this disabling disease [34, 35].

Huntington's Disease (HD)

Huntington's Disease: Genetic Origins, Clinical Manifestations, and Progressive Decline

The inherited mutation of HD in the HTT gene leads to neurodegenerative symptoms because it creates a defective huntingtin protein from excessive CAG repeats on chromosome 4. Each offspring born to HD patients has an identical 50% probability of inheriting HD through dominant inheritance. The patient population with HD typically presents clinical manifestations when they reach their 30s to 50s, but physicians have identified juvenile onset of the disease [36]. Three distinct symptomatic sets emerge in HD patients who display chorea together with dystonia additional coordination problems and mental decline resulting in depression and anxiety with psychotic signs. People with HD experience increasing functional disabilities and independence loss resulting in severe disability that usually occurs before their premature death takes place about 15–20 years after symptoms emerge. Huntington's disease relentlessly causes devastating effects on patient life quality together with demanding care obligations for their caregivers [37, 38]. Modern scientific knowledge about HD molecular research has not succeeded in developing treatments capable of either treating the disease or stopping its progression. The lack of a cure for HD requires medical staff to focus primarily on symptom-based treatments for both motor and psychiatric conditions. Scientific experts conduct continuous research to develop gene silencing methods alongside stem cell treatments combined with protective approaches for Huntington's disease pathology. Continuous research into Huntington's disease offers essential progress because it creates effective therapies to improve treatment outcomes for HD patients [39].

Amyotrophic Lateral Sclerosis (ALS)

Unraveling ALS: Mechanisms of Motor Neuron Degeneration and Emerging Therapies

The degenerative process of ALS specifically harms upper and lower motor neurons, thereby causing weak muscles that eventually grow thinner before resulting in overall body paralysis. Glutamate excitotoxicity together with oxidative stress and defective mitochondria and the neurotoxic effects of aggregated misfolded proteins such as mutant superoxide dismutase 1 (SOD1) work as destructive mechanisms against motor neurons [40]. SOD1, C9orf72, TARDBP, and FUS comprise significant genetic risk factors for ALS inheritance through family lines, while sporadic cases develop from environmental causes and RNA breakdown irregularities. The medical community considers Riluzole the main therapeutic substance for ALS patients, since it combats glutamate toxicity, whereas Edaravone defends cells from oxidative damage. The current treatment options provide minimal assistance in controlling the disease progression rate [41]. The modern research domain encompasses antisense oligonucleotides (ASOs), with applications in gene therapy, stem cell transplantation, neuroinflammation, and protein misfolding agents that provide potential ways to modify the disease. Current experimental therapies have failed to address the operational challenges that prevent their distribution as viable clinical treatment options. The identified key for improved ALS patient well-being depends on research studying molecular aspects and developing personalized treatment methods [42].

Mechanisms Underlying Neurodegeneration

Protein Aggregation in Neurodegeneration: A Core Pathology in Alzheimer's, Parkinson's, and Huntington's Diseases

A group of several severe neurodegenerative diseases form when proteins aggregate and misfold leading to the stepwise degradation of neurons until they eventually die. Multiple brain structures accumulate Alzheimer's hallmark plaques containing amyloid-β peptides and neurofibrillary tangles made of tau proteins in Alzheimer's disease patients before synapse failure and neuroinflammation occur. The pathogenesis of Parkinson's disease forms α-synuclein protein irregularities that generate Lewy compounds that destroy dopaminergic neurons and malfunction the mitochondria [43, 44]. The huntingtin protein in Huntington's disease evolves into harmful cellular clusters that prevent transcription processes, while disrupting both axonal transport and proteolytic-function mechanisms. The increased aggregation of disease-related proteins exceeds the capacity of proteasome together with autophagic pathways to remove cellular waste thus generating oxidative stress as well as extended endoplasmic reticulum stress and inflammatory response. The damage to cellular functions due to genetic mutation leads to a process of programmed cell death that causes the stepwise reduction of specific

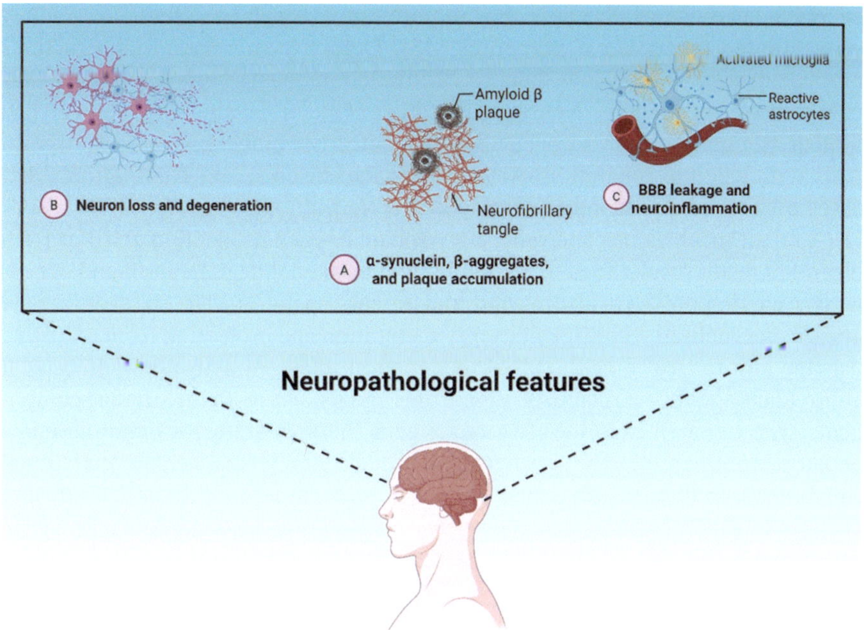

Fig. 1.2 Key pathologies of age-related neurodegeneration. The main pathologic elements of age-related brain change together with neurodegenerative diseases. Thirdly, the picture shows that blood–brain barrier (BBB) leakages produce neuroinflammation from activated microglia together with reactive astrocytes while progressive neuronal degeneration and decline as well as β-aggregates accumulation and α-synuclein deposition occur together. The various destructive neurological processes in Alzheimer's and Parkinson's disease remain strongly connected to create conditions that advance these conditions

neural cell populations [45, 46]. Medical strategies now focus on protein-handling processes and try to stabilize built-in protein structures, while developing brand-new combinations that protect cells from irregular protein assemblies. Pathogenic-protein reduction becomes feasible through the employment of antisense oligonucleotides as well as RNA-interference gene-silencing methods. Research has to continue indefinitely to track protein misfolding and aggregation, since this knowledge will generate therapeutic solutions against neurodegenerative conditions [47] (Fig. 1.2).

Mitochondrial Dysfunction and Energy Deficits

The origin of neurodegenerative processes mainly stems from defective mitochondria, which causes three principal problems that result in insufficient ATP production, elevated reactive oxygen species production, and disrupted calcium balance, leading to neuronal cell death. Neuronal cells experience unique energy-related dangers because they require massive energy to function yet they show limited regenerative properties [48]. Defective mitochondrial structures produce excessive ROS that damage cellular components before activating suicide mechanisms in cells. Neuronal alteration in electrical signaling and nervous system stimulation fails to occur because defective mitochondria create problems with calcium regulation. Alzheimer's, Parkinson's, and Huntington's diseases evolve as mitochondria display defective traits that damage DNA hinder DNA repair mechanisms and reduce phosphate oxygen production abilities [49]. Neuronal death results from the combination of axonal degeneration and synaptic dysfunction that stems from abnormal process operation. The suitable therapy approaches including biological energy restoration and protection from oxidative damage aim to treat mitochondrial dysfunction. PGC-1α activators, along with antioxidant agents and stability-supporting agents, act as the treatment components. The research examines two novel gene therapy strategies targeting mitochondria with peptides for advancing their quality maintenance and operational capabilities. Better health management of mitochondria presents promising chances to shield brain tissue while delaying the progression of neurodegenerative diseases [50, 51].

Neuroinflammation and Glial Cell Activation

Neurodegenerative processes primarily develop from faulty mitochondria, which produce three essential problems including insufficient ATP production together with elevated reactive oxygen species levels and unbalanced calcium equilibrium that ultimately leads to neuronal degradation. Energy shortages affect neurons specifically because metabolism requires high energy levels while their regeneration occurs slowly [52]. The damaged structure of mitochondria leads to overproduction of reactive oxygen species (ROS) until cellular elements get damaged and suicide processes begin. The malfunction of calcium buffering by mitochondria leads to both synaptic signal disruption as well as neuronal excitation failure. The progression of Alzheimer's, Parkinson's, and Huntington's disease strongly relates to mitochondrial disturbance marked by impaired oxidative phosphorylation and DNA mutations together with regulatory failures of mitochondria fusion and fission [53]. These processes undergo abnormalities that simultaneously activate synaptic dysfunction and generate axonal degeneration along with triggering neuronal death. Treatment strategies for mitochondrial dysfunction target two objectives, which are energy restoration of the biological system and harmful oxidation control. Medical intervention includes PGC-1α activators as well as antioxidant agents alongside compounds that maintain the stability of mitochondrial membrane potential and dynamics [54]. Research involving mitochondrial-targeted peptides and gene therapy tools is studied by scientists to develop better methods that improve mitochondrial quality control functions. The advancement of mitochondrial health opens new possibilities for both brain tissue defense and neurodegeneration rate slowing [55].

Impaired Autophagy and Lysosomal Dysfunction

The primary source of neurodegenerative processes depends on mitochondrial dysfunction, which causes three main problems including inadequate ATP generation together with elevated reactive oxygen species production and disrupted calcium regulation, leading to neuronal cell death. Neuronal cells face unique stress due to energy deprivation because they have high metabolic demands yet show low regeneration capacity [56]. Failing organelles among mitochondria produce excessive ROS molecules that destroy cellular elements while triggering cell demise mechanisms. Cellular synapses and neuronal excitatory patterns get disrupted because mitochondria cannot adequately manage calcium levels. Poor oxidative phosphorylation function, together with DNA mutations and dysregulated mitochondria fusion and fission, defines advanced stages of Alzheimer's, as well as Parkinson's and Huntington's diseases. These faulty processes simultaneously activate synaptic failure and lead to axonal destruction before causing neuronal cell death [57]. The main objective of treatments for mitochondrial dysfunction includes returning biological energy function while reducing the effects of oxidative damage. PGC-1α activators together with antioxidant agents and compounds that protect mitochondrial membrane potential and dynamics stability form the therapeutic approach. Scientific researchers currently developing two new approaches using mitochondrial-targeted peptides and genetic treatments that work to enhance both mitochondrial function and quality control mechanisms. The health improvement of mitochondria brings potential opportunities to shield brain tissue alongside rate extensions for neurodegenerative progression [58].

Prevention and Management Strategies

Lifestyle and Behavioral Interventions

Neuronal degeneration derives primarily from mitochondrial failures that produce three critical shortcomings including reduced ATP synthesis in addition to elevated reactive oxygen species and distorted calcium distribution leading to cell death. The high metabolic demands of neurons make them vulnerable to various energy deficiencies as they do not tend to generate new cells easily [59]. The defective structures of mitochondria produce excessive reactive oxygen species (ROS) that cause damage to cell components before activating cellular suicide procedures. Synaptic signals fail to transmit properly, in conjunction with neuronal excitation, because mitochondria lose the ability to buffer calcium effectively. The evolution of Alzheimer's disease, alongside Parkinson's disease and Huntington's disease, directly matches with mitochondrial deterioration that causes incomplete oxidative phosphorylation and mitochondrial DNA abnormalities as well as defective regulation of mitochondria fusion and fission processes. The abnormal functions of these processes trigger synaptic damage and axonal deterioration leading to the death of brain cells [60]. The main objective of mitochondrial dysfunction treatment methods merge biological energy improvement with oxidative damage control mechanisms. PGC-1α activators, antioxidant agents, and compounds that sustain mitochondrial membrane potential and dynamics stability are utilized as part of the therapeutic response. This research approaches mitochondria-targeted peptides alongside gene treatments and alternative strategies to advance both mitochondria function and quality control mechanisms. Elucidative research on mitochondria health enhancement provides promising prospects to safeguard brain structures and slow down neurodegenerative processes [61, 62].

Pharmacological Approaches

The source of neurodegenerative processes begins in dysfunctional mitochondria, which results in three main adverse effects involving insufficient ATP production, elevated reactive oxygen species, and cellular calcium imbalances that lead to neuronal death. Cellular energy shortage affects neurons specifically because these cells require extensive energy through metabolism, yet they have limited regenerative abilities [63]. Every dysfunctional mitochondrion produces extreme amounts of reactive oxygen species (ROS) that destroy cellular elements to trigger programmed neuronal cell death mechanisms. Neuronal excitation along with synaptic signal problems happens due to mitochondria failing to control calcium properly. Research shows that Alzheimer's disease and Parkinson's disease together with Huntington's disease progress due to mitochondrial abnormalities that involve impaired oxidative phosphorylation and mutated DNA as well as flaws in fission and fusion regulation [64]. The disruption of synaptic signals and the production of axonal degeneration leads to neuronal death through defects found in these particular processes. The main approach to treat mitochondrial dysfunction centers on oxygen reduction through antioxidation efforts along with biological energy recovery. Treatment through PGC-1α activators and antioxidant agents together with compounds supports mitochondrial membrane potential and dynamics stability. The study of two innovative approaches based on peptide targeting of mitochondria along with gene therapy evaluations aims to enhance the functional quality of mitochondria. Improved mitochondrial health creates optimistic prospects that protect brain cells and slow down neurodegenerative development [65].

Advanced Therapies and Research

Mitochondrial dysfunction normally leads to three core problems that include inadequate creation of ATP, excessive reactive oxygen species, and disordered calcium regulation, which collectively cause neuronal cell death. Neuronal cells face unique peril from energy deprivation because their active metabolism requires excessive energy consumption but they have limited self-replication ability. Cells with defective mitochondria create an excessive number of reactive oxygen species (ROS) that attack cellular contents before triggering programmed cell death [66]. Neuronal excitation with disrupted synaptic signals develops when mitochondria fail to manage calcium properly. The progression of Alzheimer's, Parkinson's, and Huntington's diseases strongly relates to mitochondrial failures that cause compromised DNA structure impaired mitochondrial fusion and fission control and reduced efficiency of oxidative phosphorylation. The cellular processes within these tissues create synaptic dysfunction cause axonal degeneration and finally result in neuronal death. Treatment plans for mitochondrial dysfunction aim to repair biological energy output while simultaneously combating oxidizing agents that cause damage [67]. The response recommends PGC-1α activators as well as antioxidant agents which maintain both mitochondrial membrane potential and dynamics stability. Research teams are actively investigating mitochondrial quality control enhancement through two possible avenues, which include mitochondrial-targeted peptides combined with gene therapy. Better health of mitochondria becomes a promising option that can protect brain tissue and slow down neurodegeneration development rates [68] (Table 1.1).

Table 1.1 This table shows how protein aggregation and misfolding contribute to the development of various neurodegenerative illnesses, such as Huntington's, Parkinson's, and Alzheimer's. To illustrate how these pathological processes lead to functional decline and disease development in the aging brain, it lists illness-specific misfolded proteins, their aggregate kinds, afflicted brain areas, and related clinical consequences

Disease	Misfolded protein(s)	Aggregate type	Clinical impact	Affected brain region(s)	Main affected function	Reference
Alzheimer's disease	Amyloid-β, tau	Amyloid plaques, neurofibrillary tangles	Memory loss, cognitive decline	Hippocampus, cortex	Cognition and memory	[69, 70]
Parkinson's disease	Alpha-synuclein	Lewy bodies	Tremors, rigidity, bradykinesia	Substantia nigra	Motor control	[71, 72]
Huntington's disease	Mutant huntingtin (mHTT)	Nuclear and cytoplasmic inclusions	Involuntary movements, cognitive impairment	Striatum, cortex	Motor and cognitive function	[73, 74]

Future Directions and Research Challenges

Advancing Diagnosis, Overcoming Treatment Barriers, and Addressing Ethical Challenges in Aging and Neurodegeneration

Neuronal destruction in degenerative conditions results from mitochondrial failure, which produces three significant adverse effects including energy deficits alongside elevated reactive oxygen species and altered calcium balance. Neuronal cells remain particularly susceptible to energy-related damage since they need high amounts of energy to function but they have limited capacity for self-renewal [75]. Structures that show mitochondrial dysfunction produce excessive reactive oxygen species that cause damage to cellular components and activate suicide programs inside cells. The blocking of synaptic signals together with disrupted neuronal excitation takes place when mitochondria fail to maintain appropriate calcium buffering. Research shows that Alzheimer's, as well as Parkinson's and Huntington's disease progression, strongly depends on abnormal mitochondrial structures and their dysfunction, including subpar oxidative phosphorylation function DNA mutations and inadequate mitochondria fusion and fission regulation [76]. These physiological process abnormalities activate synaptic dysfunction create axonal degeneration and result in neuronal death simultaneously. Therapeutic approaches for mitochondrial dysfunction need to achieve two goals: energy renewal of biological systems and protection against oxidation. PGC-1α activators along with antioxidant agents form the basis of this response along with compounds that keep mitochondrial membrane potential and dynamics stable. Scientists have established two new research types that use mitochondrial-targeted peptides as well as gene therapies to improve both mitochondrial quality control and their functional properties. Allocating medical focus on better mitochondria performance opens new possibilities for tissue protection and slowing down the rate of neurodegeneration [77, 78].

Conclusion

The main source of neurodegenerative damage leads to three principal problems originating from impaired mitochondria: insufficient energy production, reactive oxygen species elevation, and calcium balance disturbance, which jointly kill neurons. The high metabolic demand of neurons makes them vulnerable because they have limited capacity to regenerate along with needing significant quantities of energy. Mitochondrial organelles carrying damaged structures produce excessive reactive oxygen species that damage cell molecules until cellular suicide activation occurs. Neuronal excitation along with synaptic signal dysfunction occurs because mitochondria fail to control calcium fluxes properly. Research indicates that Alzheimer's, Parkinson's, and Huntington's disease progression relates strongly to mitochondrial abnormalities—primarily through mitochondrial dysfunction and DNA damage—together with broken oxidative phosphorylation and failed regulatory mechanisms of fission and fusion. The disturbed functions activate synaptic problems, as well as lead to axonal atrophy, which results in neuronal destruction. Treatment strategies for mitochondrial dysfunction focus on restoring biological energy together with minimizing rust-related damage. PGC-1α activators with antioxidant agents form part of the treatment approach along with compounds preserving mitochondrial membrane potential and stability. Two innovative gene therapy and mitochondrial-targeted peptide approaches improve mitochondrial quality control and function according to scientific investigations. Brain tissue protection along with delayed neurodegeneration rates become feasible because of improved mitochondrial health.

References

1. Maldonado E, Morales-Pison S, Urbina F, Solari A. Aging hallmarks and the role of oxidative stress. Antioxidants (Basel). 2023;12(3)
2. Kaur P, Khan H, Grewal AK, Dua K, Singh SK, Gupta G, et al. Exploring therapeutic strategies: the relationship between metabolic disorders and FOXO Signalling in Alzheimer's disease. CNS Neurol Disord Drug Targets. 2025;24(3):196–207.
3. Giorgi C, Marchi S, Simoes ICM, Ren Z, Morciano G, Perrone M, et al. Mitochondria and reactive oxygen species in aging and age-related diseases. Int Rev Cell Mol Biol. 2018;340:209–344.
4. Song J, Xiao L, Zhang Z, Wang Y, Kouis P, Rasmussen LJ, et al. Effects of reactive oxygen species and mitochondrial dysfunction on reproductive aging. Front Cell Dev Biol. 2024;12:1347286.
5. Bashir B, Gulati M, Vishwas S, Gupta G, Dhanasekaran M, Paudel KR, et al. Bridging gap in the treatment of Alzheimer's disease via postbiotics: current practices and future prospects. Ageing Res Rev. 2025;105:102689.
6. Brown RC, Lockwood AH, Sonawane BR. Neurodegenerative diseases: an overview of environmental risk factors. Environ Health Perspect. 2005;113(9):1250–6.
7. Thapa R, Ahmad Bhat A, Shahwan M, Ali H, PadmaPriya G, Bansal P, et al. Proteostasis disruption and senescence in Alzheimer's disease pathways to neurodegeneration. Brain Res. 2024;1845:149202.
8. Palanisamy CP, Pei J, Alugoju P, Anthikapalli NVA, Jayaraman S, Veeraraghavan VP, et al. New strategies of neurodegenerative disease treatment with extracellular vesicles (EVs) derived from mesenchymal stem cells (MSCs). Theranostics. 2023;13(12):4138–65.
9. Hussain MS, Altamimi ASA, Afzal M, Almalki WH, Kazmi I, Alzarea SI, et al. Kaempferol: paving the path for advanced treatments in aging-related diseases. Exp Gerontol. 2024;188:112389.
10. Dar MA, Arafah A, Bhat KA, Khan A, Khan MS, Ali A, et al. Multiomics technologies: role in disease biomarker discoveries and therapeutics. Brief Funct Genomics. 2023;22(2):76–96.
11. Bashir B, Alam S, Khandale N, Birla D, Vishwas S, Pandey NK, et al. Opening avenues for treatment of neurodegenerative disease using post-biotics: breakthroughs and bottlenecks in clinical translation. Ageing Res Rev. 2024;95:102236.
12. Cui H, Kong Y, Zhang H. Oxidative stress, mitochondrial dysfunction, and aging. J Signal Transduct. 2012;2012:646354.
13. Victorelli S, Passos JF. Telomeres and cell senescence - size matters not. EBioMedicine. 2017;21:14–20.
14. Dorf N, Maciejczyk M. Skin senescence-from basic research to clinical practice. Front Med. 2024;11:1484345.

15. Barnes RP, Fouquerel E, Opresko PL. The impact of oxidative DNA damage and stress on telomere homeostasis. Mech Ageing Dev. 2019;177:37–45.
16. Morris BJ, Willcox BJ, Donlon TA. Genetic and epigenetic regulation of human aging and longevity. Biochim Biophys Acta Mol basis Dis. 2019;1865(7):1718–44.
17. Bertogliat MJ, Morris-Blanco KC, Vemuganti R. Epigenetic mechanisms of neurodegenerative diseases and acute brain injury. Neurochem Int. 2020;133:104642.
18. Manjula R, Anuja K, Alcain FJ. SIRT1 and SIRT2 activity control in neurodegenerative diseases. Front Pharmacol. 2020;11:585821.
19. Lee J, Kim HJ. Normal aging induces changes in the brain and neurodegeneration Progress: review of the structural, biochemical, metabolic, cellular, and molecular changes. Front Aging Neurosci. 2022;14:931536.
20. Salat DH, Kaye JA, Janowsky JS. Prefrontal gray and white matter volumes in healthy aging and Alzheimer disease. Arch Neurol. 1999;56(3):338–44.
21. Siddiqui N, Sharma A, Kesharwani A, Anurag, Parihar VK. Exploring role of natural compounds in molecular alterations associated with brain ageing: a perspective towards nutrition for ageing brain. Ageing Res Rev. 2024;97:102282.
22. Jiang Q, Liu J, Huang S, Wang XY, Chen X, Liu GH, et al. Antiageing strategy for neurodegenerative diseases: from mechanisms to clinical advances. Signal Transduct Target Ther. 2025;10(1):76.
23. Franklin H, Clarke BE, Patani R. Astrocytes and microglia in neurodegenerative diseases: lessons from human in vitro models. Prog Neurobiol. 2021;200:101973.
24. Colonna M, Butovsky O. Microglia function in the central nervous system during health and neurodegeneration. Annu Rev Immunol. 2017;35:441–68.
25. Latham AS, Moreno JA, Geer CE. Biological agents and the aging brain: glial inflammation and neurotoxic signaling. Front Aging. 2023;4:1244149.
26. Adamu A, Li S, Gao F, Xue G. The role of neuroinflammation in neurodegenerative diseases: current understanding and future therapeutic targets. Front Aging Neurosci. 2024;16:1347987.
27. Al-Ghraiybah NF, Wang J, Alkhalifa AE, Roberts AB, Raj R, Yang E, et al. Glial cell-mediated neuroinflammation in Alzheimer's disease. Int J Mol Sci. 2022;23(18)
28. Zhang H, Wei W, Zhao M, Ma L, Jiang X, Pei H, et al. Interaction between Aβ and Tau in the pathogenesis of Alzheimer's disease. Int J Biol Sci. 2021;17(9):2181–92.
29. Bekris LM, Yu CE, Bird TD, Tsuang DW. Genetics of Alzheimer disease. J Geriatr Psychiatry Neurol. 2010;23(4):213–27.
30. Liu CC, Liu CC, Kanekiyo T, Xu H, Bu G. Apolipoprotein E and Alzheimer disease: risk, mechanisms and therapy. Nat Rev Neurol. 2013;9(2):106–18.
31. Di Luca M, Nutt D, Oertel W, Boyer P, Jaarsma J, Destrebecq F, et al. Towards earlier diagnosis and treatment of disorders of the brain. Bull World Health Organ. 2018;96(5):298–298A.
32. Zhou ZD, Yi LX, Wang DQ, Lim TM, Tan EK. Role of dopamine in the pathophysiology of Parkinson's disease. Transl Neurodegener. 2023;12(1):44.
33. Nuytemans K, Theuns J, Cruts M, Van Broeckhoven C. Genetic etiology of Parkinson disease associated with mutations in the SNCA, PARK2, PINK1, PARK7, and LRRK2 genes: a mutation update. Hum Mutat. 2010;31(7):763–80.
34. Vidović M, Rikalovic MG. Alpha-synuclein aggregation pathway in Parkinson's disease: current status and novel therapeutic approaches. Cells. 2022;11(11)
35. Li Y, Xia Y, Yin S, Wan F, Hu J, Kou L, et al. Targeting microglial α-synuclein/TLRs/NF-kappaB/NLRP3 inflammasome axis in Parkinson's disease. Front Immunol. 2021;12:719807.
36. Gusella JF, Persichetti F, MacDonald ME. The genetic defect causing Huntington's disease: repeated in other contexts? Mol Med. 1997;3(4):238–46.
37. Anderson KE, van Duijn E, Craufurd D, Drazinic C, Edmondson M, Goodman N, et al. Clinical management of neuropsychiatric symptoms of Huntington disease: expert-based consensus guidelines on agitation, anxiety, apathy, psychosis and sleep disorders. J Huntingtons Dis. 2018;7(3):355–66.

38. Paulsen JS, Miller AC, Hayes T, Shaw E. Cognitive and behavioral changes in Huntington disease before diagnosis. Handb Clin Neurol. 2017;144:69–91.
39. Ferguson MW, Kennedy CJ, Palpagama TH, Waldvogel HJ, Faull RLM, Kwakowsky A. Current and possible future therapeutic options for Huntington's disease. J Cent Nerv Syst Dis. 2022;14:11795735221092517.
40. Arnold FJ, Putka AF, Raychaudhuri U, Hsu S, Bedlack RS, Bennett CL, et al. Revisiting glutamate excitotoxicity in amyotrophic lateral sclerosis and age-related neurodegeneration. Int J Mol Sci. 2024;25(11)
41. Edgar S, Ellis M, Abdul-Aziz NA, Goh KJ, Shahrizaila N, Kennerson ML, et al. Mutation analysis of SOD1, C9orf72, TARDBP and FUS genes in ethnically-diverse Malaysian patients with amyotrophic lateral sclerosis (ALS). Neurobiol Aging. 2021;108:200–6.
42. Amanat M, Nemeth CL, Fine AS, Leung DG, Fatemi A. Antisense oligonucleotide therapy for the nervous system: from bench to bedside with emphasis on pediatric neurology. Pharmaceutics. 2022;14(11)
43. Barmaki H, Nourazarian A, Khaki-Khatibi F. Proteostasis and neurodegeneration: a closer look at autophagy in Alzheimer's disease. Front Aging Neurosci. 2023;15:1281338.
44. Forum on Neuroscience and Nervous System Disorders, Board on Health Sciences Policy, Institute of Medicine (2013) Neurodegeneration: exploring commonalities across diseases: workshop summary. The National Academies collection: reports funded by National Institutes of Health. National Academies Press (US), Washington (DC)
45. Tong H, Yang T, Xu S, Li X, Liu L, Zhou G, et al. Huntington's disease: complex pathogenesis and therapeutic strategies. Int J Mol Sci. 2024;25(7)
46. Koyuncu S, Fatima A, Gutierrez-Garcia R, Vilchez D. Proteostasis of huntingtin in health and disease. Int J Mol Sci. 2017;18(7)
47. Molinaro C, Martoriati A, Cailliau K. Proteins from the DNA damage response: regulation, dysfunction, and anticancer strategies. Cancers. 2021;13(15)
48. Wadan AS, Shaaban AH, El-Sadek MZ, Mostafa SA, Moshref AS, El-Hussein A, et al. Mitochondrial-based therapies for neurodegenerative diseases: a review of the current literature. Naunyn Schmiedeberg's Arch Pharmacol. 2025; https://doi.org/10.1007/s00210-025-04014-0.
49. Guo C, Sun L, Chen X, Zhang D. Oxidative stress, mitochondrial damage and neurodegenerative diseases. Neural Regen Res. 2013;8(21):2003–14.
50. Wang B, Huang M, Shang D, Yan X, Zhao B, Zhang X. Mitochondrial behavior in axon degeneration and regeneration. Front Aging Neurosci. 2021;13:650038.
51. Bustamante-Barrientos FA, Luque-Campos N, Araya MJ, Lara-Barba E, de Solminihac J, Pradenas C, et al. Mitochondrial dysfunction in neurodegenerative disorders: potential therapeutic application of mitochondrial transfer to central nervous system-residing cells. J Transl Med. 2023;21(1):613.
52. Chen W, Zhao H, Li Y. Mitochondrial dynamics in health and disease: mechanisms and potential targets. Signal Transduct Target Ther. 2023;8(1):333.
53. Baev AY, Vinokurov AY, Novikova IN, Dremin VV, Potapova EV, Abramov AY. Interaction of mitochondrial calcium and ROS in neurodegeneration. Cells. 2022;11(4)
54. Wen H, Deng H, Li B, Chen J, Zhu J, Zhang X, et al. Mitochondrial diseases: from molecular mechanisms to therapeutic advances. Signal Transduct Target Ther. 2025;10(1):9.
55. Hong WL, Huang H, Zeng X, Duan CY. Targeting mitochondrial quality control: new therapeutic strategies for major diseases. Mil Med Res. 2024;11(1):59.
56. Song N, Mei S, Wang X, Hu G, Lu M. Focusing on mitochondria in the brain: from biology to therapeutics. Transl Neurodegener. 2024;13(1):23.
57. Di Meo S, Reed TT, Venditti P, Victor VM. Role of ROS and RNS sources in physiological and pathological conditions. Oxidative Med Cell Longev. 2016;2016:1245049.
58. Rius-Pérez S, Torres-Cuevas I, Millán I, Ortega ÁL, Pérez S. PGC-1α, inflammation, and oxidative stress: an integrative view in metabolism. Oxidative Med Cell Longev. 2020;2020:1452696.
59. Kodavati M, Wang H, Hegde ML. Altered mitochondrial dynamics in motor neuron disease: an emerging perspective. Cells. 2020;9(4)

60. Zong Y, Li H, Liao P, Chen L, Pan Y, Zheng Y, et al. Mitochondrial dysfunction: mechanisms and advances in therapy. Signal Transduct Target Ther. 2024;9(1):124.
61. Zhou H, Wu C, Jin Y, Wu O, Chen L, Guo Z, et al. Role of oxidative stress in mitochondrial dysfunction and their implications in intervertebral disc degeneration: mechanisms and therapeutic strategies. J Orthop Transl. 2024;49:181–206.
62. Sivitz WI, Yorek MA. Mitochondrial dysfunction in diabetes: from molecular mechanisms to functional significance and therapeutic opportunities. Antioxid Redox Signal. 2010;12(4):537–77.
63. Clemente-Suárez VJ, Redondo-Flórez L, Beltrán-Velasco AI, Ramos-Campo DJ, Belinchón-deMiguel P, Martinez-Guardado I, et al. Mitochondria and brain disease: a comprehensive review of pathological mechanisms and therapeutic opportunities. Biomedicines. 2023;11(9)
64. Zorov DB, Juhaszova M, Sollott SJ. Mitochondrial reactive oxygen species (ROS) and ROS-induced ROS release. Physiol Rev. 2014;94(3):909–50.
65. Golpich M, Amini E, Mohamed Z, Azman Ali R, Mohamed Ibrahim N, Ahmadiani A. Mitochondrial dysfunction and biogenesis in neurodegenerative diseases: pathogenesis and treatment. CNS Neurosci Ther. 2017;23(1):5–22.
66. Moon HE, Paek SH. Mitochondrial dysfunction in Parkinson's disease. Exp Neurobiol. 2015;24(2):103–16.
67. Guo L, Tian J, Du H. Mitochondrial dysfunction and synaptic transmission failure in Alzheimer's disease. J Alzheimers Dis. 2017;57(4):1071–86.
68. Chen L, Qin Y, Liu B, Gao M, Li A, Li X, et al. PGC-1α-mediated mitochondrial quality control: molecular mechanisms and implications for heart failure. Front Cell Dev Biol. 2022;10:871357.
69. Nelson PT, Alafuzoff I, Bigio EH, Bouras C, Braak H, Cairns NJ, et al. Correlation of Alzheimer disease neuropathologic changes with cognitive status: a review of the literature. J Neuropathol Exp Neurol. 2012;71(5):362–81.
70. Kamatham PT, Shukla R, Khatri DK, Vora LK. Pathogenesis, diagnostics, and therapeutics for Alzheimer's disease: breaking the memory barrier. Ageing Res Rev. 2024;101:102481.
71. Dickson DW. Neuropathology of Parkinson disease. Parkinsonism Relat Disord. 2018;46:S30–S3.
72. Maiti P, Manna J, Dunbar GL. Current understanding of the molecular mechanisms in Parkinson's disease: targets for potential treatments. Transl Neurodegener. 2017;6:28.
73. Frank S. Treatment of Huntington's disease. Neurotherapeutics. 2014;11(1):153–60.
74. Irfan Z, Khanam S, Karmakar V, Firdous SM, El Khier B, Khan I, et al. Pathogenesis of Huntington's disease: an emphasis on molecular pathways and prevention by natural remedies. Brain Sci. 2022;12(10)
75. Norat P, Soldozy S, Sokolowski JD, Gorick CM, Kumar JS, Chae Y, et al. Mitochondrial dysfunction in neurological disorders: exploring mitochondrial transplantation. NPJ Regen Med. 2020;5(1):22.
76. Choi EH, Kim MH, Park SJ. Targeting mitochondrial dysfunction and reactive oxygen species for neurodegenerative disease treatment. Int J Mol Sci. 2024;25(14)
77. Klemmensen MM, Borrowman SH, Pearce C, Pyles B, Chandra B. Mitochondrial dysfunction in neurodegenerative disorders. Neurotherapeutics. 2024;21(1):e00292.
78. Song T, Song X, Zhu C, Patrick R, Skurla M, Santangelo I, et al. Mitochondrial dysfunction, oxidative stress, neuroinflammation, and metabolic alterations in the progression of Alzheimer's disease: a meta-analysis of in vivo magnetic resonance spectroscopy studies. Ageing Res Rev. 2021;72:101503.

Introduction to Cellular Senescence

2

Neeraj Kumar Fuloria, Shivkanya Fuloria, Mahendran Sekar, Mohammed Tahir Ansari, Goot Heah Khor, Anupam Biswas, Kamal Narain, Sangita Biswas, and Sumita Bhatia

Introduction

The biological process of cellular senescence causes cells to stop their cell cycle permanently after enduring DNA damage, oncogenic signaling, and telomere shortening. Cellular senescence leads to two main characteristics, including a growing cell size and elevated lysosomal expression of SA-β-gal and morphological changes that affect chromatin structure while also generating SAPS secretions with inflammatory cytokines, chemokines, growth factors, and proteases [1, 2]. Maintaining cell cycle arrest through tissue changes leads to environmental modifications that activate immune responses until senescence reaches a balance where protection

N. K. Fuloria (✉) · S. Fuloria
Faculty of Pharmacy, AIMST University, Bedong, Kedah, Malaysia
e-mail: nfuloria@aimst.edu.my; shivkanya_fuloria@aimst.edu.my

M. Sekar
School of Pharmacy, Monash University Malaysia, Bandar Sunway, Subang Jaya, Selangor, Malaysia
e-mail: mahendran.sekar@monash.edu

M. T. Ansari
School of Pharmacy, Faculty of Science and Engineering, University of Nottingham Malaysia, Jalan Broga, Semenyih, Selangor, Malaysia
e-mail: tahir.ansari@nottingham.edu.my

G. H. Khor
Centre of Preclinical Science Studies & Oral and Maxillofacial Disease Research Group, Faculty of Dentistry, Universiti Teknologi MARA, Sungai Buloh Campus, Jalan Hospital, Sungai Buloh, Selangor, Malaysia
e-mail: gootheah@uitm.edu.my

G. Gupta (ed.), *Cellular Senescence and Brain Aging*,
https://doi.org/10.1007/978-981-96-8873-9_2

meets damaging effects in multicellular organisms. In the early 1960s, Leonard Hayflick and Paul Moorhead published the first findings that demonstrated human fibroblasts had restricted proliferation abilities when cultured. Modern scientists define this discovery as the Hayflick limit [3, 4]. The groundbreaking research announced that cultured cells only survive a finite time, debunking the belief in immortal cells to define cellular lifespan regulations. Through his discovery of how population doubling accumulation causes final growth stoppage, Hayflick provided essential knowledge about how cellular aging works through telomere-based genome-defense mechanisms [5]. Research in senescence moved toward understanding both stress and oncogene-induced cellular events besides replicative exhaustion to prove this phenomenon extended across many cellular influences [6, 7].

The medical field and biological research currently acknowledge cellular senescence as a core foundational concept that directs disease health treatments and therapeutic breakthroughs. The defense against genomic integrity needs this potent tumor-suppressive mechanism to prevent harmful or cancerous cell proliferation [8]. Senescent cell persistent activation generates tissue injuries, while the SASP causes lasting inflammation, leading to continuous tissue degeneration and medical condition development. Research teams created two types of drugs: senolytics for deleting specific senescent cells and senomorphics to adjust SASP so the cells would experience less harm than removal [9–11]. Research indicates that eliminating senescent cells using available methods holds promise to treat age-related diseases, improve cancer treatment results, and control fibrotic conditions. Human physiology experiences opposing effects from cellular senescence depending on which condition prevails [12]. The aging process of fibroblasts plays a vital role in embryonic development because it helps eliminate additional structures needed for tissue pattern formation and morphological changes. Wounds reach healing and repair states by activating immune and stem cells through paracrine signals from senescent cells [13, 14]. Multiple age-related health problems, such as osteoarthritis, atherosclerosis, and pulmonary fibrosis, emerge from tissue-wide senescent cell accumulation that results in the breakdown of immune system function and reduced tissue repair potential. Cellular senescence protects against cancer by stopping cell cycle progression, but its SASP inflammatory response and matrix destruction may support tumor development [15, 16]. Current research demonstrates how neuron

A. Biswas
Faculty of Medicine, Department of Physiology, AIMST University, Bedong, Kedah, Malaysia
e-mail: anupam@aimst.edu.my

K. Narain · S. Bhatia
Faculty of Medicine, AIMST University, Bedong, Kedah, Malaysia
e-mail: kamal@aimst.edu.my

S. Biswas
Department of Preclinical Sciences, Faculty of Dentistry, MAHSA University, Jenjarom, Selangor, Malaysia
e-mail: sangita@mahsa.edu.my

and glial cells with senescence behaviour cause extensive inflammation-related damage to neurons in patients experiencing Alzheimer's and Parkinson's [17]. Cellular senescence functions as a fundamental protecting mechanism in biology because it creates connections between developmental sequences and aging patterns and progressions that continue to be the primary focus for new therapy research [18].

Hallmarks and Characteristics of Cellular Senescence

A strict regulatory system establishes cellular senescence through sustained permanent cell cycle arrest following different stress factors including telomere shortening, DNA damage, and oncogenic signaling. The permanent cell cycle arrest develops mainly through tumor suppressor pathways, including p53/p21CIP1 and p16INK4a/Rb, which prevent cellular cyclin-dependent kinases from promoting growth arrest [19, 20]. DNA maintenance proteins at damage sites constitute the only exception to typical DDR activation, which characterizes senescent cells by creating structural alterations including enlarged cell dimensions with flattened shape and increased granulation in addition to modified nuclear structure patterns that signify massive cellular metabolic and structural changes [21]. Scientific investigations and experimental research heavily rely on elevated SA-β-Gal activity at pH 6.0 as the specific senescence biomarker that arises from enlarged lysosomes [22]. Persistent activation of γ-H2AX foci and other repair proteins at DNA lesions through DDR creates a diagnostic trait that defines both the establishment and the preservation of a senescent condition [23]. When SAHF develops, cellular chromatin remodels to accomplish genome silencing and cellular cycle arrest, stability, and proliferation control. Senescent cells develop senescence-associated secretory phenotype (SASP) when cellular homeostasis breaks down, releasing inflammatory cytokines in combination with chemokines, growth factors, and proteases [24, 25]. The cellular process of SASP enables self-maintenance of cellular senescence inside cells while simultaneously altering tissue signals that affect tissue repair, immunological responses, and tumor development [26]. The interconnected activity of these cellular components maintains the senescence status as an independent cellular condition compared to quiescence or terminal differentiation. The hallmarks reveal how tissue protection from cancer co-occurs as aging-related tissue dysfunctions develop due to senescent cell accumulation [27, 28].

Molecular Mechanisms Driving Senescence

Multiple cellular networks produce the senescence mechanism through which intrinsic or extrinsic stimuli trigger permanent cell cycle arrest. The p53/p21CIP1 and p16INK4a/Rb serve as main cellular senescence regulators that restrict cyclin-dependent kinase activity to stabilize Rb in its hypophosphorylated state, thereby preventing G1 phase advancement [29]. After telomere shortening in replicative senescence cells, cells experience DNA damage response (DDR) that stabilizes p53 while inducing p21 expression to halt cell division permanently. Cells develop p16INK4a signaling independently when experiencing constant growth signals and specific changes to chromatin that occur without telomere erosion [30]. Oncogene activation of RAS or BRAF causes damaged cells to receive faulty signals, leading to DNA damage responses that force cells into terminal proliferation arrest or cancerous mutation pathways [31]. Activating stress-induced premature senescence (SIPS) through p53 and p16INK4a axes operation occurs before genotoxic damage prompts telomere erosion. The dysfunctional mitochondria initiate an inflammatory response through ROS production, resulting in DNA damage, SASP, and autocrine and paracrine signaling pathway activation [32]. The global network serves as an effective defense system that removes damaged or abnormally functioning cells and precancerous cells from entering the proliferative cycle [33]. Researchers currently study molecular aging networks that employ various senescence triggers for two purposes. Both goals assist with understanding dual antitumoral responses alongside age-associated tissue harm mechanisms, and they enable the creation of new clinical treatments that focus on identifying and destroying senescent cells [34] (Fig. 2.1).

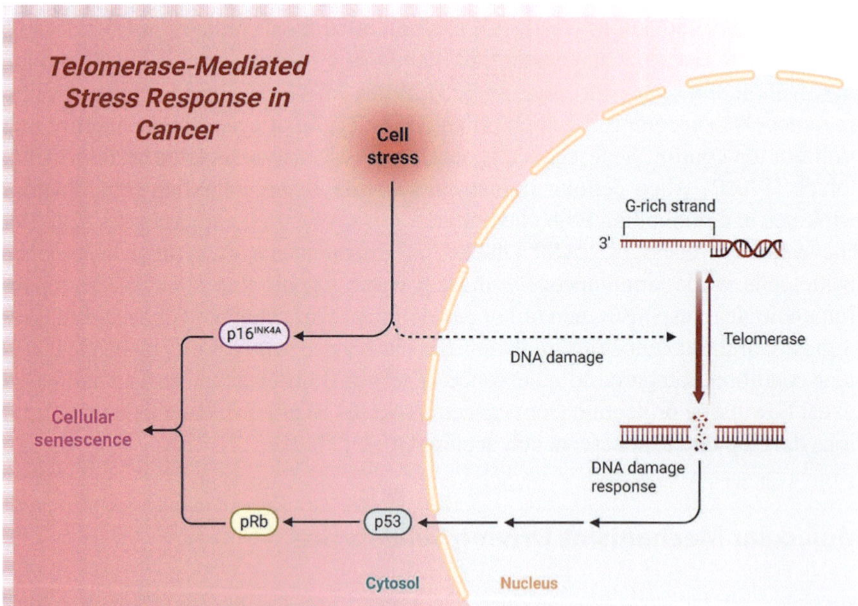

Fig. 2.1 Telomere-Mediated Senescence Pathway. The telomerase-mediated stress response in cancer is seen in the figure. Cellular stress causes DNA damage, activating the p16INK4A and p53-pRb pathways. A shorter G-rich strand causes telomere malfunction, which triggers telomerase and starts a DNA-damaging reaction. Together, these molecular processes encourage cellular senescence as a defense against cancerous transformation

Types of Cellular Senescence

Replicative Senescence

The inception of replicative senescence emerged when Leonard Hayflick studied human fibroblasts in his experiments. When repeated cell cycles damage telomeres until they reach their length threshold, the cells activate a DNA damage response. Cell cycle arrest becomes permanent through p53/p21 activation and the control of p16^INK4a^/Rb axes [35]. Replicative senescence is a robust anticancer defense because it determines the maximum reproductive capacity of somatic cells. Telomere shortening affects organism aging in tissues, which cannot be restored. Cell replication and telomere dysfunction lead to age-related conditions such as idiopathic pulmonary fibrosis and specific haematological disorders [36]. Over the last few years, scientific research has established that cancer cells gain immunity from replicative senescence through telomerase reactivation. The process of replicative senescence both protects genomic integrity and serves as a danger factor leading to degenerative disease development [37].

Stress-Induced Premature Senescence (SIPS)

Environmental and intracellular stress leads to stress-activated premature senescence (SIPS) independent of telomere reduction. SIPS results from four primary stress triggers: oxidative stress, ionizing radiation, genotoxic drugs, and mitochondrial dysfunction. Biochemical stressors activate broad DNA damage responses through which p53 and p16INK4a pathways become active, eventually carrying out cell cycle arrest functions [38]. Rapid SIPS development can be observed in cells that did not demonstrate substantial growth expansion. The protecting system SIPS is a vital checkpoint to control tissue cell expansion post-damage execution to prevent cancer development. The cells that undergo SIPS subsequently produce SASP substances that sustain chronic inflammation while damaging cellular architecture [39]. The medical community acknowledges SIPS operations as essential for clinical pathology formation in neurodegenerative diseases, cardiovascular dysfunction, and metabolic diseases. Therapeutic research about combining antioxidants with DDR-modulating agents became necessary because some evidence demonstrates SIPS could potentially reverse [40].

Oncogene-Induced Senescence (OIS)

An abnormal activation of oncogenic signaling leads to the protective cellular process called oncogene-induced senescence (OIS). The DNA damage buildup and blocked replication forks occur when oncogenes mutate or the genes RAS, BRAF, or MYC overexpress, leading to hyperproliferative stress [41]. The DDR triggers permanent cell growth cessation with the p53/p21 pathway and the p16IN4a/Rb tumor suppressor pathway. OIS resistance maintains a robust prevention mechanism against cancer progression, which medical professionals identify as precancerous tissue in cases of melanocytic nevi and pancreatic intraepithelial neoplasias [42]. Tumor suppression pathways fail to stop cancer formation by losing their protective effect against tumors, either through unstable senescence or pathway deactivation. The secretion of SASP substances by OIS cells modifies cellular immune function and the tissue microenvironment for tumor prevention or promotion actions. Scientists today focus their research on cancer therapy applications of OIS to derive tumor suppression benefits without causing destructive inflammatory responses [43, 44].

Therapy-Induced Senescence (TIS)

Therapy-induced senescence is established when DNA damage reaches the point of repair because of chemotherapy or radiation therapy chemotherapeutic agents. Using these methods, the initial cancer killing objective causes cells to perform stable cycle arrest by activating DDR, followed by p53, p21, and p16INK4a. Oncology faces two simultaneous challenges due to TIS because the protective mechanism stops tumor progression but produces undesirable effects related to SASP [45]. Resistance to treatment and immune suppression emerge together with disease relapse because pro-inflammatory factors that stem from senescent tumor or stromal cells cause tissue remodeling. Tissue complications from cancer treatment in the later stages of life prove to have origins from the tissue-integrity system. Research investigates using chemotherapy with senolytic drugs to enhance therapeutic effects by reducing unfavourable toxicities derived from senescent cells [46].

Developmental Senescence

Developmental senescence is a vital cell development process through its planned temporary mechanisms during embryonic development. Developmental senescence operates through distinct regulatory processes that do not result from DNA damage and proceed apart from telomere deterioration. The p21CIP1 pathway and TGF-β signaling are significant regulators in developmental senescence, even though p16INK4a and embryonic development rely on developmental senescence to build bodily structures by eliminating temporary cell elements, which allows tissue organization as well as preparing organs for growth. The sequence includes short-lived activation of SA-β-gal and p21 senescence markers that the immune system eliminates from tissues [47]. New research evidence has compelled scientists to question prevailing thoughts that describe cellular senescence as only harmful to biological systems. The aging pattern in development showcases how the senescence system operates differently as it maintains a naturally preserved function beyond its role in aging and disease management. Insights into developmental senescence offer valuable clues for regenerative medicine and tissue engineering [27, 48].

Paracrine and Bystander Senescence

Senescence traits originating from damaged cells propagate between healthy cells through signaling molecules and cellular contacts, resulting in two types of senescence: paracrine and bystander. When senescent cells activate SASP functions, they induce secretion of multiple inflammatory substances that drive cellular senescence in surrounding cells [49]. Persistent inflammatory conditions and tumors demonstrate the spread of senescence because they worsen cellular dysfunctions in both cases. NF-κB signaling, DDR signaling pathways, and ROS production use three cellular mechanisms to activate paracrine senescence. The spread of bystander senescence happens through gap junction-signaling pathways and extracellular vesicle transfer [43, 50]. The stress reaction within cells protects against excessive cell multiplication in damaged tissues. However, prolonged cellular spread leads to structural issues that impair tissue healing and worsen age-related diseases. Scientific research has identified ways to stop tissue transmission during cellular aging through targeted interventions of signaling substances, IL-6 and IL-8, and SASP while mitigating disease development in aging and disease-prone situations [51].

Biological Functions of Senescence

Cellular senescence creates oppositely beneficial and harmful results through its biological activation, each with different effects at various stages. The vital purpose of tumor suppression exists in senescence since it uses p53/p21 and p16INK4a/Rb pathways to establish permanent cell growth blockades for cells with DNA damage, telomere degradation, or oncologic stress. This protective mechanism stops the cellular spread of possible malignancies to prevent cancer formation inside the body [52, 53]. The process of embryonic senescence enables the creation of tissue patterns as it shapes embryonic body development without causing any detrimental effects to embryonic growth. The temporary appearance of senescent cells at developmental phases seems to initiate structural changes during tissue clearance procedures and organogenesis, together with cellular element restructuring leading to proper complex structural formation [54]. Adult healing requires the senescence-associated secretory phenotype (SASP) to pull immune cells, then drive vessel generation for rebuilding tissue structures and matrix restoration. Scientific evidence shows that diminished senescent cell clearance capacity in aged individuals produces uncontrolled SASP triggers leading to tissue inflammation, which damages normal homeostasis levels [55]. The ongoing activity of the SASP results in age-related tissue dysfunction that leads to regeneration problems and fibrosis development while weakening tissue resistance to degenerative diseases. The protective mechanisms of cellular senescence support tumor suppression, developmental processes, and tissue healing, but persistent cellular senescence presence causes aging along with related medical conditions [56, 57] (Fig. 2.2).

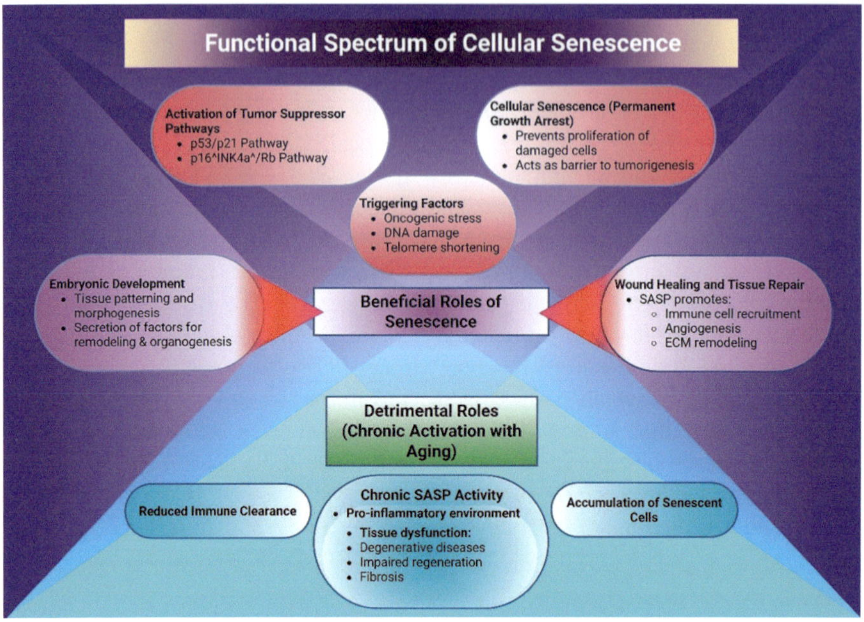

Fig. 2.2 Dual Roles of Cellular Senescence in Health and Aging. Cellular senescence is triggered by stressors like DNA damage or telomere shortening, activating p53/p21 and p16INK4a/Rb pathways to halt cell proliferation. It is beneficial in tumor suppression, embryonic development, and wound healing via SASP. However, with aging, senescent cells accumulate due to poor clearance, causing chronic inflammation, tissue dysfunction, and disease, highlighting its dual role in health and aging

Cellular Senescence in Aging and Disease

The aging process leads cells toward an undisputable cessation of their dividing capabilities, and this causes both tissue aging and organ damage. At the start, senescent cell benefits turn into an imbalanced condition that initiates multiple age-related diseases. The inflammatory processes in atherosclerosis match the progression of osteoarthritis, with senescent chondrocytes blocking cartilage-healing capabilities and producing inflammatory markers [58]. Neurodegenerative disease inflammation worsens because glial cells develop senescence after experiencing damage. The main functioning element of senescent cells provides inflammatory mediators through a phenomenon called SASP, which releases cytokines and chemokines with proteases to modify the surrounding tissue. The protective aspect of cellular senescence prevents early cancer development through cell growth restriction. Yet, the secreted SASP factors create an inflammatory condition that speeds up tumor advancement by stimulating invasion and angiogenesis while blocking immune defenses [59]. Human health and disease events display dual properties from senescence, ranging from protective to harmful effects. New therapeutic approaches must identify why senescent cells build up and how SASP functions to combat aging diseases, and develop precise methods to eliminate toxic but protectively essential cell groups [60].

Detection and Biomarkers of Senescent Cells

Researchers require methods for detecting senescent cell populations and their attributes to study their functions in tissue preservation and age-related disease formation. Senior cell detection heavily depends on the expression of unique molecular markers that scientists primarily use to detect these cells. SA-β-galactosidase is the leading marker for senescence because it shows lysosomal functionality in histochemical detection assays [61, 62]. The growth arrest in senescent cells functions through p16INK4a and p21CIP1/WAF1 inhibition, and the cells display DNA damage signals through γ-H2AX activity. Complex identification of senescent cells in living tissues faces difficulty because markers exhibit tissue-specific expression patterns that interact with changing marker activities and show a wide range of cellular traits [63]. The distinction between these cells becomes difficult because biomarkers detecting cellular aging exist in all three cellular states. New technological solutions attempt to enhance the current methods for detecting cellular senescence by resolving existing limitations. New research develops two assessment tools that unite protein analysis methods with transcriptomic data, metabolic measurements, and single-cell RNA sequencing to track senescent cellular groups [64, 65]. With fluorescence lifetime imaging and in vivo reporters as advanced imaging methods, the researchers study senescent cells' locations and time patterns. An improved detection mechanism achieved through these methods shows promise for clinical significance when studying senescence in laboratories [66].

Therapeutic Targeting of Senescent Cells

Senolytics

The therapeutic agents known as senolytics activate selective apoptosis mechanisms in age-compounded senescent cells, which causes chronic inflammation and organ dysfunction. Thus, therapeutic agents focus on survival pathway elements, BCL-2 and BCL-xL, and PI3K/AKT to attack senescent cells. Health durations extend and tissue healing improves due to senolytic therapy, which removes aged cells [67]. The prominent drug characterizing these agents is dasatinib, among the three compounds comprising navitoclax and quercetin. Earlier preclinical research demonstrated the effectiveness of these treatments by using models of osteoarthritis, along with models of pulmonary fibrosis and atherosclerosis, heading toward future use as age-related disease and aging improvement human therapies [68].

Senomorphics

Therapeutic compounds known as senomorphics function by stopping the apoptosis process of senescent cells. Therapeutic compounds called senomorphics help lower the senescence-associated secretory phenotype (SASP), which generates inflammatory machinery and remodeling tissue processes that cause age-related inflammation and cancer expansion while worsening tissues [69]. The injected drugs restrict pro-inflammatory factors, IL-6 and IL-8 production, and tissue-digesting MMPs, but keep protective mechanisms of aged cells intact. The combination of rapamycin, metformin, and JAK inhibitors demonstrates their ability to undergo senomorphic functions. Therapeutic compounds demonstrate promising safety features for medical use because they cause minimal harm to cells without interrupting normal tissue health [70].

Clinical Trials and Potential Interventions

Medical professionals now advance senotherapy studies from preclinical research into first-stage clinical trials because of their increasing interest in treating age-related diseases through senescent cell therapies. As part of their clinical evaluation, UBX1325 and UBX0101 drugs are being assessed to treat diabetic macular edema and osteoarthritis diseases [54]. Medical experts evaluate how the fisetin, in addition to dasatinib-quercetin medication, affects functional changes in elderly patients during ongoing research studies. Researchers investigate the age-slowing properties of metformin through the TAME project, but they are conducting those investigations exclusively. The scientific community shows evidence of clinical worth in life-span extension strategies for disease therapy and improved human lifespan duration [71].

Experimental Models to Study Cellular Senescence

Complete recognition of cellular senescence demands experimental models that perform physical characterization. Cell culture systems using fibroblasts and epithelial cells allow scientists to investigate senescence onset phenomena and measure the establishment of cell cycle arrest and secretory signals that become evident during cellular aging through SASP [72]. Scientists operating within these laboratory models can modify separate stress elements by changing oxidative damage, DNA damage and oncogene activation conditions. Physiological research into senescent cell effects on aging mechanisms and their impact on tissue repair functions and cancer development requires genetic modification of laboratory mice for live-animal study [73]. The research models allow scientists to analyze the interactions between senescent cells and their immune system components. Research scientists track senescent cells through p16INK4a and p21CIP1 promoters, which activate fluorescent or luminescent reporters in combination with SA-β-gal reporters in and outside the body [74]. The proper tools need to be used for tracing cellular ancestry, studying initial points of cellular senescence development, measuring therapy effectiveness, and measuring senescent cell response. An effective research system develops from experimental techniques, enabling better comprehension of senescence origins and advanced therapeutic options, including senolytics and senomorphics [75].

Emerging Concepts and Future Directions

The dynamic aspect of cellular senescence emerges from ideas that demonstrate diverse characteristics across different cell types and their related areas and triggering situations. The cellular context and microenvironmental signals shape various phenotypes constituting senescence instead of representing a single goal [76]. The senescent cell interaction with immune components affects inflammation, tissue remodeling, and immune surveillance through the key pathway known as senescence-associated secretory phenotype (SASP). Research indicates a better understanding of this two-way communication, which impacts how senescent cells survive and continue to exist. The new scientific evidence shows that senescent cells possess a reversible potential through cellular reprogramming techniques that develop restorative possibilities for ancient tissue repair [77]. The current understanding that senescence produces irreversible effects has encountered significant challenges through this discovery. System-level biological studies and combined omics procedures comprising transcriptomics, proteomics, metabolomics, and single-cell research techniques now help scientists detect fresh biomarkers along with regulatory inputs and therapeutic targets for senescence networks. These recent discoveries present fresh regulatory concepts for senescence in cancer cells, age-related conditions, and regenerative medicine while advancing senescence research knowledge [78, 79] (Table 2.1).

Table 2.1 Overview of cellular senescence: types, mechanisms, biomarkers, functions, and clinical implications

Type of senescence	Inducing mechanisms	Key biomarkers/ oathways	Biological roles	Clinical relevance	Reference
Replicative senescence	Telomere shortening from repeated cell division	Telomere dysfunction, DDR, p53/p21, p16INK4a/Rb, SA-β-Gal	Tumor suppression limits proliferative lifespan	Aging, idiopathic pulmonary fibrosis, hematologic disorders	[76, 80]
Stress-induced premature senescence (SIPS)	Oxidative stress, ionizing radiation, chemotherapeutic/ genotoxic agents	DDR activation, p53, p16INK4a, SA-β-Gal	Protection from damaged cells prevents malignant transformation	Neurodegeneration, cardiovascular, and metabolic diseases	[81, 82]
Oncogene-induced senescence (OIS)	Aberrant oncogene activation (e.g., RAS, BRAF) → replication stress	DDR, p53/p21, p16INK4a/Rb, SASP factors (IL-6, IL-8)	Prevents early tumorigenesis	Pre-malignant lesions (e.g., nevi), cancer therapy targets	[83]
Therapy-induced senescence (TIS)	DNA-damaging chemotherapy, radiation therapy	p53, p21, p16INK4a, persistent DDR, SASP	Arrests tumor cells, a side effect of treatment	Contributes to therapy resistance, relapse, and tissue dysfunction	[84]
Developmental senescence	Programmed senescence during embryogenesis	p21, TGF-β (not p16 or DDR), transient SA-β-Gal	Morphogenesis, tissue remodeling	Essential for embryonic development and patterning	[85]
Paracrine/bystander senescence	SASP factors, ROS, NF-κB, gap junctions, extracellular vesicles	Secondary DDR activation, cytokine signaling (IL-6, IL-8)	Amplifies senescence, maintains tissue integrity	Promotes chronic inflammation and age-related tissue damage	[86]
Common pathways across all types	Various stressors (intrinsic/ extrinsic) → p53/p21, p16INK4a/Rb, DDR	SA-β-Gal, γ-H2AX, SAHF, chromatin remodeling, SASP	Cell cycle arrest, immune surveillance, tissue repair	Dual role in cancer and aging-related diseases	[83, 87]

Conclusion

Due to its complicated nature, the biological process of cellular senescence appeared exclusively as a cell-terminating event in early research. This chapter presented fundamental concepts about senescence alongside three essential characteristics involving cell morphology changes, constant cell division pause, and bioactive substance release called SASP. Research investigated fundamental biological processes which included p53/p21 and p16INK4a/Rb pathways, DNA damage responses, telomere attrition, oxidative stress, and replicative stress-induced and oncogene-induced forms of senescence. Senescent cells demonstrated two-sided capabilities during developmental stages and tissue repair. Still, they accumulated over time to create issues leading to aging symptoms while triggering chronic inflammation in addition to promoting cancer advancement, according to the chapter. Identifying senescent cells remains complicated since multiple biomarkers alongside modern omics technologies are required to perform the detection. Current research demonstrates that senescence shows variable and reversible patterns and involves an advanced understanding of the connection between cells and their surrounding tissue environment. The correct utilization of senescent cells in therapy requires a complete knowledge of the regulatory systems at work. The scientific field currently explores fresh specific methods for combating age-related damage and extending wellness duration by repurposing detrimental senescent cells or altering their signaling capacity or senescent cell depletion using medicines. A deep comprehension of cellular senescence has therapeutic and biological significance in human health development.

References

1. Campisi J. Aging, cellular senescence, and cancer. Annu Rev Physiol. 2013;75:685–705.
2. Thapa R, Marianesan AB, Rekha A, Ganesan S, Kumari M, Bhat AA, et al. Hypoxia-inducible factor and cellular senescence in pulmonary aging and disease. Biogerontology. 2025;26(2):64.
3. Ou HL, Hoffmann R, González-López C, Doherty GJ, Korkola JE, Muñoz-Espín D. Cellular senescence in cancer: from mechanisms to detection. Mol Oncol. 2021;15(10):2634–71.
4. Pandey SN, Afzal M, Uikey J, Ganesan S, Mishra S, Bansal P, et al. ATM and p53 in aging and cancer: a double-edged sword in genomic integrity. Biogerontology. 2025;26(3):102.
5. Kaur P, Khan H, Grewal AK, Dua K, Singh SK, Gupta G, et al. Exploring therapeutic strategies: the relationship between metabolic disorders and FOXO signalling in Alzheimer's disease. CNS Neurol Disord Drug Targets. 2025;24(3):196–207.
6. Hayflick L. A brief history of the mortality and immortality of cultured cells. Keio J Med. 1998;47(3):174–82.
7. Gupta S, Afzal M, Agrawal N, Almalki WH, Rana M, Gangola S, et al. Harnessing the FOXO-SIRT1 axis: insights into cellular stress, metabolism, and aging. Biogerontology. 2025;26(2):65.
8. Goyal A, Afzal M, Khan NH, Goyal K, Srinivasamurthy SK, Gupta G, et al. Targeting p53-p21 signaling to enhance mesenchymal stem cell regenerative potential. Regen Ther. 2025;29:352–63.

9. Prasanna PG, Citrin DE, Hildesheim J, Ahmed MM, Venkatachalam S, Riscuta G, et al. Therapy-induced senescence: opportunities to improve anticancer therapy. J Natl Cancer Inst. 2021;113(10):1285–98.
10. Calcinotto A, Kohli J, Zagato E, Pellegrini L, Demaria M, Alimonti A. Cellular senescence: aging, cancer, and injury. Physiol Rev. 2019;99(2):1047–78.
11. Bashir B, Gulati M, Vishwas S, Gupta G, Dhanasekaran M, Paudel KR, et al. Bridging gap in the treatment of Alzheimer's disease via postbiotics: current practices and future prospects. Ageing Res Rev. 2025;105:102689.
12. Balaraman AK, Altamimi ASA, Babu MA, Goyal K, PadmaPriya G, Bansal P, et al. The interplay of senescence and MMPs in myocardial infarction: implications for cardiac aging and therapeutics. Biogerontology. 2025;26(1):46.
13. Liao Z, Yeo HL, Wong SW, Zhao Y. Cellular senescence: mechanisms and therapeutic potential. Biomedicines. 2021;9(12)
14. Calabrò A, Accardi G, Aiello A, Caruso C, Galimberti D, Candore G. Senotherapeutics to counteract senescent cells are prominent topics in the context of anti-ageing strategies. Int J Mol Sci. 2024;25(3)
15. McHugh D, Gil J. Senescence and aging: causes, consequences, and therapeutic avenues. J Cell Biol. 2018;217(1):65–77.
16. Babu MA, Jyothi SR, Kaur I, Kumar S, Sharma N, Kumar MR, et al. The role of GATA4 in mesenchymal stem cell senescence: a new frontier in regenerative medicine. Regen Ther. 2025;28:214–26.
17. Thapa R, Moglad E, Afzal M, Gupta G, Bhat AA, Hassan Almalki W, et al. The role of sirtuin 1 in ageing and neurodegenerative disease: a molecular perspective. Ageing Res Rev. 2024;102:102545.
18. Wang Y, Kuca K, You L, Nepovimova E, Heger Z, Valko M, et al. The role of cellular senescence in neurodegenerative diseases. Arch Toxicol. 2024;98(8):2393–408.
19. Reimann M, Lee S, Schmitt CA. Cellular senescence: neither irreversible nor reversible. J Exp Med. 2024;221(4)
20. Thapa R, Gupta S, Gupta G, Bhat AA, Smriti SM, et al. Epithelial-mesenchymal transition to mitigate age-related progression in lung cancer. Ageing Res Rev. 2024;102:102576.
21. Thapa R, Ahmad Bhat A, Shahwan M, Ali H, PadmaPriya G, Bansal P, et al. Proteostasis disruption and senescence in Alzheimer's disease pathways to neurodegeneration. Brain Res. 2024;1845:149202.
22. Wu X, Zhou X, Wang S, Mao G. DNA damage response (DDR): a link between cellular senescence and human cytomegalovirus. Virol J. 2023;20(1):250.
23. Goyal K, Afzal M, Altamimi ASA, Babu MA, Ballal S, Kaur I, et al. Chronic kidney disease and aging: dissecting the p53/p21 pathway as a therapeutic target. Biogerontology. 2024;26(1):32.
24. Podhorecka M, Skladanowski A, Bozko P. H2AX phosphorylation: its role in DNA damage response and cancer therapy. J Nucleic Acids. 2010;2010:920161.
25. Si Z, Sun L, Wang X. Evidence and perspectives of cell senescence in neurodegenerative diseases. Biomed Pharmacother. 2021;137:111327.
26. El Sherkawi T, Bani Saeid A, Yeung S, Chellappan DK, Mohamad S, Kokkinis S, et al. Therapeutic potential of 18-β-glycyrrhetinic acid-loaded poly (lactic-co-glycolic acid) nanoparticles on cigarette smoke-induced in-vitro model of COPD. Pathol Res Pract. 2024;263:155629.
27. Childs BG, Durik M, Baker DJ, van Deursen JM. Cellular senescence in aging and age-related disease: from mechanisms to therapy. Nat Med. 2015;21(12):1424–35.
28. Qin Y, Liu H, Wu H. Cellular senescence in health, disease, and lens aging. Pharmaceuticals (Basel, Switzerland). 2025;18(2)
29. Lee S, Lee JS. Cellular senescence: a promising strategy for cancer therapy. BMB Rep. 2019;52(1):35–41.
30. Bhat AA, Moglad E, Goyal A, Afzal M, Thapa R, Almalki WH, et al. Nrf2 pathways in neuroprotection: alleviating mitochondrial dysfunction and cognitive impairment in aging. Life Sci. 2024;357:123056.

31. Shreeya T, Ansari MS, Kumar P, Saifi M, Shati AA, Alfaifi MY, et al. Senescence: a DNA damage response and its role in aging and neurodegenerative diseases. Front Aging. 2023;4:1292053.
32. Mirzayans R, Andrais B, Hansen G, Murray D. Role of p16(INK4A) in replicative senescence and DNA damage-induced premature senescence in p53-deficient human cells. Biochem Res Int. 2012;2012:951574.
33. Bhat AA, Moglad E, Afzal M, Thapa R, Almalki WH, Kazmi I, et al. Therapeutic approaches targeting aging and cellular senescence in Huntington's disease. CNS Neurosci Ther. 2024;30(10):e70053.
34. Lee KS, Lin S, Copland DA, Dick AD, Liu J. Cellular senescence in the aging retina and developments of senotherapies for age-related macular degeneration. J Neuroinflammation. 2021;18(1):32.
35. Chen JH, Hales CN, Ozanne SE. DNA damage, cellular senescence and organismal ageing: causal or correlative? Nucleic Acids Res. 2007;35(22):7417–28.
36. Razgonova MP, Zakharenko AM, Golokhvast KS, Thanasoula M, Sarandi E, Nikolouzakis K, et al. Telomerase and telomeres in aging theory and chronographic aging theory (review). Mol Med Rep. 2020;22(3):1679–94.
37. Sorrenti V, Buriani A, Fortinguerra S, Davinelli S, Scapagnini G, Cassidy A, et al. Cell survival, death, and proliferation in senescent and cancer cells: the role of (poly)phenols. Adv Nutr. 2023;14(5):1111–30.
38. Toussaint O, Medrano EE, von Zglinicki T. Cellular and molecular mechanisms of stress-induced premature senescence (SIPS) of human diploid fibroblasts and melanocytes. Exp Gerontol. 2000;35(8):927–45.
39. Walen KH. Cell cycle stress in normal human cells: a route to "first cells" (with/without fitness gain) and cancer-like cell-shape changes. Semin Cancer Biol. 2022;81:73–82.
40. Zhu J, Wu C, Yang L. Cellular senescence in Alzheimer's disease: from physiology to pathology. Transl Neurodegener. 2024;13(1):55.
41. Gorgoulis VG, Halazonetis TD. Oncogene-induced senescence: the bright and dark side of the response. Curr Opin Cell Biol. 2010;22(6):816–27.
42. Engeland K. Cell cycle regulation: p53-p21-RB signaling. Cell Death Differ. 2022;29(5):946–60.
43. Dong Z, Luo Y, Yuan Z, Tian Y, Jin T, Xu F. Cellular senescence and SASP in tumor progression and therapeutic opportunities. Mol Cancer. 2024;23(1):181.
44. Yang Y, Fan L, Li M, Wang Z. Immune senescence: a key player in cancer biology. Semin Cancer Biol. 2025;108:71–82.
45. Saleh T, Bloukh S, Carpenter VJ, Alwohoush E, Bakeer J, Darwish S, et al. Therapy-induced senescence: an "old" friend becomes the enemy. Cancers. 2020;12(4)
46. Liu H, Zhao H, Sun Y. Tumor microenvironment and cellular senescence: understanding therapeutic resistance and harnessing strategies. Semin Cancer Biol. 2022;86:769–81.
47. Muñoz-Espín D, Cañamero M, Maraver A, Gómez-López G, Contreras J, Murillo-Cuesta S, et al. Programmed cell senescence during mammalian embryonic development. Cell. 2013;155(5):1104–18.
48. Trubitsyn AG. The mechanism of programmed aging: the way to create a real remedy for senescence. Curr Aging Sci. 2020;13(1):31–41.
49. Soto-Gamez A, Demaria M. Therapeutic interventions for aging: the case of cellular senescence. Drug Discov Today. 2017;22(5):786–95.
50. Baechle JJ, Chen N, Makhijani P, Winer S, Furman D, Winer DA. Chronic inflammation and the hallmarks of aging. Mol Metab. 2023;74:101755.
51. Guo J, Huang X, Dou L, Yan M, Shen T, Tang W, et al. Aging and aging-related diseases: from molecular mechanisms to interventions and treatments. Signal Transduct Target Ther. 2022;7(1):391.
52. Takahashi A, Ohtani N, Hara E. Irreversibility of cellular senescence: dual roles of p16INK4a/Rb-pathway in cell cycle control. Cell Div. 2007;2:10.

53. Yang J, Liu M, Hong D, Zeng M, Zhang X. The paradoxical role of cellular senescence in cancer. Front Cell Develo Biol. 2021;9:722205.
54. Khalil R, Diab-Assaf M, Lemaitre JM. Emerging therapeutic approaches to target the dark side of senescent cells: new hopes to treat aging as a disease and to delay age-related pathologies. Cells. 2023;12(6)
55. Coppé JP, Desprez PY, Krtolica A, Campisi J. The senescence-associated secretory phenotype: the dark side of tumor suppression. Annu Rev Pathol. 2010;5:99–118.
56. Tenchov R, Sasso JM, Wang X, Zhou QA. Aging hallmarks and progression and age-related diseases: a landscape view of research advancement. ACS Chem Neurosci. 2024;15(1):1–30.
57. Tchkonia T, Zhu Y, van Deursen J, Campisi J, Kirkland JL. Cellular senescence and the senescent secretory phenotype: therapeutic opportunities. J Clin Invest. 2013;123(3):966–72.
58. Li Y, Tian X, Luo J, Bao T, Wang S, Wu X. Molecular mechanisms of aging and anti-aging strategies. Cell Commun Signal. 2024;22(1):285.
59. Melo dos Santos LS, Trombetta-Lima M, Eggen BJL, Demaria M. Cellular senescence in brain aging and neurodegeneration. Ageing Res Rev. 2024;93:102141.
60. Mansfield L, Ramponi V, Gupta K, Stevenson T, Mathew AB, Barinda AJ, et al. Emerging insights in senescence: pathways from preclinical models to therapeutic innovations. npj Aging. 2024;10(1):53.
61. Faragher RGA. Simple detection methods for senescent cells: opportunities and challenges. Front Aging. 2021;2:686382.
62. González-Gualda E, Baker AG, Fruk L, Muñoz-Espín D. A guide to assessing cellular senescence in vitro and in vivo. FEBS J. 2021;288(1):56–80.
63. Coppé JP, Rodier F, Patil CK, Freund A, Desprez PY, Campisi J. Tumor suppressor and aging biomarker p16(INK4a) induces cellular senescence without the associated inflammatory secretory phenotype. J Biol Chem. 2011;286(42):36396–403.
64. Muthamil S, Kim HY, Jang HJ, Lyu JH, Shin UC, Go Y, et al. Biomarkers of cellular senescence and aging: current state-of-the-art, challenges and future perspectives. Adv Biol. 2024;8(9):e2400079.
65. Zhou D, Borsa M, Simon AK. Hallmarks and detection techniques of cellular senescence and cellular ageing in immune cells. Aging Cell. 2021;20(2):e13316.
66. Hassan M, Riley J, Chernomordik V, Smith P, Pursley R, Lee SB, et al. Fluorescence lifetime imaging system for in vivo studies. Mol Imaging. 2007;6(4):229–36.
67. Robbins PD, Jurk D, Khosla S, Kirkland JL, LeBrasseur NK, Miller JD, et al. Senolytic drugs: reducing senescent cell viability to extend health span. Annu Rev Pharmacol Toxicol. 2021;61:779–803.
68. Suda M, Paul KH, Tripathi U, Minamino T, Tchkonia T, Kirkland JL. Targeting cell senescence and senolytics: novel interventions for age-related endocrine dysfunction. Endocr Rev. 2024;45(5):655–75.
69. Prašnikar E, Borišek J, Perdih A. Senescent cells as promising targets to tackle age-related diseases. Ageing Res Rev. 2021;66:101251.
70. Moiseeva O, Deschênes-Simard X, St-Germain E, Igelmann S, Huot G, Cadar AE, et al. Metformin inhibits the senescence-associated secretory phenotype by interfering with IKK/NF-κB activation. Aging Cell. 2013;12(3):489–98.
71. Lee E, Carreras-Gallo N, Lopez L, Turner L, Lin A, Mendez TL, et al. Exploring the effects of Dasatinib, Quercetin, and Fisetin on DNA methylation clocks: a longitudinal study on senolytic interventions. Aging. 2024;16(4):3088–106.
72. Ogrodnik M, Carlos Acosta J, Adams PD, d'Adda di Fagagna F, Baker DJ, Bishop CL, et al. Guidelines for minimal information on cellular senescence experimentation in vivo. Cell. 2024;187(16):4150–75.
73. Di Carlo E, Sorrentino C. Oxidative stress and age-related tumors. Antioxidants (Basel, Switzerland). 2024;13(9)
74. Burd CE, Sorrentino JA, Clark KS, Darr DB, Krishnamurthy J, Deal AM, et al. Monitoring tumorigenesis and senescence in vivo with a p16(INK4a)-luciferase model. Cell. 2013;152(1–2):340–51.

75. Andrade AM, Sun M, Gasek NS, Hargis GR, Sharafieh R, Xu M. Role of senescent cells in cutaneous wound healing. Biology. 2022;11(12)
76. Roger L, Tomas F, Gire V. Mechanisms and regulation of cellular senescence. Int J Mol Sci. 2021;22(23)
77. Tripathi U, Misra A, Tchkonia T, Kirkland JL. Impact of senescent cell subtypes on tissue dysfunction and repair: importance and research questions. Mech Ageing Dev. 2021;198:111548.
78. Tiburcio R, Rappaport J, Palmer C. Chapter 10. Omics metabolism tools in antiaging drug discovery. In: Ryan TAJ, O'Neill LAJ, Zaslona Z, editors. Metabolites as signals in immunity and inflammation. Academic Press; 2025. p. 209–25.
79. Lorusso JS, Sviderskiy OA, Labunskyy VM. Emerging omics approaches in aging research. Antioxid Redox Signal. 2018;29(10):985–1002.
80. Deng Y, Chan SS, Chang S. Telomere dysfunction and tumour suppression: the senescence connection. Nat Rev Cancer. 2008;8(6):450–8.
81. Ngoi NY, Liew AQ, Chong SJF, Davids MS, Clement MV, Pervaiz S. The redox-senescence axis and its therapeutic targeting. Redox Biol. 2021;45:102032.
82. Mijit M, Caracciolo V, Melillo A, Amicarelli F, Giordano A. Role of p53 in the regulation of cellular senescence. Biomol Ther. 2020;10(3)
83. Kumari R, Jat P. Mechanisms of cellular senescence: cell cycle arrest and senescence associated secretory phenotype. Front Cell Dev Biol. 2021;9:645593.
84. Liu Y, Lomeli I, Kron SJ. Therapy-induced cellular senescence: potentiating tumor elimination or driving cancer resistance and recurrence? Cells. 2024;13(15)
85. Childs BG, Baker DJ, Kirkland JL, Campisi J, van Deursen JM. Senescence and apoptosis: dueling or complementary cell fates? EMBO Rep. 2014;15(11):1139–53.
86. Nelson G, Kucheryavenko O, Wordsworth J, von Zglinicki T. The senescent bystander effect is caused by ROS-activated NF-κB signalling. Mech Ageing Dev. 2018;170:30–6.
87. Kudlova N, De Sanctis JB, Hajduch M. Cellular senescence: molecular targets, biomarkers, and senolytic drugs. Int J Mol Sci. 2022;23(8)

Introduction to Senescence-Associated Secretory Phenotype (SASP) 3

Muhammad Afzal, Mustafa Zeyadi, Salman Hosawi,
Imran Kazmi, and Gaurav Gupta

Introduction to Cellular Senescence and SASP

Several stresses include DNA damage, shortening of the telomeres, and activation of oncogenes that lead to cellular senescence, an irreversible state of the cell cycle that stops [1]. This process helps maintain tissue homeostasis as a protective response to prevent abnormal/cancerous cell growth. However, when one gets older or responds to chronic stress, accumulating senescent cells impair tissue function and cause age-related diseases [2]. Since severely shortened telomeres trigger DDR tissue pathways that would end up in cell cycle arrest, shortened telomeres, as a side product of the frequent cellular division, are a major cause of senescence [3]. Moreover, accumulation of reactive oxygen species (ROS) may cause oxidative damage that promotes cell death and increases the pace of senescence. This process is the senescence, which is activated by oncogene activation, often associated with

M. Afzal (✉)
Department of Pharmaceutical Sciences, Pharmacy Program, Batterjee Medical College, Jeddah, Saudi Arabia
e-mail: mohmmad.afzal@bmc.edu.sa

M. Zeyadi · S. Hosawi · I. Kazmi
Department of Biochemistry, Faculty of Science, King Abdulaziz University, Jeddah, Saudi Arabia
e-mail: ikazmi@kau.edu.sa

G. Gupta
Centre for Research Impact & Outcome, Chitkara College of Pharmacy, Chitkara University, Rajpura, Punjab, India
e-mail: Gaurav.gupta@chitkara.edu.in

G. Gupta (ed.), *Cellular Senescence and Brain Aging*,
https://doi.org/10.1007/978-981-96-8873-9_3

changes in genes responsible for the tumor suppressor [4]. Nonetheless, in some cases, this process becomes maladaptive. Cellular senescence is characterized by the secretion of a complicated set of bioactive chemicals known as the SASP. From the initial identification as a senescence by-product, SASP has become an essential regulator of senescent cells' interaction with their environment [5, 6]. Molecules that are parts of growth factors, cytokines, chemokines, extracellular matrix (ECM) remodeling enzymes, and metabolites are all encompassed in the SASP, and they all affect tissue function for good or bad. SASP has tissue repair, immunity-gardening, and tumor suppression abilities in physiological conditions. Once chronically activated, it leads to fibrosis, chronic inflammation, and the growth of tumors. SASP's ambiguous nature reflects its complex involvement in the process of aging and diseases because chronic activation plays a role in the pathophysiology of diseases ranging from cancer to neurodegeneration [7].

In contrast, brief activation can induce tissue regeneration [8, 9]. The current chapter aims to test the molecular components and regulatory systems that can govern SASP regarding its impact on the cellular microenvironment and treatment. The chapter will discuss the role of SASP and aging, and how it leads to disease. It will also cover approaches to interventions for senescence, such as senolytic and senomorphic interventions, to mitigate the adverse effects of SASP and improve health outcomes in age-related conditions [10, 11].

Molecular and Functional Composition of SASP

SASP is an assortment of bioactive chemicals that cooperate to regulate autocrine and paracrine signaling within the microenvironment of senescent cells. Pro-inflammatory cytokines that are important for SASP are interleukin-6 (IL-6) and interleukin-1α/β (IL-1α/β), which enhance growth arrest and senescence-associated inflammation through the NF-κB and C/EBPβ signaling pathways [12]. Chemokine such as CCL2 and CXCL8/IL-8 enhances the immigration of the immune cells. It supports immunopathology in chronic settings as well as immunosurveillance [13]. Growth factors like hepatocyte growth factor (HGF) and vascular endothelial growth factor (VEGF) modify the balance of proliferation and stroma at the local level, causing remodeling of the tissues, angiogenesis, and possibly carcinogenesis. Components of the extracellular matrix (ECM) are digested by proteolytic enzymes like matrix metalloproteinases (MMPs) and serine protease inhibitors (SERPINs); both modify structural properties and facilitate cell migration [14]. These ECM remodeling proteins, which continue to influence matrix degradation and the structure of tissues, are complemented by fibronectin and tissue inhibitors of metalloproteinases (TIMPs). Reactive oxygen species (ROS) (bioactive lipids) and metabolites are other components of SASP that promote the spread of senescence signal and aggravate oxidative stress besides the spread of senescence proteins [15, 16].

Exosomes and extracellular vesicles also facilitate long-range intercellular communications and systemic inflammation by taking on the role of transporters of proteins, lipids, and nucleic acids produced from SASP [17]. As a whole, these combinations of SASPs lead to a range of detrimental outcomes, including fibrosis, neurodegeneration, cancer progression, and chronic inflammation, while supporting positive processes including tissue regeneration and wound healing upon transient senescence. Context-dependent functions and functional diversity exhibited by the components of SASP indicate the significance of their role as therapeutic targets and biomarkers for diseases connected with aging and cancers [18].

Regulation of SASP Expression: Central Signaling Pathways

Important signaling pathways, namely NF-κB, C/EBPβ, p38 MAPK, and mTOR, control the SASP. They preserve stress responses and facilitate the expression of SASP genes by synergy. The sustained DNA harm facilitates the induction of SASP by stabilizing GATA4 and inducing DDR [19]. Enhancer accessibility is affected by epigenetic mechanisms, including chromatin remodeling and histone modifications, while transcription regulations modulate SASP output. Importantly, SASP is highly variable depending on the context, representing transcriptional programs for various cell types, as well as the inputs of the microenvironment [20, 21]. Mitogen-activated protein kinases (MAPKs), including p38 MAPK and ERK, control SASP. They activate transcription factors such as NF-κB and C/EBPβ. Therefore, they change the SASP and promote the release of chemokines, pro-inflammatory cytokines, and matrix-remodeling enzymes that affect senescent cells' secretome [21, 22]. Ren et al. demonstrated that in HEK293 cells, Cd induced oxidative stress and inflammation of cells, which led to the inhibition of MMP, cell death, and activation of the MAPK and NF-κB pathways. Cadmium exposure also triggered the SASP production and DNA damage [23]. Limiting the signaling pathways, by NF-κB BAY11–7082 inhibitor and the antioxidant NAC, decreased these effects [24]. In response to stress, the crucial tumor suppressor gene TP53 is responsible for cell cycle arrest that regulates cell senescence. TP53 controls the widespread expression of secretory factors in the context of SASP, often suppressing pro-inflammatory factors. It inhibits chronic inflammation and damage to tissues that result from SASP and overcomes the tumor-suppressive forces of senescence [25]. Zhou et al. proved that Jianpi Qingre Tongluo prescription decreases the SASP in OA by modifying the STAG1/TP53/P21 signaling pathway. According to histology and diminished SASP fragments (IL-1β, IL-6, MMP 13, etc.), HQC therapy reduced the degrees of OA. HQC decelerated cartilage destruction and returned the balance to the extracellular matrix due to the attention to critical combining (STAG1, TP53, P21, and P16) [26]. Galectin-1 regulates Rho GTPase signaling, leading to changes in cell shape and cytoskeleton movements. Galectin-1-driven Rho activation in conditions of senescence-associated-secretory phenotype, or SASP, changes existing vesicle trafficking and secretion schemes, raising the quantity of pro-inflammatory SASP components discharged into the tissue microenvironment of senescent cells [27]. The study by Setayesh et al. revealed that Rho signaling, ECM remodeling, and SASP pathways were strongly activated in MASH-HCC and metabolic dysfunction-associated fatty liver disease (MAFLD) induced by Western diet (WD), particularly at the tumor areas and margins within the MASH-HCC mice model. In the n-Cadherin positive cells, the activation of Galectin-1 (Gal-1) helped induce epithelial-to-mesenchymal transition (EMT), whereas the modulation of expression of the Gal-1 showed its role in regulating Rho signaling and SASP in MASH-HCC [28]. One of the kinases switched on by stress that is required for regulating the phenomenon of SASP is p38 MAPK. It alters the synthesis of the pro-inflammatory cytokines and other SASP components by activating transcription factors such as NF-κB and C/EBPβ and is involved in the tissue-remodeling and inflammatory aspects of senescence [29]. According to Liu et al., puerarin successfully suppresses the aging pace in mice by decreasing age-related indicators (SA-β-gal, p53, p21, p16) and SASP markers. The process slows down the cellular senescence in the case of mammary epithelial cells by inhibiting the p38MAPK signaling pathway, repairing mitochondrial damage, and regulating the gut microbiome that leads to beneficial metabolites [30]. The cytokine signaling is powered by the intracellular tyrosine kinases, Janus kinases (JAKs), particularly through the JAK/STAT pathway. JAKs amplify SASP activity and are involved in chronic inflammation and dysfunction of tissues in the disease and old age by regulating the expression of significant SASP constituents in senescent cells, namely the pro-inflammatory cytokines such as IL-6 and IL-8 [31]. Chen et al. demonstrated that the SASP that results from the activation of the JAK/STAT signaling pathway is responsible for the overexpression of PD-L1 under the condition of the bleomycin (BON)-induced senescence in the human cancer cells. Measured using the senescence tests, SASP detection, and specific inhibitors, the study confirmed PD-L1 increase in NCI-H460 and MDA-MB-231 cells, bringing information for enhancement of bleomycin-mediated treatments [32].

SASP and the Cellular Microenvironment

The SASP via autocrine and paracrine signaling has a major effect on the cellular microenvironment, promoting the sustenance of senescence and propagating it to adjacent cells [33]. Securing the removal of senescent cells or, on the contrary, the chronic inflammation, SASP components capture immune effectors like T lymphocytes, NK cells, and macrophages. SASP also regulates tissue regeneration, fibrosis, and wound healing by regulating the dynamics of the stem cell niche, activation of fibroblasts, and epithelial integrity. SASP participates in the two-faced relationships with the tissue homeostasis and pathogenesis upon the aging and disease terrains as supported by the context-dependent SASP's tumor-suppressive and tumor-promoting activities [8, 34]. Wang et al. studied the kinetics, content, and diversity of the SASP, where they focused on the mechanisms that control its formation. Other than using SASP inhibitors as senomorphic treatment of cancer and age-related diseases, the study looked at the beneficial and adverse effects of SASP in various physiological and pathological settings and its effects on the health span and potential as biomarkers [35]. Involving the enzyme phosphoglucose dehydrogenase (PGD), the pentose phosphate pathway produces ribose-5-phosphate and NADPH. PGD is involved in SASP-related metabolic reprogramming of senescent cells, which results in inflammation and oxidative stress in an SASP, the two significant characteristics of SASP that impact the tumor microenvironment [36, 37].

Glutaryl-CoA dehydrogenase (GCDH) inhibits the growth of hepatocellular carcinoma (HCC) and forms an antitumor environment, as studied by Lao et al. The GCDH overexpression opposed the effect of GCDH depletion, which enhanced the growth and metastasis of HCC. GCDH-induced crotonylation of important metabolic enzymes, including PGD, TKT, and ALDOC, inhibited the pentose phosphate pathway and glycolysis, causing senescence and immune cell infiltration through the SASP by blocking the usability of the metabolic enzymes [38]. In the GCDH-low group, the anti-PD-1 treatment showed better response patterns [39]. These brain tumors are referred to as pediatric low-grade gliomas (pLGGs), and they are of a good prognosis and moderate development. The SASP of pLGGs can interfere with the microenvironment and tumor development. Some cytokines and growth factors, among the SASP constituents, can alter immune response, tumor cell survival, and recurrence [40, 41]. Sigaud et al. studied the MAPK-OIS-SASP-TME axis (MOST) in primary pediatric low-grade glioma (pLGG) and emphasized its significant role in tumor creation and response to therapy. In the given work, the MOST axis is presented as pivotal for the future treatment development and patient care; the necessity of sophisticated preclinical models for the study of new therapeutic targets is highlighted along with an all-embracing approach in pLGG research [40]. Preneoplastic cells are the aberrant cells that can potentially transform into cancer in their initial stages. Senescent cells in SASP secrete compounds that can enhance growth arrest in such cells and prevent their transformation into cancers [42]. However, with specificities of the environment, long SASP can paradoxically promote preneoplastic cells, inflammation, and tumors [43, 44]. In their study of the roles of the immune cells and senescent cancer in the cervical cancer tumor milieu, Yijiang He et al. underscored that the considered senescent cells and their SASP remodel the microenvironment, affecting antitumor immunity, tumor start, and growth. The study also shows the potential of senotherapeutics in suppressing the growth of tumors, promoting the antitumor immunity, and killing the senescent cells [45] (Fig. 3.1).

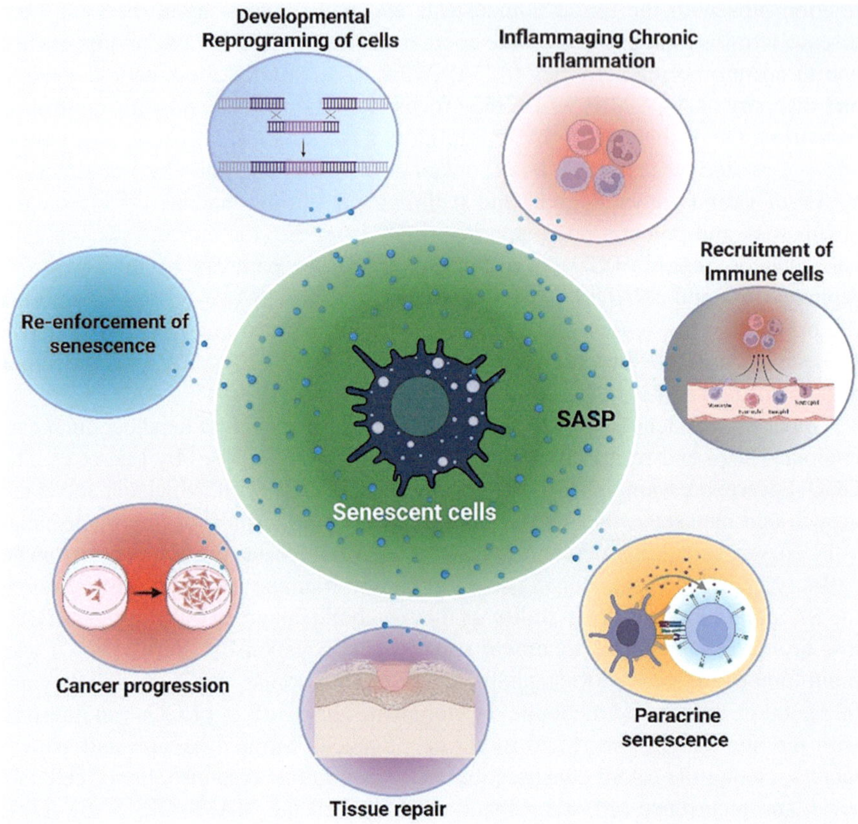

Fig. 3.1 Multifaceted roles of the senescence-associated secretory phenotype (SASP). The various impacts of the senescence-associated secretory phenotype (SASP), which is generated by senescent cells, are depicted in this figure. SASP factors play a role in immune cell recruitment, tissue healing, and developmental reprogramming. Nevertheless, they also encourage inflammaging, chronic inflammation, and the development of cancer, reinforce senescence, and cause paracrine senescence in nearby cells. These conflicting results highlight SASP's intricate function in physiology and diseases

SASP in Health and Disease

SASP is an important agent in aging and inflammation since it enhances low-grade chronic inflammation that accelerates age-related diseases and tissue malfunction. In pro-inflammatory conditions, SASP is a pro-tumor suppressor and surveillance of senescence biomarker in the context of cancer [46]. Yet, simultaneously, it activates the therapeutic circumvention and formation of tumors. With the help of immune cell recruitment, collagen synthesis stimulation, and tissue regeneration inhibition, SASP enhances pathological remodeling in all fibrotic diseases, like liver, lung, and kidney fibrosis, worsening fibrosis [47]. Similarly, in neurodegenerative diseases such as Parkinson's and Alzheimer's, SASP aggravates the disease symptoms, promotes neuroinflammation, and interferes with neuronal function. The long-term activation of SASP is triggered by the mechanisms of DDR, NF-κB, and p38 MAPK signaling pathways governing the release of cytokines, chemokines, and matrix remodeling enzymes [48]. Consequently, since with senolytics the senescent cells are getting eliminated and with the senomorphics changes being made to the harmful impact of the SASP factors without cell death, clinically, targeting SASP has become a feasible therapeutic option [49]. Based on recent studies, SASP-modulating drugs such as rapamycin and JAK inhibitors can mitigate neuroinflammation, enhance tumor response to treatments, and alleviate fibrosis. However, this issue remains challenging in developing a specific treatment that balances between the SASP's positive and negative impacts. Focusing on SASP has potential for new therapy for aging diseases and regenerative medicine as studies develop [11].

Therapeutic Targeting of SASP

Senolytics such as navitoclax and dasatinib, in combination with quercetin, for the specific killing of senescent cells and senomorphics such as rapamycin, metformin, and JAK inhibitors that prevent SASP without cell death are examples of therapeutics that interrogate S drug-induced inhibition of upstream regulators, including NF-κB and mTOR [50]. Immunotherapeutic procedures to enhance immune-mediated clearance of senescent cells are other strategies. While encouraging results, specific targeting, systemic toxicity, and inadvertent disruption of beneficial senescence responses remain serious challenges to clinical translation and need more study [51, 52]. One of the central DDR mediators, checkpoint kinase 2 (CHK2), regulates DNA repair and cell cycle arrest. Since CHK2 activation leads to an increase in the synthesis of the growth factors and pro-inflammatory cytokines, it affects the tumor microenvironment and the aging processes, further stimulating the expression of the SASP in senescence context [53]. Cellular senescence (CS), the same phenomenon as the first cancer prevention in Dong et al. study, before the occurrence of tumors, senescent cells can promote the growth of tumor by transforming the tumor microenvironment (TME) via senescence-associated secretory phenotypes (SASPs). Despite possible enhancements of immunosurveillance in some cases, the net effect of SASPs on cancer development inspires investigation of the treatment strategies that affect CS and SASPs, including senolytics and senomorphics [54]. H2 AX is a histone variant, which turns into γH2AX after phosphorylation at serine 139 when there is DNA damage. It is important for the reaction and repair of damage to DNA. During cellular senescence, the accumulation of γH2AX identifies DNA damage and leads to SASP, modulating inflammation and tissue remodeling [55]. The recent advancements in therapy methods against aging treatment that focus on the signaling pathways associated with DDR and SASP were reviewed by Nadeem et al. The study provides wide insight into the existing signaling networks, highlighting the known and potential new targets for antiaging therapies. Such results are useful for generating focused antiaging treatment plans [20]. The tumor microenvironment (TME) composition includes extracellular matrix, cells that surround it, and signaling chemicals that influence tumor development and progression [56]. By releasing pro-inflammatory cytokines, growth factors, and extracellular matrix components, an SASP helps create the TME. These substances may influence cancer development by triggering carcinogenesis, immunosuppression, and tissue remodeling [57]. Liu et al. studied the regulatory mechanisms of cellular senescence in the growth and treatment of tumours. The paper discusses new treatment prospects while new approaches to modulating senescence are addressed. A deeper understanding of such pathways is critical to enhance clinical outcomes and make precision medicines. These discoveries will help create cancer treatments that focus on the process of senescence [58]. The ABT-263 drug, which is commonly known as navitoclax, inhibits the Bcl-2 family proteins, namely Bcl-2, Bcl-xl, and Bcl-w, very strongly. Reducing the pro-inflammatory cytokines secreted by senescent cells induces the dying of these cells. It alters the SASP, possibly reducing the inflammation and disease created by SASP in the tissues [59]. Gold et al. found that acetaldehyde and alcohol use result in the aging of liver cells. Cells of the senescent hepatocytes and stellate cells were eliminated due to the senolytic drug, ABT263, increasing plasma levels of triglycerides and alanine aminotransferase. On the other hand, nicotinamide (NAM) blocked alcohol-induced damage to the liver because it suppressed senescence and SASP. RNA-sequencing indicated that NAM enhanced Sirt1 and dampened the indicators of DNA damage, which implies the safer therapeutic potential of NAM in ALD, while ABT263 downregulated adipogenesis genes and activated the complement system [60]. The chemotherapy drug doxorubicin induces senescence in cancer cells through DNA damage and the activation of SASP. Since it affects the tumor microenvironment and promotes inflammation, recruitment of immune cells, and remodeling of extracellular matrix, the SASP precipitated is involved in both tumor suppression and development [58]. Xia et al. hypothesized that the Bcl-2 inhibitor ABT-199 decreases Dox-induced cardiotoxicity by eliminating SASP senescent cardiomyocytes. In vitro and in vivo studies established that when Dox is administered, the cardiomyocytes go through an SASP, thus causing an upregulation in Bcl-2. ABT-199 eliminated partial cardiotoxicity by selectively eliminating SASP senescent cardiomyocytes. According to the results, targeting the SASP senescence is one of the possible approaches to treating Dox-related cardiac damage [61]. Stressors such as DNA damage or oncogene activation can cause cellular senescence, an irreversible cell cycle stop. The result of this condition is the SASP, which is brought on by the release of growth factors, cytokines, and enzymes that affect the tissue microenvironment and aid in tissue repair and disease development [62]. Liu et al. reviewed new advancements in the processes and SASP-focused treatment strategies in the milieu of intervertebral disc degeneration. Apart from proposing the therapeutic techniques, their results reveal information about the regulatory mechanisms for disc cell senescence. The knowledge of the connection between disc degeneration and cellular senescence may guide the development of effective medicines for the regulation of senescence for better results [63].

Emerging Directions and Future Perspectives

Current studies bring to attention the variability of the SASP, which depends on different cell types, tissue contexts, and stimuli. There is growing awareness that noncoding RNAs, rewiring of the metabolism, and signals propagated from the microbiome play significant roles as modulators of SASP composition and dynamics. In the aging and diseases continuum, some of the variables of SASP are promising as biomarkers for the diagnosis and prognosis of diseases. Besides, regenerative medicine can use the transitory activity of SASP to enhance the tissue repair process. But, the resolution of temporal dynamics, specific cell-based regulatory networks, and therapeutic windows is still a prime goal in creating methods to assist in targeting SASP [8, 13]. Oxidative stress damages the cells by upsetting the antioxidant defences and reactive oxygen species (ROS) balance. DNA is destroyed by oxidative stress, which triggers cellular stress responses, causing SASP or senescence-associated secretory phenotype.

Also, SASP influences aging and the course of the disease by promoting tissue remodeling and sustained inflammation [64]. Despite the SASP association with carcinogenesis, Li et al. reported that the oxidative stress-induced cellular senescence acts as a preventive barrier for cancer transformation. Genes or pathways that induce oxidative stress-induced senescence are viable targets for inhibiting cancer development. The need for additional studies on components of SASP and functions of senescent cells in different tumor stages is evidenced by favourable outcomes of new therapeutics that promote the oxidative stress-induced senescence [65]. Eczema is a chronic skin inflammation characterized by swelling, redness, and itching. The immune system's dysregulation leads to inflammation and skin cell damage. With the maintenance of fibrosis, recruitment of the immune cells, and sustained inflammation, the SASP would worsen failure in the skin barrier and hasten the emergence of eczema [66]. Nurkolis et al. explore the possibility of the skin and salivary microbiomes as bio-detectors of aging-associated illnesses. They highlight how aging, affected by the environments and lifestyles, makes people more likely to be affected by neurological, musculoskeletal, and cardiovascular diseases. The work sheds light on new diagnostic and treatment approaches, as the role of the inflammation that SASP facilitates, and the microbiome's contribution to the aging-related disorders are also presented [67]. Cells with genetic defects or mutations that increase the probability are called cancer-prone cells. These cells undergo a growth arrest in case of senescence due to oncogenic stress or DNA damage. These cells secrete a SASP, influencing the tumor microenvironment and causing fibrosis, inflammation, and perhaps tumorigenesis [68, 69]. Takasugi et al. investigated the role of cellular senescence and the SASP in the tumor microenvironment. They underlined that SASP can create an environment favourable for cancer, while senescence hinders tumorization via growth arrest. Apart from providing outlines on prospective therapeutic tactics, such as senotherapy, directed to senescent cells treatment in cancer, the paper discusses the mechanisms of SASP's context-dependent effects on carcinogenesis [70] (Fig. 3.2; Table 3.1).

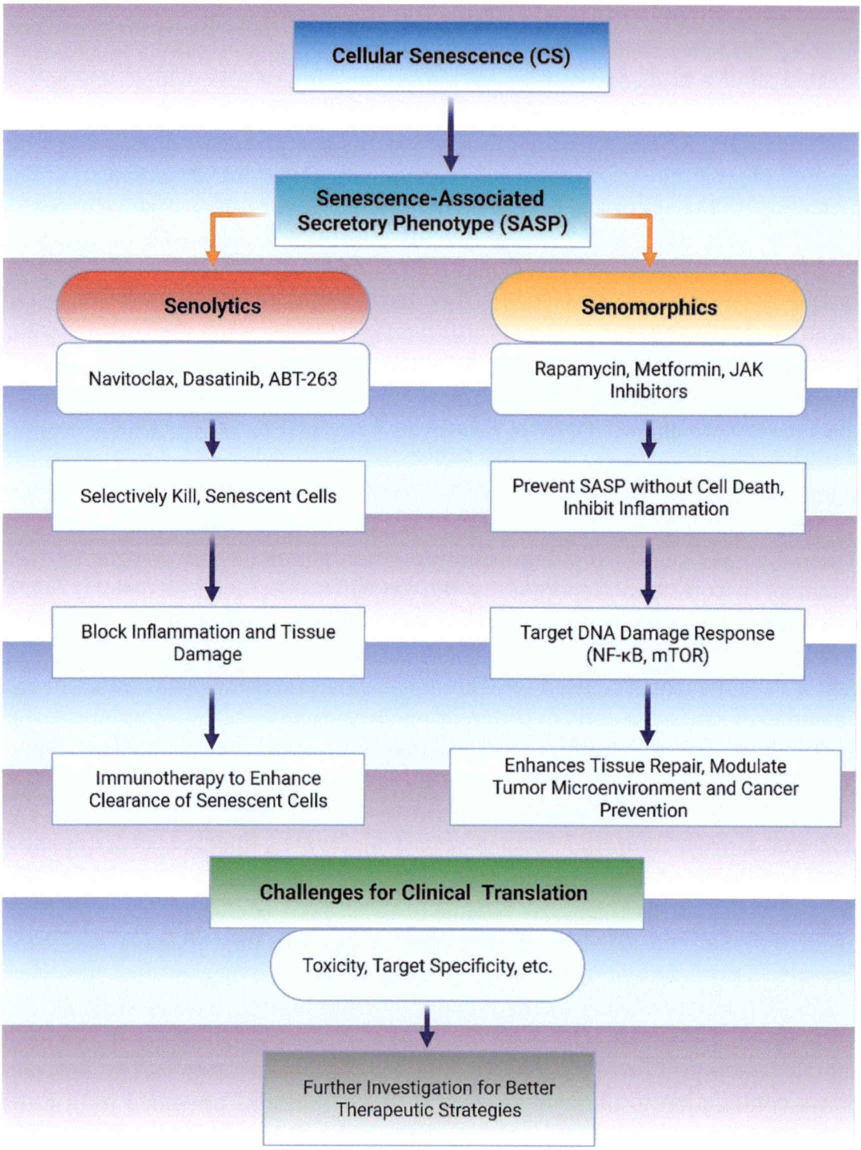

Fig. 3.2 Therapeutic approaches targeting cellular senescence and senescence-associated secretory phenotype (SASP). Therapeutic approaches that target cellular senescence and the senescence-associated secretory phenotype (SASP) are depicted in the figure. While senomorphics, such as metformin and rapamycin, inhibit SASP without causing cell death, senolytics, such as dasatinib and navitoclax, kill senescent cells. These treatments seek to improve tissue healing, alter the tumor microenvironment, and lessen inflammation. Other tactics include using immunotherapy to increase the removal of senescent cells and focusing on upstream regulators such as NF-κB and mTOR. Clinical translation is hampered by issues including toxicity and the loss of advantageous senescence effects, despite advancements

Table 3.1 Molecular landscape and therapeutic relevance of senescence-associated secretory phenotype (SASP) components

SASP component class	Representative molecules	Molecular pathways involved	Biological functions	Pathophysiological implications	Therapeutic implications	Reference
Pro-inflammatory cytokines	IL-6, IL-1α/β, TNF-α	NF-κB, C/EBPβ, p38 MAPK	Promote inflammation, growth arrest	Chronic inflammation, tissue damage, and tumor progression	Senomorphics to modulate SASP, NF-κB inhibitors	[71, 72]
Chemokines	CCL2, CXCL8/IL-8	NF-κB, JAK/STAT	Recruit immune cells (e.g., macrophages, NK cells)	Immunopathology, immuno-senescence	JAK inhibitors, immune modulation	[73, 74]
Growth factors	VEGF, HGF, TGF-β	ERK, MAPK, PI3K/AKT	Angiogenesis, tissue remodeling	Fibrosis, cancer progression	Targeted anti-VEGF therapy, TGF-β inhibitors	[75, 76]
ECM remodeling enzymes	MMPs, SERPINs, TIMPs	MAPK, Galectin-1/rho/rho GTPase	Matrix degradation, cell migration	Tissue stiffness, metastasis, and OA	MMP inhibitors, HQC therapy	[77]
Reactive oxygen species (ROS) and metabolites	ROS, PGD, GCDH-regulated metabolites	DDR, MAPK, PPP reprogramming	Oxidative damage, metabolic stress	Senescence propagation, cancer support	Antioxidants (e.g., NAC), GCDH activation	[78, 79]
Extracellular vesicles (EVs)	Exosomes containing SASP proteins, lipids, and RNAs	Galectin-1, rho signaling, NF-κB	Intercellular communication, systemic inflammation	Neurodegeneration, tumorigenesis	EV-targeted therapies, senescence-targeted delivery	[80, 81]
Epigenetic and transcriptional regulators	TP53, p16, p21, GATA4	TP53, STAG1, chromatin remodeling, DDR	Regulate SASP gene expression, senescence entry	Imbalanced SASP output, aging	Senolytics (e.g., ABT263), TP53 modulators	[82, 83]

Conclusion

SASP is another feature of cellular senescence; this is characterized by dynamic and controlled release of a wide variety of pro-inflammatory cytokines, chemokines, growth factors, proteases, bioactive lipids, and extracellular vesicles. SASP has a double role for tissue homeostasis and diseases, as an amplifier and modulator of intercellular communication. Hypothermia induces good effects in acute circumstances, including tumor suppression, immune-mediated elimination of senescent or damaged cells, and tissue restoration. Conversely, sustained SASP activation culminates in tissue architectural disruption, a pro-inflammatory environment, and pathophysiology of many age-related diseases, including cancer, fibrotic diseases, and neurodegenerative diseases. Complicated regulatory networks responsible for regulating SASP composition, its intensity, and temporal dynamics, including DDR signaling, transcription factors such as NF-κB and C/EBPβ, p38 MAPK and mTOR kinases, epigenetic roadblocks, among others, are responsible for the regulation of this bipolar behavior. SASP is an attractive subject for drug-based intervention due to its critical participation in age-related disease. On one hand, senomorphic medicines attempt to decrease bad SASP factors, but preserve cell viability.

On the other hand, senolytic agents aim at selectively destroying senescent cells. Despite the promising preclinical and preliminary clinical results, the major inhibitors are the insurmountable barriers with significant variability in their SASP profiles, lack of available biomarkers, cell-type selective toxicity, and ability to compromise the physiological function of senescence. To effectively exploit the therapeutic potential that SASP presents, we are going to have to continue to work on its context dependent regulation, the understanding of how SASP interacts with noncoding RNAs, metabolic signals, and the modulation of immunology and the development of low-invasive techniques that can reg In the end, the SASP precise targeting is a game-changing opportunity to diminish age-related diseases, assist tissue regeneration, and prolong the life expectancy, thus, making it an important element of the nearing translational strategies in geroscience and regenerative medicine.

References

1. Thapa R, Marianesan AB, Rekha A, Ganesan S, Kumari M, Bhat AA, et al. Hypoxia-inducible factor and cellular senescence in pulmonary aging and disease. Biogerontology. 2025;26(2):64.
2. Pandey SN, Afzal M, Uikey J, Ganesan S, Mishra S, Bansal P, et al. ATM and p53 in aging and cancer: a double-edged sword in genomic integrity. Biogerontology. 2025;26(3):102.
3. Ohtani N, Mann DJ, Hara E. Cellular senescence: its role in tumor suppression and aging. Cancer Sci. 2009;100(5):792–7.
4. Kaur P, Khan H, Grewal AK, Dua K, Singh SK, Gupta G, et al. Exploring therapeutic strategies: the relationship between metabolic disorders and FOXO Signalling in Alzheimer's disease. CNS Neurol Disord Drug Targets. 2025;24(3):196–207.
5. Ogrunc M, Di Micco R, Liontos M, Bombardelli L, Mione M, Fumagalli M, et al. Oncogene-induced reactive oxygen species fuel hyperproliferation and DNA damage response activation. Cell Death Differ. 2014;21(6):998–1012.

6. Di Carlo E, Sorrentino C. Oxidative stress and age-related tumors. Antioxidants (Basel, Switzerland). 2024;13(9)
7. Gupta S, Afzal M, Agrawal N, Almalki WH, Rana M, Gangola S, et al. Harnessing the FOXO-SIRT1 axis: insights into cellular stress, metabolism, and aging. Biogerontology. 2025;26(2):65.
8. Ohtani N. The roles and mechanisms of senescence-associated secretory phenotype (SASP): can it be controlled by senolysis? Inflamm Regen. 2022;42(1):11.
9. Yue Z, Nie L, Zhao P, Ji N, Liao G, Wang Q. Senescence-associated secretory phenotype and its impact on oral immune homeostasis. Front Immunol. 2022;13:1019313.
10. Khavinson V, Linkova N, Dyatlova A, Kantemirova R, Kozlov K. Senescence-associated secretory phenotype of cardiovascular system cells and inflammaging: perspectives of peptide regulation. Cells. 2022;12(1)
11. Alqahtani S, Alqahtani T, Venkatesan K, Sivadasan D, Ahmed R, Sirag N, et al. SASP modulation for cellular rejuvenation and tissue homeostasis: therapeutic strategies and molecular insights. Cells. 2025;14(8)
12. Bashir B, Gulati M, Vishwas S, Gupta G, Dhanasekaran M, Paudel KR, et al. Bridging gap in the treatment of Alzheimer's disease via postbiotics: current practices and future prospects. Ageing Res Rev. 2025;105:102689.
13. Coppé JP, Desprez PY, Krtolica A, Campisi J. The senescence-associated secretory phenotype: the dark side of tumor suppression. Annu Rev Pathol. 2010;5:99–118.
14. Balaraman AK, Altamimi ASA, Babu MA, Goyal K, PadmaPriya G, Bansal P, et al. The interplay of senescence and MMPs in myocardial infarction: implications for cardiac aging and therapeutics. Biogerontology. 2025;26(1):46.
15. Radisky ES. Extracellular proteolysis in cancer: proteases, substrates, and mechanisms in tumor progression and metastasis. J Biol Chem. 2024;300(6):107347.
16. Xin X, Yang S, Ingle G, Zlot C, Rangell L, Kowalski J, et al. Hepatocyte growth factor enhances vascular endothelial growth factor-induced angiogenesis in vitro and in vivo. Am J Pathol. 2001;158(3):1111–20.
17. Babu MA, Jyothi SR, Kaur I, Kumar S, Sharma N, Kumar MR, et al. The role of GATA4 in mesenchymal stem cell senescence: a new frontier in regenerative medicine. Regen Ther. 2025;28:214–26.
18. Di Bella MA. Overview and update on extracellular vesicles: considerations on exosomes and their application in modern medicine. Biology. 2022;11(6)
19. Thapa R, Moglad E, Afzal M, Gupta G, Bhat AA, Hassan Almalki W, et al. The role of sirtuin 1 in ageing and neurodegenerative disease: a molecular perspective. Ageing Res Rev. 2024;102:102545.
20. Nadeem J, Sultana R, Parveen A, Kim SY. Recent advances in anti-aging therapeutic strategies targeting DNA damage response and senescence-associated secretory phenotype-linked signaling cascade. Cell Biochem Funct. 2025;43(3):e70046.
21. Xu Y, Li N, Xiang R, Sun P. Emerging roles of the p38 MAPK and PI3K/AKT/mTOR pathways in oncogene-induced senescence. Trends Biochem Sci. 2014;39(6):268–76.
22. Wang T, Liu C, Hu X, Yang N, Qiu C. Senescent macrophages in cancer: roles in tumor progression and treatment opportunities. Cancer Biol Med. 2025; https://doi.org/10.20892/j.issn.2095-3941.2024.0589.
23. Thapa R, Gupta S, Gupta G, Bhat AA, Smriti, Singla M, et al. Epithelial-mesenchymal transition to mitigate age-related progression in lung cancer. Ageing Res Rev. 2024;102:102576.
24. Ren Z, Zheng Y, Liu J, Liu Z, Chen J, Liu H, et al. Cadmium induces the secretion of SASP factors regulated by MAPK and NF-κB signaling pathways in HEK293 cells: a possible mechanism of acute kidney damage induced by cadmium. Toxicology. 2025;515:154166.
25. Clark JS, Kayed R, Abate G, Uberti D, Kinnon P, Piccirella S. Post-translational modifications of the p53 protein and the impact in Alzheimer's disease: a review of the literature. Front Aging Neurosci. 2022;14:835288.

26. Zhou Q, Liu J, Qi Y, Hu Y, Li Y, Cong C, et al. Jianpi qingre tongluo prescription alleviates the senescence-associated secretory phenotype with osteoarthritis by regulating STAG1/TP53/P21 signaling pathway. J Ethnopharmacol. 2025;337:118953.
27. Camby I, Belot N, Lefranc F, Sadeghi N, de Launoit Y, Kaltner H, et al. Galectin-1 modulates human glioblastoma cell migration into the brain through modifications to the actin cytoskeleton and levels of expression of small GTPases. J Neuropathol Exp Neurol. 2002;61(7):585–96.
28. Setayesh T, Hu Y, Vaziri F, Wei D, Wan YY. The spatial impact of a Western diet in enriching Galectin-1-regulated Rho, ECM, and SASP signaling in a novel MASH-HCC mouse model. Biomark Res. 2024;12(1):122.
29. Freund A, Patil CK, Campisi J. p38MAPK is a novel DNA damage response-independent regulator of the senescence-associated secretory phenotype. EMBO J. 2011;30(8):1536–48.
30. Liu S, Hu H, Zhang M, Zhang Y, Geng R, Jin Y, et al. Puerarin delays mammary gland aging by regulating gut microbiota and inhibiting the p38MAPK signaling pathway. J Agric Food Chem. 2024;72(19):10879–96.
31. Sarapultsev A, Gusev E, Komelkova M, Utepova I, Luo S, Hu D. JAK-STAT signaling in inflammation and stress-related diseases: implications for therapeutic interventions. Mol Biomed. 2023;4(1):40.
32. Chen Y, Shen J, Zhao X, He Q, Zhang J. The up-regulation of PD-L1 during boningmycin-induced senescence in human cancer cells depends on the activation of the JAK/STAT signaling pathway mediated by SASP. Immunol Cell Biol. 2024;102(9):847–59.
33. Thapa R, Ahmad Bhat A, Shahwan M, Ali H, PadmaPriya G, Bansal P, et al. Proteostasis disruption and senescence in Alzheimer's disease pathways to neurodegeneration. Brain Res. 2024;1845:149202.
34. Venkataraman A, Kordic I, Li J, Zhang N, Bharadwaj NS, Fang Z, et al. Decoding senescence of aging single cells at the nexus of biomaterials, microfluidics, and spatial omics. npj Aging. 2024;10(1):57.
35. Wang B, Han J, Elisseeff JH, Demaria M. The senescence-associated secretory phenotype and its physiological and pathological implications. Nat Rev Mol Cell Biol. 2024;25(12):958–78.
36. Polat IH, Tarrado-Castellarnau M, Bharat R, Perarnau J, Benito A, Cortés R, et al. Oxidative pentose phosphate pathway enzyme 6-phosphogluconate dehydrogenase plays a key role in breast cancer metabolism. Biology. 2021;10(2)
37. Qiao J, Yu Z, Zhou H, Wang W, Wu H, Ye J. The pentose phosphate pathway: from mechanisms to implications for gastrointestinal cancers. Int J Mol Sci. 2025;26(2)
38. Jha SK, De Rubis G, Devkota SR, Zhang Y, Adhikari R, Jha LA, et al. Cellular senescence in lung cancer: molecular mechanisms and therapeutic interventions. Ageing Res Rev. 2024;97:102315.
39. Lao Y, Cui X, Xu Z, Yan H, Zhang Z, Zhang Z, et al. Glutaryl-CoA dehydrogenase suppresses tumor progression and shapes an anti-tumor microenvironment in hepatocellular carcinoma. J Hepatol. 2024;81(5):847–61.
40. Sigaud R, Brummer T, Kocher D, Milde T, Selt F. MOST wanted: navigating the MAPK-OIS-SASP-tumor microenvironment axis in primary pediatric low-grade glioma and preclinical models. Childs Nerv Syst. 2024;40(10):3209–21.
41. Yvone GM, Breunig JJ. Pediatric low-grade glioma models: advances and ongoing challenges. Front Oncol. 2023;13:1346949.
42. Hussain MS, Altamimi ASA, Afzal M, Almalki WH, Kazmi I, Alzarea SI, et al. Kaempferol: paving the path for advanced treatments in aging-related diseases. Exp Gerontol. 2024;188:112389.
43. Aswani BS, Sajeev A, Hegde M, Mishra A, Abbas M, Vayalpurayil T, et al. Exosomal dynamics: bridging the gap between cellular senescence and cancer therapy. Mech Ageing Dev. 2025;225:112045.
44. Zaczek A, Kalenik S, Rodacka A. Mechanisms of senescence in cancer: positive and negative aspects of cancer cells senescence. Cell Physiol Biochem. 2023;57(6):478–511.
45. He Y, Qiu Y, Yang X, Lu G, Zhao S-S. Remodeling of tumor microenvironment by cellular senescence and immunosenescence in cervical cancer. Semin Cancer Biol. 2025;108:17–32.

46. Goyal K, Afzal M, Altamimi ASA, Babu MA, Ballal S, Kaur I, et al. Chronic kidney disease and aging: dissecting the p53/p21 pathway as a therapeutic target. Biogerontology. 2024;26(1):32.
47. Wang B, Kohli J, Demaria M. Senescent cells in cancer therapy: friends or foes? Trends Cancer. 2020;6(10):838–57.
48. El Sherkawi T, Bani Saeid A, Yeung S, Chellappan DK, Mohamad S, Kokkinis S, et al. Therapeutic potential of 18-β-glycyrrhetinic acid-loaded poly (lactic-co-glycolic acid) nanoparticles on cigarette smoke-induced in-vitro model of COPD. Pathol Res Pract. 2024;263:155629.
49. Wang Y, Kuca K, You L, Nepovimova E, Heger Z, Valko M, et al. The role of cellular senescence in neurodegenerative diseases. Arch Toxicol. 2024;98(8):2393–408.
50. Bhat AA, Moglad E, Goyal A, Afzal M, Thapa R, Almalki WH, et al. Nrf2 pathways in neuroprotection: alleviating mitochondrial dysfunction and cognitive impairment in aging. Life Sci. 2024;357:123056.
51. Kirkland JL, Tchkonia T. Senolytic drugs: from discovery to translation. J Intern Med. 2020;288(5):518–36.
52. de Godoy MCX, Macedo JA, Gambero A. Researching new drug combinations with senolytic activity using senescent human lung fibroblasts MRC-5 cell line. Pharmaceuticals (Basel, Switzerland). 2024;17(1)
53. Huang RX, Zhou PK. DNA damage response signaling pathways and targets for radiotherapy sensitization in cancer. Signal Transduct Target Ther. 2020;5(1):60.
54. Dong Z, Luo Y, Yuan Z, Tian Y, Jin T, Xu F. Cellular senescence and SASP in tumor progression and therapeutic opportunities. Mol Cancer. 2024;23(1):181.
55. Rybak P, Hoang A, Bujnowicz L, Bernas T, Berniak K, Zarębski M, et al. Low level phosphorylation of histone H2AX on serine 139 (γH2AX) is not associated with DNA double-strand breaks. Oncotarget. 2016;7(31):49574–87.
56. Bhat AA, Moglad E, Afzal M, Thapa R, Almalki WH, Kazmi I, et al. Therapeutic approaches targeting aging and cellular senescence in Huntington's disease. CNS Neurosci Ther. 2024;30(10):e70053.
57. Brassart-Pasco S, Brézillon S, Brassart B, Ramont L, Oudart JB, Monboisse JC. Tumor microenvironment: extracellular matrix alterations influence tumor progression. Front Oncol. 2020;10:397.
58. Liu B, Peng Z, Zhang H, Zhang N, Liu Z, Xia Z, et al. Regulation of cellular senescence in tumor progression and therapeutic targeting: mechanisms and pathways. Mol Cancer. 2025;24(1):106.
59. Wilson WH, O'Connor OA, Czuczman MS, LaCasce AS, Gerecitano JF, Leonard JP, et al. Navitoclax, a targeted high-affinity inhibitor of BCL-2, in lymphoid malignancies: a phase 1 dose-escalation study of safety, pharmacokinetics, pharmacodynamics, and antitumour activity. Lancet Oncol. 2010;11(12):1149–59.
60. Gold NM, Ding Q, Yang Y, Pu S, Cao W, Ge X, et al. Therapeutic potential of nicotinamide and ABT263 in alcohol-associated liver disease through targeting cellular senescence. MedComm. 2025;6(2):e70086.
61. Xia W, Chen H, Yang H, Zhu L, Xie C, Hou M. Depletion of SASP senescent cardiomyocytes with senolytic drugs confers therapeutic effects in doxorubicin-related cardiotoxicity. FEBS J. 2024;291(18):4029–42.
62. Kumari R, Jat P. Mechanisms of cellular senescence: cell cycle arrest and senescence associated secretory phenotype. Front Cell Dev Biol. 2021;9:645593.
63. Liu Y, Dou Y, Sun X, Yang Q. Mechanisms and therapeutic strategies for senescence-associated secretory phenotype in the intervertebral disc degeneration microenvironment. J Orthop Translat. 2024;45:56–65.
64. Pizzino G, Irrera N, Cucinotta M, Pallio G, Mannino F, Arcoraci V, et al. Oxidative stress: harms and benefits for human health. Oxidative Med Cell Longev. 2017;2017:8416763.
65. Li D, Yu Q, Wu R, Tuo Z, Wang J, Ye L, et al. Interactions between oxidative stress and senescence in cancer: mechanisms, therapeutic implications, and future perspectives. Redox Biol. 2024;73:103208.

66. Cianciulli A, Calvello R, Porro C, Lofrumento DD, Panaro MA. Inflammatory skin diseases: focus on the role of suppressors of cytokine signaling (SOCS) proteins. Cells. 2024;13(6)
67. Nurkolis F, Utami TW, Alatas AI, Wicaksono D, Kurniawan R, Ratmandhika SR, et al. Can salivary and skin microbiome become a biodetector for aging-associated diseases? Current insights and future perspectives. Front Aging. 2024;5:1462569.
68. Schosserer M, Grillari J, Breitenbach M. The dual role of cellular senescence in developing tumors and their response to cancer therapy. Front Oncol. 2017;7:278.
69. He S, Sharpless NE. Senescence in health and disease. Cell. 2017;169(6):1000–11.
70. Takasugi M, Yoshida Y, Hara E, Ohtani N. The role of cellular senescence and SASP in tumour microenvironment. FEBS J. 2023;290(5):1348–61.
71. Huang W, Hickson LJ, Eirin A, Kirkland JL, Lerman LO. Cellular senescence: the good, the bad and the unknown. Nat Rev Nephrol. 2022;18(10):611–27.
72. Landskron G, De la Fuente M, Thuwajit P, Thuwajit C, Hermoso MA. Chronic inflammation and cytokines in the tumor microenvironment. J Immunol Res. 2014;2014:149185.
73. Zhao H, Wu L, Yan G, Chen Y, Zhou M, Wu Y, et al. Inflammation and tumor progression: signaling pathways and targeted intervention. Signal Transduct Target Ther. 2021;6(1):263.
74. Cambier S, Gouwy M, Proost P. The chemokines CXCL8 and CXCL12: molecular and functional properties, role in disease and efforts towards pharmacological intervention. Cell Mol Immunol. 2023;20(3):217–51.
75. Peng D, Fu M, Wang M, Wei Y, Wei X. Targeting TGF-β signal transduction for fibrosis and cancer therapy. Mol Cancer. 2022;21(1):104.
76. Sheikh KA, Amjad M, Irfan MT, Anjum S, Majeed T, Riaz MU, et al. Exploring TGF-β signaling in cancer progression: prospects and therapeutic strategies. Onco Targets Ther. 2025;18:233–62.
77. Yuan Z, Li Y, Zhang S, Wang X, Dou H, Yu X, et al. Extracellular matrix remodeling in tumor progression and immune escape: from mechanisms to treatments. Mol Cancer. 2023;22(1):48.
78. Forrester SJ, Kikuchi DS, Hernandes MS, Xu Q, Griendling KK. Reactive oxygen species in metabolic and inflammatory signaling. Circ Res. 2018;122(6):877–902.
79. Snezhkina AV, Kudryavtseva AV, Kardymon OL, Savvateeva MV, Melnikova NV, Krasnov GS, et al. ROS generation and antioxidant defense systems in normal and malignant cells. Oxidative Med Cell Longev. 2019;2019:6175804.
80. Wallis R, Mizen H, Bishop CL. The bright and dark side of extracellular vesicles in the senescence-associated secretory phenotype. Mech Ageing Dev. 2020;189:111263.
81. Liu YJ, Wang C. A review of the regulatory mechanisms of extracellular vesicles-mediated intercellular communication. Cell Commun Signal. 2023;21(1):77.
82. Mijit M, Caracciolo V, Melillo A, Amicarelli F, Giordano A. Role of p53 in the regulation of cellular senescence. Biomol Ther. 2020;10(3)
83. Kudlova N, De Sanctis JB, Hajduch M. Cellular senescence: molecular targets, biomarkers, and senolytic drugs. Int J Mol Sci. 2022;23(8)

Cellular Senescence in the Aging Brain

4

Riya Mishra

Introduction

Cellular senescence is a state of stable cell-cycle arrest that cells enter in response to various stressors: telomere shortening, DNA damage, oxidative stress, oncogenic signaling, mitochondrial dysfunction, etc. First described by Hayflick and Moorhead in 1961, senescence was initially viewed as a proliferative limit in cultured fibroblasts [1, 2]. Over the past two decades, however, it has become clear that senescence is an active, multifaceted program that impacts tissue homeostasis, regeneration, and aging. In peripheral tissues, the accumulation of senescent cells contributes causally to physiological decline and age-related pathologies; emerging evidence now implicates senescence as a central mechanism in brain aging and neurodegeneration [3, 4].

Senescent cells are not merely passive bystanders. They acquire a complex secretory profile known as the senescence-associated secretory phenotype (SASP), characterized by pro-inflammatory cytokines (e.g., IL-6, IL-1β), chemokines, growth factors, and proteases. In the brain, the SASP can disrupt the tightly regulated immune milieu, exacerbate neuroinflammation, impair synaptic function, and compromise blood–brain barrier integrity. While transient senescence may serve beneficial roles in wound healing, limiting aberrant cell proliferation or clearing damaged cells, chronic accumulation of senescent cells in neural and glial populations fosters a maladaptive environment that underlies cognitive decline and vulnerability to neurodegenerative diseases [5–7].

The concept of senescence in the central nervous system has evolved considerably. Historically, postmitotic neurons were deemed incapable of senescence due to their terminally differentiated state. Yet recent studies have revealed that neurons

R. Mishra (✉)
Department of Pharmacy, Vivekanand Global University, Jaipur, Rajasthan, India

G. Gupta (ed.), *Cellular Senescence and Brain Aging*,
https://doi.org/10.1007/978-981-96-8873-9_4

can exhibit a senescence-like phenotype, marked by DNA damage foci, cell-cycle reentry signals, mitochondrial dysfunction, and SASP factor expression [8, 9]. Glial cells—astrocytes, microglia, and oligodendrocyte precursor cells—have emerged as primary reservoirs of senescent cells in the aging brain. Astrocyte senescence impairs neurotrophic support and glutamate recycling; senescent microglia adopt a pro-inflammatory, phagocytosis-deficient state; and oligodendrocyte lineage senescence disrupts myelin maintenance, collectively undermining neuronal health and circuit integrity [10–12].

Understanding the drivers and consequences of cellular senescence in the brain holds translational promise. Preclinical studies using genetically engineered "suicide" models and small-molecule senolytics (agents that selectively eliminate senescent cells) have demonstrated improved cognitive performance, reduced neuroinflammation, and attenuated pathology in mouse models of Alzheimer's disease and tauopathy [13, 14]. Senomorphics compounds that modulate the SASP without killing cells also effectively restore a more youthful tissue environment. Yet significant challenges remain: identifying robust, cell type–specific senescence biomarkers; delivering senolytics across the blood–brain barrier with minimal off-target toxicity; and determining optimal treatment windows to prevent tipping the balance from beneficial, acute senescence to detrimental chronic accumulation [15–17].

This chapter will explore the molecular underpinnings of senescence, survey how different brain cell types adopt senescent phenotypes, and examine the tools currently used to detect cellular senescence in vivo and in vitro. We will discuss the functional ramifications of senescent cell buildup in neural tissue, focusing on neuroinflammatory cascades, synaptic dysfunction, and blood–brain barrier breakdown. Particular attention will be paid to the interplay between senescence and hallmark neurodegenerative diseases, Alzheimer's disease, Parkinson's disease, and amyotrophic lateral sclerosis, highlighting mechanistic links and potential intervention points. Finally, we will assess emerging therapeutic avenues targeting senescence, from senolytic and senomorphic drugs to gene-editing approaches and lifestyle interventions, and outline key challenges and priorities for future research. By integrating current knowledge across molecular, cellular, and translational domains, we aim to provide a comprehensive framework for understanding how cellular senescence shapes the aging brain and devising strategies to mitigate its deleterious effects.

Molecular and Cellular Mechanisms of Senescence

Cellular senescence is orchestrated by a network of interconnected stress-response pathways that converge on stable cell-cycle arrest, metabolic rewiring, and a distinctive secretory profile. Here, we review the key molecular triggers: DNA damage and telomere attrition, oxidative stress and mitochondrial dysfunction, epigenetic alterations, and the effector program known as the senescence-associated secretory phenotype (SASP) [18, 19].

DNA Damage Response and Telomere Attrition

Persistent DNA damage is the canonical trigger for cellular senescence, and telomere attrition represents a physiologically relevant source of such damage in aging cells. Each cell division erodes telomeric repeats due to the end-replication problem. Once telomeres reach a critical length or become dysfunctional, for example, through oxidative base modifications or replication stress, they are recognized as double-strand breaks by the DNA damage response (DDR) machinery [20, 21]. Central to this surveillance is the ATM (ataxia-telangiectasia mutated) kinase, which phosphorylates histone H2AX (forming γ-H2AX foci) around damaged sites and activates checkpoint kinases CHK2 and p53. Activated p53 transcriptionally upregulates the cyclin-dependent kinase inhibitor p21CIP1, enforcing a G_1/S arrest bolstered by the p16INK4a–RB axis as damage persists [22, 23].

Beyond replicative exhaustion, many genotoxic insults, including ionizing radiation, chemotherapeutic agents, and reactive oxygen species (ROS), can induce irreparable DNA lesions. Rodier et al. demonstrated that unrepaired double-strand breaks coalesce into persistent DDR foci that maintain p53 and NF-κB signaling, thus locking cells into a senescent state. Critically, these foci resist resolution even when upstream damage subsides, a phenomenon termed "irreparable damage domains." In neural cells, both dividing glia and postmitotic neurons accrue DNA damage with age; neurons display persistent γ-H2AX and 53BP1 foci at sites of oxidative base lesions, indicating that even noncycling cells can mount a senescence-like arrest in response to DNA damage [24, 25].

Telomere dysfunction also activates the shelterin complex's kinase TRF2-ATM axis, feeding into the broader DDR network. Studies in mice deficient for telomerase RNA component show accelerated neurodegeneration, impaired hippocampus neurogenesis, and cellular senescence features in both neural stem cells and astrocytes. Importantly, forced expression of telomerase reverse transcriptase (TERT) in aged rodents attenuates telomere attrition, reduces markers of DNA damage, and improves cognitive performance, underscoring the causal link between telomere integrity, DNA damage signaling, and brain aging [26, 27].

Oxidative Stress and Mitochondrial Dysfunction

Oxidative stress serves both as an upstream inducer of senescence and a downstream amplifier of the senescent state. Mitochondria, the cell's primary source of ROS through leakage of electrons at complexes I and III, become progressively dysfunctional with age. In senescent cells, mitochondrial DNA (mtDNA) damage, altered dynamics (fission/fusion imbalance), and impaired mitophagy result in damaged mitochondria accumulating excess ROS [28, 29]. Passos et al. showed that this creates a positive feedback loop in fibroblasts. ROS drive further DNA damage and DDR activation, promoting mitochondrial biogenesis via p53 and AMPK signaling, thereby exacerbating oxidative stress [30, 31].

In the central nervous system, neurons are especially vulnerable to mitochondrial dysfunction due to their high metabolic demands and postmitotic nature. Age-related decline in mitochondrial respiratory chain activity has been documented in hippocampal and cortical neurons, where increased superoxide production correlates with senescence markers, including SA-β-gal positivity and elevated p16INK4a expression. Moreover, glial cells display similar mitochondrial derangements: astrocytes exhibit reduced mitochondrial membrane potential and heightened ROS levels, impairing their capacity for glutamate clearance and neurotrophic support; microglia show mitochondrial fragmentation that links to a pro-inflammatory activation state, further reinforcing tissue inflammation [32, 33].

Key regulators of this axis include p66Shc, an adaptor protein whose phosphorylation promotes mitochondrial ROS generation and senescence induction. Knockout of p66Shc in mice reduces oxidative damage, extends lifespan, and alleviates age-associated cognitive decline, demonstrating the centrality of mitochondrial ROS in brain aging [34]. Concurrently, energy-sensing pathways AMPK and SIRT1 modulate mitochondrial quality control. SIRT1 deacetylates PGC-1α, the master regulator of mitochondrial biogenesis, and promotes the expression of antioxidant enzymes like SOD2; however, SIRT1 activity diminishes with age, tilting the balance toward ROS accumulation [35].

Epigenetic Alterations and Chromatin Remodeling

Senescence is cemented by sweeping changes to the epigenetic landscape that reinforce cell-cycle arrest and reprogram gene expression. A hallmark of senescent cells is the formation of senescence-associated heterochromatic foci (SAHF) distinct nuclear domains of compacted chromatin enriched for repressive marks such as H3K9me3 and macroH2A [36]. Narita et al. first described SAHF in oncogene-induced senescence models, showing that these foci silence proliferation-promoting genes (E2F targets) by sequestering them into transcriptionally inert regions. Concurrently, genome-wide DNA methylation undergoes bidirectional remodeling: global hypomethylation at intergenic and repetitive elements, which may derepress transposable elements, and focal hypermethylation at CpG islands, potentially locking down tumor suppressor and cell-cycle genes [37, 38].

Histone modifications also shift: loss of histone H3 and H4 acetylation at proliferation gene promoters reinforces silencing, while increased H4K20 methylation stabilizes the condensed chromatin state. The histone chaperone HIRA and the histone variant H3.3 are recruited to genomic sites undergoing SAHF assembly, suggesting an active remodeling process. Beyond heterochromatin, "DNA damage response–induced chromatin alterations" (DISCAs) involve the ubiquitin ligase RNF168 and the polycomb repressive complex, linking DDR factors directly to chromatin architecture changes [39, 40].

In neural cells, epigenetic drift with age mirrors these senescence-associated signatures. Aged neurons exhibit increased DNA methylation entropy and loss of lamin B1, a nuclear lamina component whose downregulation is a senescence

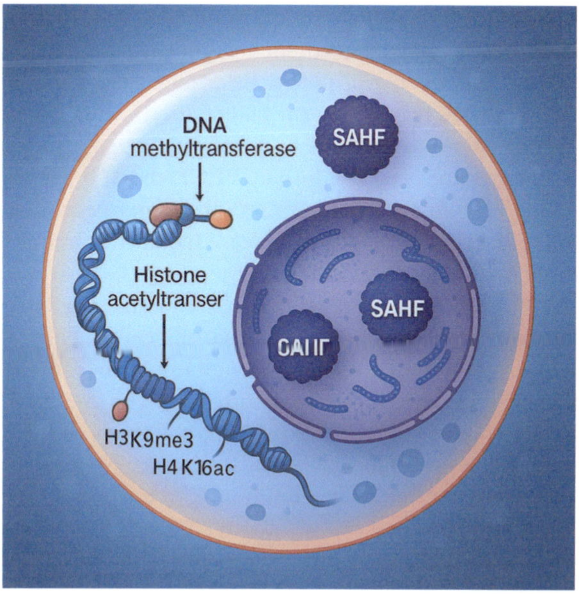

Fig. 4.1 Epigenetic remodeling and SAHF formation in senescent cells. This illustration depicts how senescent cells reorganize their chromatin into senescence-associated heterochromatic foci (SAHF) within the nucleus, creating compact, transcriptionally silent domains. DNA methyltransferase and histone acetyltransferase enzymes are shown depositing the repressive H3K9me3 mark and the activating H4K16ac mark to reinforce the stable epigenetic state characteristic of senescence

marker. Astrocytes isolated from older individuals show elevated SAHF markers and chromatin changes that correlate with impaired support functions. Significantly, pharmacological modulation of epigenetic regulators, such as inhibition of the histone methyltransferase SUV39H1 or activation of SIRT1, can attenuate senescence-associated chromatin changes, reduce SASP expression, and improve cellular phenotypes in vitro [41, 42] (Fig. 4.1).

Senescence-Associated Secretory Phenotype (SASP)

The senescence-associated secretory phenotype (SASP) is a multifaceted secretome comprising cytokines (IL-6, IL-8), chemokines (MCP-1/CCL2), growth factors (VEGF, TGF-β), proteases (MMP-1, MMP-3), and extracellular matrix components. Coppè et al. characterized the SASP in human fibroblasts, demonstrating that NF-κB and C/EBPβ transcription factors orchestrate its expression downstream of DDR kinases. Activation of innate immune sensors such as the cytosolic DNA sensor cGAS and its adaptor STING further amplifies SASP via IRF3 and NF-κB, linking nuclear DNA damage to inflammatory cytokine production [43, 44].

SASP factors have profound paracrine and systemic effects in the brain's specialized milieu. As shown in hippocampal slice studies, IL-6 and IL-1β secreted by senescent astrocytes and microglia potentiate a chronic inflammatory state that impairs synaptic plasticity and long-term potentiation. MMPs degrade basement membrane components of the blood–brain barrier, increasing permeability and allowing peripheral immune cells to infiltrate. VEGF and TGF-β dysregulate angiogenesis and extracellular matrix composition, disrupting neurovascular coupling. Senescent oligodendrocyte precursor cells release factors that inhibit remyelination, exacerbating white matter degeneration [45, 46].

Moreover, SASP perpetuates senescence via "bystander senescence," wherein secreted cytokines and ROS induce DNA damage and cell-cycle arrest in neighboring cells, expanding the senescent niche. Longitudinal rodent studies show that accumulation of SASP in the hippocampus correlates with age-associated memory deficits and vulnerability to Alzheimer's pathology. Genetic ablation of NF-κB in astrocytes or pharmacological blockade of IL-6 signaling mitigates SASP's detrimental effects, improving cognitive outcomes in aged mice [47, 48].

Senescence in Different Brain Cell Types

Cellular senescence within the aging brain does not affect all populations equally. Neurons, astrocytes, microglia, and oligodendrocyte lineage cells each adopt unique senescence-associated features that impair their specialized functions and contribute to tissue-level decline. Below, we examine how these four cell types display senescence-like phenotypes, the molecular signatures they share and diverge upon, and the consequences for neural circuitry and homeostasis [49, 50].

Neurons

Although mature neurons no longer divide, they can mount a senescence-like response when confronted with chronic stress. DNA damage foci marked by persistent γ-H2AX and 53BP1 accumulations appear in aged neurons, reflecting unresolved oxidative lesions and replication-independent damage. Mitochondrial dysfunction in these cells leads to energy deficits and excess reactive oxygen species, compromising DNA integrity. In parallel, neurons up-regulate classical senescence markers such as p21CIP1 and p16INK4a, even in the absence of cell-cycle reentry, suggesting that these cyclin-dependent kinase inhibitors also function to enforce a durable stress response in postmitotic contexts [51, 52].

Functionally, senescent neurons secrete pro-inflammatory cytokines and chemokine elements of a neuronal SASP that disrupt synaptic transmission and plasticity. Studies in aged hippocampal cultures show that this secretome diminishes long-term potentiation and impairs dendritic spine maintenance, undermining learning and memory. Moreover, distressed neurons release damage-associated molecular patterns (DAMPs) that activate neighboring glia, propagating inflammation across the neural network. Intriguingly, interventions that bolster mitochondrial quality control, such as pharmacological activation of PGC-1α or restoration of mitophagy, can reduce neuronal SASP signals and partially rescue synaptic function in aged rodent models [53, 54].

Astrocytes

Astrocytes are vital in neurotransmitter clearance, metabolic support, and blood–brain barrier maintenance. A subset of these cells enters senescence with age, undermining their homeostatic functions. Senescent astrocytes display enlarged, flattened morphologies, an up-regulation of SA-β-gal activity, and increased expression of p16INK4a. Epigenetic profiling reveals heightened DNA methylation at glutamate transporters and neurotrophic factor production genes, diminishing the cells' capacity to regulate excitatory signaling and support neuronal health [55, 56].

Crucially, astrocytic SASP factors including IL-6, LCN2, and CCL2, promote a pro-inflammatory milieu that impairs synaptic pruning and potentiates microglial activation. In coculture assays, conditioned medium from aged astrocytes reduces neuronal survival and interferes with the maturation of oligodendrocyte precursor cells, highlighting astrocyte senescence as a nexus for multicellular dysfunction. Restoration of SIRT1 activity in these cells via NAD^+ precursors or small-molecule activators attenuates SASP output. It improves glutamate uptake, hinting at epigenetic and metabolic levers to reverse astrocyte-senescence phenotypes [57, 58].

Microglia

As the brain's resident immune sentinels, microglia respond rapidly to injury and infection. In aging, however, microglial cells can transition into a dysfunctional, senescent-like state. Key features include reduced phagocytic capacity, fragmented mitochondrial networks, and sustained NF-κB activation driving chronic cytokine release. Senescent microglia often exhibit increased autofluorescence from lipofuscin accumulation, a hallmark of impaired lysosomal clearance. This altered phenotype dampens the clearance of amyloid-β and other protein aggregates, exacerbating proteotoxic stress in neurodegenerative disease contexts [59]. Furthermore, senescent microglia secrete complement proteins (e.g., C1q) and pro-inflammatory mediators that tag synapses for elimination, contributing to synaptic loss and network disintegration. Interventions that enhance autophagy or inhibit the cGAS–STING pathway in aged microglia can restore phagocytic function and reduce inflammatory signaling in mouse models, proving that microglial senescence is malleable [60].

Oligodendrocyte Lineage Cells

Oligodendrocyte precursor cells (OPCs) and mature oligodendrocytes ensure proper myelination of axons, which is critical for rapid electrical conduction. With aging, OPCs accumulate DNA damage and adopt senescence markers, including p21CIP1 expression and SA-β-gal positivity, which impairs their proliferative and differentiation potential. Epigenetic changes such as increased H3K27me3 on myelination gene promoters further inhibit remyelination capacity [61].

Mature oligodendrocytes also display a stress-associated secretome that includes MMPs and inflammatory cytokines, disturbing extracellular matrix composition and axon–glial interactions. In demyelinating models, clearance of senescent OPCs via senolytic compounds rejuvenates the progenitor pool and enhances remyelination, illustrating the direct link between oligodendrocyte senescence and white-matter integrity [62] (Table 4.1).

Table 4.1 Summary of senescence in brain cell types

Cell type	Key senescence markers	Functional consequences	Intervention insights
Neurons	γ-H2AX foci, p21CIP1, p16INK4a	Impaired synaptic plasticity, secretion of neuronal SASP, DAMP release	Enhance mitophagy (PGC-1α activators), ROS scavengers
Astrocytes	SA-β-gal, p16INK4a, DNA methylation changes	Reduced glutamate uptake, loss of neurotrophic support, pro-inflammatory SASP	SIRT1 activation (NAD^+ precursors), epigenetic modulators
Microglia	Lipofuscin accumulation, NF-κB activation, mitochondrial fragmentation	↓ Phagocytosis of debris/aggregates, chronic cytokine/complement release	Autophagy enhancers, cGAS–STING inhibitors
Oligodendrocyte Lineage	SA-β-gal, p21CIP1, H3K27me3 enrichment	Impaired OPC proliferation/differentiation, myelin disruption, MMP secretion	Senolytics to clear OPCs, remyelination promoters

Detection and Biomarkers of Brain Senescence

Accurate identification of senescent cells in the brain hinges on combining multiple complementary markers and techniques. No single assay suffices, so researchers typically employ a battery of molecular, enzymatic, transcriptomic, and imaging-based readouts to confirm the presence of senescence in neural tissues. The following subsections discuss classical histochemical markers, emerging "omics" signatures, and in vivo imaging strategies that provide a robust toolkit for detecting brain senescence [63, 64].

Classic Markers (p16INK4a, p21CIP1, SA-β-Gal)

The foundational assays for cellular senescence rely on the expression of cell-cycle inhibitors and a histochemical stain for lysosomal β-galactosidase activity at pH 6.0. Upregulation of p16INK4a and p21CIP1 arrests cells in G_1 phase: antibodies against these proteins, applied to fixed brain sections or dissociated cells, reveal their localization and abundance. In aged rodent hippocampi, immunohistochemistry shows that p16INK4a levels rise sharply in astrocytes and microglia, whereas p21CIP1 accumulates even in postmitotic neurons that harbour DNA damage foci [65, 66].

Senescence-associated β-galactosidase (SA-β-gal) staining, though developed initially in fibroblasts, adapts to brain slices and primary glial cultures with caveats. Due to the brain's intrinsically high lysosomal content, careful pH adjustment and parallel negative controls are essential to avoid false positives. When properly calibrated, SA-β-gal reveals enlarged, flattened glial cells and occasional neurons with intense blue precipitate in aged samples. Coupling SA-β-gal histochemistry with immunostaining for cell-type markers (e.g., Iba1 for microglia, GFAP for astrocytes) clarifies which populations exhibit lysosomal hyperactivity [67, 68].

Despite their widespread use, these classic markers have limitations. p16INK4a and p21CIP1 can be induced by stress without full senescence, and SA-β-gal activity sometimes overlaps with autophagy or lysosomal storage disorders. Therefore, researchers often corroborate these findings with additional readouts, such as accumulation of DNA damage foci or loss of nuclear lamin B1, to strengthen the senescence diagnosis in brain cells [69, 70].

Transcriptomic and Proteomic Signatures

High-throughput analyses have expanded the senescence toolbox by revealing gene and protein expression patterns that distinguish senescent from healthy cells. Bulk and single-cell RNA sequencing of aging brain tissue uncover upregulation of CDKN2A (encoding p16INK4a), inflammatory cytokines (IL-6, CCL2), and matrix metalloproteinases (MMP-3, MMP-9). Single-cell datasets further resolve which cell types of astrocytes, microglia, neurons, or oligodendrocyte precursors exhibit these transcriptomic shifts. In one study, the aged mouse cortex showed a unique "senescence signature" in astrocyte clusters, marked by elevated SASP transcripts and decreased metabolic gene expression [71, 72].

Proteomic profiling of conditioned media from brain cell cultures identifies secreted senescence biomarkers, such as lipocalin-2 (LCN2) from astrocytes and osteopontin (SPP1) from microglia. Mass spectrometry of cerebrospinal fluid (CSF) from aged rodents and human donors has revealed increased extracellular HMGB1 and fibronectin fragments factors linked to chronic inflammation and extracellular matrix remodeling. These soluble markers hold the potential for minimally invasive detection of brain senescence in clinical samples [73, 74].

Imaging and In Vivo Detection Strategies

Visualizing senescent cells within living brains is challenging but increasingly feasible thanks to innovative probes and reporter models. Transgenic mice expressing a fluorescent or bioluminescent reporter under the control of the CDKN2A promoter allow real-time monitoring of p16INK4a activation in neural tissues. In one example, aged p16LUC mice emit luminescence in the hippocampus that correlates with cognitive deficits and histological senescence markers [75, 76].

Small-molecule probes have also been developed. Senescence-associated β-galactosidase activatable fluorophores accumulate in cells with high lysosomal β-galactosidase activity; when injected intravenously, these dyes cross the blood–brain barrier and highlight senescent glia and neurons via two-photon microscopy. Other approaches employ radiolabeled tracers targeting SASP components such as IL-6 or MMPs that can be imaged by positron emission tomography (PET) to map neuroinflammation hotspots associated with senescence [77, 78].

Finally, advanced MRI techniques capture tissue properties altered by senescence. Diffusion tensor imaging detects microstructural changes in white matter tracts affected by senescent oligodendrocytes, while iron-sensitive sequences reveal microglial iron accumulation linked to lipofuscin and oxidative stress. Although still in preclinical stages, these modalities promise non-invasive assessments of cellular senescence landscapes in the aging brain, opening avenues for monitoring disease progression and therapeutic response [79, 80] (Fig. 4.2).

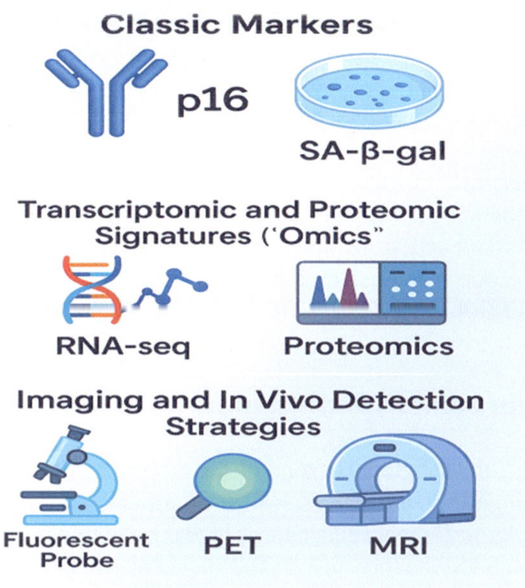

Fig. 4.2 Integrated approaches for detecting brain senescence. This schematic combines three complementary detection strategies: histochemical markers, high-throughput omics, and in vivo imaging, providing a unified framework for identifying senescent cells in neural tissue. Classic assays like p16INK4a immunostaining and SA-β-gal activity highlight individual cells undergoing arrest, while transcriptomic and proteomic profiling capture broader molecular signatures of senescence across cell populations. Advanced imaging modalities, including fluorescent probes, PET tracers, and MRI techniques, enable noninvasive mapping of senescence-associated changes in living brains, facilitating longitudinal studies and therapeutic monitoring

Consequences of Senescence in the Aging Brain

Cellular senescence in the central nervous system affects tissue health and function. Beyond halting cell proliferation, senescent cells secrete factors that reshape the local environment, disrupt cell–cell communication, and compromise vital barriers. This section explores three primary downstream outcomes of accumulating senescent cells: enhanced neuroinflammatory tone, loss of plasticity and new neuron formation, and weakening of blood–brain barrier integrity [81, 82].

Altered Neuroinflammation

Senescent glia and neurons unleash a chronic inflammatory cascade through their senescence-associated secretory phenotype (SASP). Key SASP cytokines interleukin-6, interleukin-1β, tumour necrosis factor-α and chemokines like CCL2 create a sustained pro-inflammatory milieu. Unlike the acute, resolving inflammation that follows injury, this low-grade, persistent inflammatory state erodes homeostatic control: microglia remain locked in an activated, phagocytosis-deficient configuration, and astrocytes adopt a reactive, neurotoxic profile [83, 84].

Furthermore, SASP-derived matrix metalloproteinases remodel the extracellular matrix, promoting infiltration of peripheral immune cells across a compromised barrier. Together, these changes amplify oxidative stress and DNA damage in neighboring cells, generating a feed-forward loop that accelerates senescence spread. The net result is a brain environment predisposed to degeneration, in which inflammatory mediators erode synapses, impair neuronal signaling, and predispose circuits to malfunction [85, 86].

Impaired Neurogenesis and Synaptic Plasticity

In young brains, the subgranular zone of the hippocampus and the subventricular zone sustain lifelong neurogenesis, which is critical for memory formation and adaptive learning. Senescent cells in these niches blunt this regenerative capacity in two ways. First, SASP factors such as TGF-β and interleukin-8 suppress the proliferation and differentiation of neural progenitor cells. Second, local oxidative stress and epigenetic alterations within progenitors reduce their responsiveness to growth cues [87, 88].

Synaptic plasticity, the physiological basis of learning and memory, also suffers. Neurons exposed to a senescent secretome exhibit diminished long-term potentiation and retract dendritic spines, a phenomenon linked to the downregulation of BDNF signaling and elevated reactive oxygen species. Consequently, aged animals show slower acquisition of new tasks and difficulty consolidating spatial memories. Restoring plasticity in model systems by pharmacologically clearing senescent cells or neutralizing specific SASP cytokines reawakens progenitor activity and rescues synaptic strength, underscoring the reversible nature of these impairments [89, 90].

Blood–Brain Barrier Integrity

The blood–brain barrier (BBB) is an essential gatekeeper, maintaining CNS homeostasis by tightly regulating molecular and cellular traffic. Senescence undermines this barrier on multiple fronts. Endothelial cells lining cerebral capillaries can themselves become senescent, marked by p16INK4a upregulation and reduced tight-junction protein expression (claudin-5, occludin). Concurrently, SASP-derived MMPs degrade basement membrane components, and inflammatory cytokines loosen endothelial junctions [91, 92].

These alterations increase vascular permeability, allowing plasma proteins and peripheral immune cells to enter the parenchyma. The breach heightens local inflammation and generates oedema, further stressing neurons and glia. In models of Alzheimer's disease, BBB breakdown correlates with early cognitive deficits and accelerates amyloid-β deposition. Therapeutic strategies that preserve BBB integrity, such as MMP inhibitors or senolytic clearance of vascular senescent cells, have shown promise in reestablishing tight-junction continuity and reducing neuroinflammation in aged rodents [93, 94].

Interplay Between Senescence and Neurodegenerative Diseases

Cellular senescence shapes normal aging and weaves into the pathophysiology of major neurodegenerative disorders. By altering the microenvironment, senescent cells create fertile ground for protein aggregation, neuroinflammation, and synaptic failure that underlie clinical syndromes such as Alzheimer's, Parkinson's, and amyotrophic lateral sclerosis. Here, we explore how senescence amplifies disease-specific processes and how, in turn, pathological proteins may reinforce the senescent state [95, 96].

Alzheimer's Disease

In Alzheimer's disease (AD), the accumulation of amyloid-β plaques and tau tangles coincides with a marked rise in senescence markers across multiple cell types. Astrocytes and microglia surrounding amyloid deposits exhibit elevated p16INK4a expression and secrete an inflammatory mix including IL-1β, IL-6, and MMPs, further compromising neuronal health. Senescent neurons bearing DNA damage foci and mitochondrial defects tend to accumulate hyperphosphorylated tau, suggesting that the loss of proteostasis inherent to senescence accelerates tangle formation [97, 98]. Conversely, amyloid-β oligomers can induce oxidative stress and DNA damage in nearby glia, pushing them into a senescent-like arrest. This bidirectional interplay fuels a self-reinforcing cycle: senescent glia exacerbate plaque toxicity, while protein aggregates deepen cellular stress, expanding the senescent cell pool. Interventions that clear senescent cells in AD mouse models reduce plaque burden, dampen neuroinflammation, and rescue memory performance, underscoring the therapeutic potential of breaking this vicious loop [99, 100].

Parkinson's Disease

Parkinson's disease (PD) is characterized by the loss of dopaminergic neurons in the substantia nigra and the accumulation of α-synuclein aggregates. Senescent astrocytes in PD models show impaired glutamate uptake and secrete pro-inflammatory cytokines that heighten oxidative damage in vulnerable neurons. Microglia exposed to α-synuclein fibrils adopt a senescent phenotype, marked by lysosomal dysfunction and chronic NF-κB activation, which reduces their ability to phagocytose toxic protein species. Moreover, dopaminergic neurons display signs of senescence, including upregulated p21CIP1 and senescence-like SASP, which can propagate stress to neighboring cells [101]. By reinforcing both protein aggregation and inflammatory circuits, senescence deepens the motor deficits and neuronal loss characteristic of PD. Early pharmacological clearance of senescent glia in rodent PD models preserves dopaminergic cell numbers and improves motor coordination, highlighting senescence as a key mediator of disease progression [102].

Amyotrophic Lateral Sclerosis

Amyotrophic lateral sclerosis (ALS) involves the degeneration of motor neurons in the spinal cord and cortex, often accompanied by mutated SOD1 or TDP-43 proteinopathies. In ALS, reactive astrocytes acquire a senescent signature, with elevated SA-β-gal activity and neurotoxic mediators such as TNF-α and C1q secretion. Senescent microglia lose phagocytic efficiency and release complement components that tag synapses for elimination, contributing to early synaptic withdrawal. Motor neurons under chronic ER stress and oxidative insult upregulate p16INK4a and secrete SASP factors that feedback on glia, exacerbating inflammation. Animal studies demonstrate that targeting senescent astrocytes prolongs motor neuron survival and extends lifespan, suggesting that interrupting senescence-driven toxicity may slow ALS progression [103, 104].

Other Tauopathies and Dementias

Beyond Alzheimer's, a spectrum of tauopathies, including frontotemporal dementia, progressive supranuclear palsy, and corticobasal syndrome, share senescence-related features. In these disorders, cortical and subcortical neurons accumulate hyperphosphorylated tau parallel with senescence markers such as lamin B1 loss and SA-β-gal positivity. Senescent oligodendrocyte precursor cells in white-matter tracts fail to remyelinate damaged axons, amplifying tau-induced transmission deficits. Microglial senescence, characterized by lipofuscin buildup and chronic complement release, accelerates synaptic pruning and network collapse. These common threads suggest that senescence is a convergent pathway across diverse tauopathies, opening the door for senolytic and senomorphic strategies to have broad efficacy in dementia beyond Alzheimer's (Fig. 4.3).

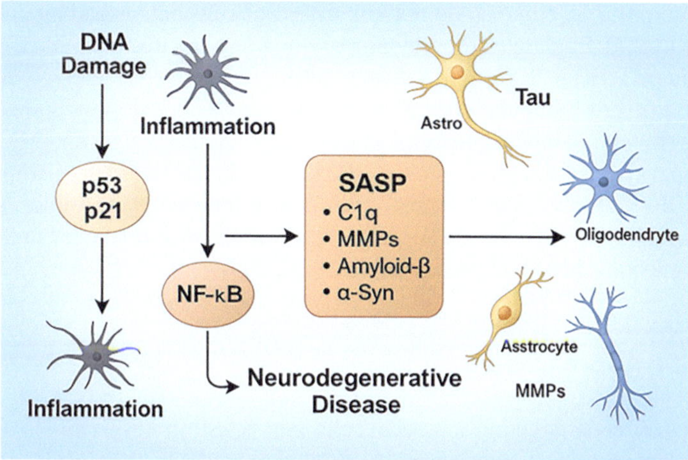

Fig. 4.3 Signaling pathways linking senescence to neurodegeneration. This diagram illustrates how DNA damage and inflammatory cues converge on p53/p21 and NF-κB activation, triggering a robust SASP that releases factors such as MMPs, amyloid-β, α-synuclein, and complement proteins. These secreted molecules act on neurons, astrocytes, oligodendrocytes, and microglia to promote synaptic dysfunction, protein aggregation, and barrier disruption. Together, these interconnected signaling cascades create a pro-degenerative milieu that accelerates the onset and progression of Alzheimer's, Parkinson's, and other neurodegenerative disorders

Therapeutic Strategies Targeting Senescence

Various interventions have emerged to counteract the harmful effects of senescent cells in the brain. Broadly, these approaches fall into four categories: senolytic compounds that selectively eliminate senescent cells, senomorphic agents that suppress their deleterious secretions, gene- and cell-based methods to correct or replace dysfunctional cells, and lifestyle or metabolic modulators that reduce senescence burden [105].

Senolytics and Senomorphics

Senolytics are small molecules designed to exploit vulnerabilities in senescent cells, such as their reliance on antiapoptotic BCL-2 family proteins to induce selective apoptosis. Agents like navitoclax (ABT-263) and the flavonoid quercetin, often combined with the tyrosine kinase inhibitor dasatinib, have effectively eliminated senescent glia and improved cognitive measures in aged rodents. These drugs blunt chronic inflammation and restore tissue homeostasis by clearing the senescent cell pool [106].

Senomorphics, in contrast, do not kill senescent cells but instead modulate their secretory profile. Rapamycin and metformin suppress the mammalian target of the rapamycin (mTOR) pathway and activate AMPK, respectively, thereby reducing the production of pro-inflammatory cytokines and matrix metalloproteinases. In preclinical Alzheimer's models, rapamycin treatment attenuates IL-6 and MMP-9 levels in hippocampal tissue, improving synaptic function without disrupting cell viability. Both senolytics and senomorphics carry translational promise, but challenges remain in optimizing dosing regimens, minimizing off-target toxicity, and ensuring adequate blood–brain barrier penetration [107].

Anti-SASP Approaches

Rather than directly targeting senescent cells, anti-SASP strategies aim to neutralize or inhibit key secretome components. Neutralizing antibodies against IL-1β or IL-6 receptors dampen the inflammatory cascade and have demonstrated cognitive benefits in aged mice. Small-molecule inhibitors of the cGAS–STING pathway blocks the cytosolic DNA sensing that amplifies SASP via type I interferon and NF-κB signaling, reducing glial activation and preserving synaptic integrity [84].

Matrix metalloproteinase inhibitors (e.g., GM6001) protect blood–brain barrier components by preventing basement membrane degradation, while complement cascade modulators (such as C1q-blocking peptides) avert excessive synaptic pruning by microglia. These interventions sidestep the need to eliminate senescent cells but require sustained administration to control continuous SASP production, raising questions about long-term safety and immune suppression [72].

Gene- and Cell-Based Interventions

Genetic engineering offers precision tools to ablate or reprogram senescent cells. "Suicide gene" models employ promoters from senescence-associated genes (e.g., p16INK4a) to drive expression of inducible caspases; administration of a prodrug then triggers apoptosis in p16-expressing cells. While powerful in rodent studies, translating this approach to humans demands safe, cell type–specific gene delivery vectors and stringent off-target controls [95].

Cell-based therapies focus on replenishing lost or dysfunctional neural populations. Transplantation of young neural stem cells can both replace damaged cells and secrete factors that rejuvenate endogenous niches. CRISPR-based epigenome editing has been used in vitro to downregulate CDKN2A expression in aged glial cultures, restoring proliferative and support functions. These personalized strategies hold significant potential but face hurdles, including immunogenicity, ethical considerations, and delivery across the blood–brain barrier [88].

Lifestyle and Metabolic Modulators

Non-pharmacological interventions also exert anti-senescence effects. Caloric restriction and intermittent fasting activate SIRT1 and AMPK, enhancing mitochondrial biogenesis and autophagy, which reduce oxidative damage and delay senescence onset. Exercise promotes the release of neurotrophic factors (BDNF, IGF-1) that counteract SASP-mediated inflammation and support neurogenesis [108].

Nutraceuticals such as resveratrol, nicotinamide riboside, and omega-3 fatty acids boost NAD^+ levels or provide antioxidant capacity, mitigating DNA damage and epigenetic drift. Early clinical trials of NAD^+ precursors report improved cognitive scores in older adults, though larger randomized studies are needed. While lifestyle modifications may not fully reverse established senescence, they represent accessible, low-risk strategies that complement pharmacological and advanced biotechnological therapies [48] (Fig. 4.4).

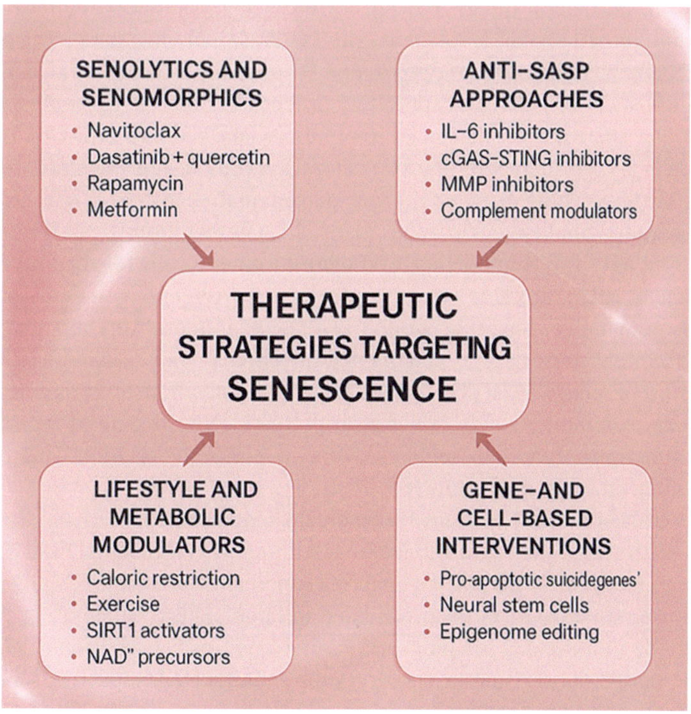

Fig. 4.4 Therapeutic strategies targeting cellular senescence. This infographic organizes four major intervention categories: senolytics and senomorphics, anti-SASP approaches, gene- and cell-based methods, and lifestyle/metabolic modulators around a central theme of combating senescence in the aging brain. Each box lists key examples, from navitoclax and rapamycin to IL-6 inhibitors, proapoptotic "suicide" genes, and caloric restriction, highlighting how diverse tactics converge on reducing senescent cell burden or neutralizing their harmful effects

Challenges and Future Directions

Despite remarkable advances in understanding how cellular senescence drives brain aging and neurodegeneration, several key hurdles must be overcome before senescence-based therapies can be reliably translated to the clinic. First, the heterogeneity of senescent cells in the central nervous system poses a major obstacle for targeted interventions. Neurons, astrocytes, microglia, oligodendrocyte precursors, and endothelial cells adopt distinct senescent phenotypes and secretory profiles. A therapy that efficiently clears p16INK4a-positive astrocytes may miss neurons exhibiting p21CIP1-driven arrest or vice versa. Developing high-resolution, cell type–specific biomarkers and coupling them with delivery systems such as antibody-drug conjugates or viral vectors tuned to individual cell populations will be essential to maximize on-target efficacy while sparing healthy tissue [55, 57, 62].

Second, the blood–brain barrier remains a formidable pharmacological barrier. Many senolytic small molecules and biologics demonstrate promising activity in vitro or in peripheral organs but fail to reach therapeutic concentrations in the brain. Future efforts should leverage carrier systems, including nanoparticle formulations, receptor-mediated transcytosis, and focused ultrasound to transiently open tight junctions and deliver agents precisely to senescent niches. In parallel, optimizing molecular scaffolds for enhanced lipophilicity and reduced efflux by transporters such as P-glycoprotein will improve central nervous system penetration [75, 78].

Third, distinguishing beneficial from detrimental senescence is a critical yet underappreciated challenge. An acute, transient senescent arrest can limit aberrant proliferation, aid tissue remodeling, and recruit immune cells for repair. Wholesale elimination of all p16INK4a-expressing cells risks disrupting these homeostatic functions, potentially impairing wound healing or immune surveillance. Moving forward, the field must define "senescence thresholds" that discriminate adaptive, short-lived senescence from chronic, pathological states. This will require longitudinal studies combining single-cell transcriptomics, live imaging of reporter models, and functional readouts such as synaptic plasticity or blood–brain barrier integrity after timed interventions [65, 70, 82].

Finally, robust clinical measures to evaluate senescence burden and therapeutic response in human patients are still lacking. Although cerebrospinal fluid assays for SASP components and PET tracers targeting senescence-associated β-galactosidase hold promise, none have yet been validated in large cohorts. Integrating fluid biomarkers with multimodal imaging and digital cognitive assessments will create composite endpoints capable of capturing the multifaceted impact of senescence-modulating treatments. Equally important will be designing early-phase trials with adaptive protocols that can swiftly de-risk novel agents and optimize dosing windows.

Looking ahead, several avenues warrant particular attention. The convergence of spatial "omics" technologies with in situ imaging will map senescence across brain regions at single-cell resolution, revealing microenvironmental cues that modulate secretory phenotypes. Advances in synthetic biology may enable programmable "smart" senolytic constructs that activate only in the presence of multiple senescence signatures, dramatically improving specificity. Finally, combining senescence-targeted drugs with established neuroprotective strategies such as anti-amyloid immunotherapy or BDNF agonists may yield synergistic benefits, arresting disease progression on multiple fronts.

The field can build a rigorous translational pipeline by addressing these challenges through interdisciplinary collaboration and bringing together chemical engineers, neuroscientists, immunologists, and clinicians. Ultimately, harnessing the power to precisely detect, modulate, and eliminate deleterious senescent cells offers the best hope for delaying cognitive decline and forestalling the devastating impact of neurodegenerative diseases.

Conclusions

Cellular senescence has emerged as a central driver of brain aging and a key contributor to neurodegenerative disease processes. Through durable cell-cycle arrest, extensive chromatin remodeling, and a complex pro-inflammatory secretome, senescent neurons and glia reshape their surroundings to undermine synaptic integrity, neurogenesis, and vascular health. Importantly, this multifaceted program reflects accumulated cellular damage and actively propagates dysfunction, creating a self-amplifying cycle of inflammation, proteostasis collapse, and barrier breakdown.

Therapeutic strategies aimed at intercepting senescence from targeted clearance of senescent cells to modulation of the SASP have demonstrated proof-of-concept efficacy in preclinical models of Alzheimer's, Parkinson's, ALS, and other tauopathies. Lifestyle and metabolic interventions further complement pharmacological approaches, suggesting that multi-modal regimens may offer the most robust benefits. Yet translating these findings into safe and effective human treatments demands solutions to outstanding challenges: cell-type specificity, blood–brain barrier delivery, preservation of beneficial senescence, and validated clinical biomarkers.

The integration of single-cell and spatial "omics" technologies with advanced imaging will deepen our understanding of senescence heterogeneity and its context-dependent roles. Synthetic biology approaches hold promise for crafting "smart" senolytic agents that activate only upon recognition of precise senescence signatures. Moreover, combining senescence-targeted therapies with existing neuroprotective modalities may yield synergistic effects that arrest or even reverse early disease stages.

By uniting expertise across neuroscience, immunology, chemical engineering, and clinical disciplines, the field is poised to translate mechanistic insights into tangible interventions. Success in this endeavour could redefine how we treat age-related cognitive decline and neurodegenerative disorders, shifting the paradigm from symptomatic management toward fundamental restoration of brain health.

References

1. Tamatta R, Pai V, Jaiswal C, Singh I, Singh AK. Neuroinflammaging and the immune landscape: the role of autophagy and senescence in aging brain. Biogerontology. 2025;26(2):52.
2. Ma Y, Erb ML, Moore DJ. Aging, cellular senescence and Parkinson's disease. J Parkinsons Dis. 2025;15(2):239–54.
3. Franco AC, Martini H, Victorelli S, Lagnado AB, Wyles SP, Rowsey JL, et al. Senescent cell transplantation into the skin induces age-related peripheral dysfunction and cognitive decline. Aging Cell. 2025;24(1):e14340.
4. Dubal DB, Murphy CT, Suh Y, Benayoun BA. Biological sex matters in brain aging. Neuron. 2025;113(1):2–6.
5. Zhang W, Sun HS, Wang X, Dumont AS, Liu Q. Cellular senescence, DNA damage, and neuroinflammation in the aging brain. Trends Neurosci. 2024;47(6):461–74.
6. Wu D, Sun JK, Chow KH. Neuronal cell cycle reentry events in the aging brain are more prevalent in neurodegeneration and lead to cellular senescence. PLoS Biol. 2024;22(4):e3002559.
7. Wu C, Tu T, Xie M, Wang Y, Yan B, Gong Y, et al. Spatially resolved transcriptome of the aging mouse brain. Aging Cell. 2024;23(5):e14109.
8. Vecchiarelli HA, Lopes LT, Paolicelli RC, Stevens B, Wake H, Tremblay M. Synapse regulation. Adv Neurobiol. 2024;37:179–208.
9. Stojiljkovic MR, Schmeer C, Witte OW. Senescence and aging differentially alter key metabolic pathways in murine brain microglia. Neurosci Lett. 2024;828:137751.
10. Singh M, Ali H, Renuka Jyothi S, Kaur I, Kumar S, Sharma N, et al. Tau proteins and senescent cells: targeting aging pathways in Alzheimer's disease. Brain Res. 2024;1844:149165.
11. Shaulson ED, Cohen AA, Picard M. The brain-body energy conservation model of aging. Nat Aging. 2024;4(10):1354–71.
12. Ritzel RM, Jiang D, McCullough LD. CAMs in command: aging brain macrophages fine-tune stroke immune responses. Trends Neurosci. 2024;47(12):965–7.
13. Rim C, You MJ, Nahm M, Kwon MS. Emerging role of senescent microglia in brain aging-related neurodegenerative diseases. Transl Neurodegener. 2024;13(1):10.
14. Real MGC, Falcione SR, Boghozian R, Clarke M, Todoran R, St Pierre A, et al. Endothelial cell senescence effect on the blood-brain barrier in stroke and cognitive impairment. Neurology. 2024;103(11):e210063.
15. Rachmian N, Medina S, Cherqui U, Akiva H, Deitch D, Edilbi D, et al. Identification of senescent, TREM2-expressing microglia in aging and Alzheimer's disease model mouse brain. Nat Neurosci. 2024;27(6):1116–24.
16. Melo Dos Santos LS, Trombetta-Lima M, Eggen B, Demaria M. Cellular senescence in brain aging and neurodegeneration. Ageing Res Rev. 2024;93:102141.
17. Maeda C, Tsuruta F. Molecular basis of neuronal and microglial states in the aging brain and impact on cerebral blood vessels. Int J Mol Sci. 2024;25(8)
18. Kumar D. Sleep-memory relationships during brain aging. Aging Brain. 2024;6:100127.
19. Katsuumi G, Shimizu I, Suda M, Yoshida Y, Furihata T, Joki Y, et al. SGLT2 inhibition eliminates senescent cells and alleviates pathological aging. Nat Aging. 2024;4(7):926–38.
20. Karabag D, Heneka MT, Ising C. The putative contribution of cellular senescence to driving tauopathies. Trends Immunol. 2024;45(10):837–48.
21. Herdy JR, Mertens J, Gage FH. Neuronal senescence may drive brain aging. Science. 2024;384(6703):1404–6.
22. Gulej R, Csik B, Faakye J, Tarantini S, Shanmugarama S, Chandragiri SS, et al. Endothelial deficiency of insulin-like growth factor-1 receptor leads to blood-brain barrier disruption and accelerated endothelial senescence in mice, mimicking aspects of the brain aging phenotype. Microcirculation. 2024;31(2):e12840.
23. Fjell AM. Aging brain from a lifespan perspective. Curr Top Behav Neurosci. 2024;68:349–70.

24. Ferrer I. Amyloid-β pathology is the common nominator proteinopathy of the primate brain aging. J Alzheimers Dis. 2024;100(s1):S153–s64.
25. Dong H, Wang S, Hu C, Wang M, Zhou T, Zhou Y. Neuroprotective effects of intermittent fasting in the aging brain. Ann Nutr Metab. 2024;80(4):175–85.
26. Byrns CN, Perlegos AE, Miller KN, Jin Z, Carranza FR, Manchandra P, et al. Senescent glia link mitochondrial dysfunction and lipid accumulation. Nature. 2024;630(8016):475–83.
27. Bondy SC, Wu M. The critical role of autophagy and phagocytosis in the aging brain. Int J Mol Sci. 2024;26(1)
28. Zhang H, Li J, Yu Y, Ren J, Liu Q, Bao Z, et al. Nuclear lamina erosion-induced resurrection of endogenous retroviruses underlies neuronal aging. Cell Rep. 2023;42(6):112593.
29. Zeng W, Zhang W, Tse EHY, Liu J, Dong A, Lam KSW, et al. Restoration of CPEB4 prevents muscle stem cell senescence during aging. Dev Cell. 2023;58(15):1383–98.e6.
30. Wong GC, Chow KH. DNA damage response-associated cell cycle re-entry and neuronal senescence in brain aging and Alzheimer's disease. J Alzheimers Dis. 2023;94(s1):S429–s51.
31. Ng PY, McNeely TL, Baker DJ. Untangling senescent and damage-associated microglia in the aging and diseased brain. FEBS J. 2023;290(5):1326–39.
32. Nair P, Prasad K, Balasundaram P, Vibha D, Nand Dwivedi S, Gaikwad SB, et al. Multimodal imaging of the aging brain: baseline findings of the LoCARPoN study. Aging Brain. 2023;3:100075.
33. Malvaso A, Gatti A, Negro G, Calatozzolo C, Medici V, Poloni TE. Microglial senescence and activation in healthy aging and Alzheimer's disease: systematic review and neuropathological scoring. Cells. 2023;12(24)
34. Lu Y, Jarrahi A, Moore N, Bartoli M, Brann DW, Baban B, et al. Inflammaging, cellular senescence, and cognitive aging after traumatic brain injury. Neurobiol Dis. 2023;180:106090.
35. Liu X, Liu Z, Wu Z, Ren J, Fan Y, Sun L, et al. Resurrection of endogenous retroviruses during aging reinforces senescence. Cell. 2023;186(2):287–304.e26.
36. Li M, An H, Wang W, Wei D. Biomolecular markers of brain aging. Adv Exp Med Biol. 2023;1419:111–26.
37. Craik FIM. Memory, aging and the brain: old findings and current issues. Aging Brain. 2023;4:100096.
38. Budamagunta V, Kumar A, Rani A, Bean L, Manohar-Sindhu S, Yang Y, et al. Effect of peripheral cellular senescence on brain aging and cognitive decline. Aging Cell. 2023;22(5):e13817.
39. Akbulut KG, Keskin-Aktan A, Abgarmi SA, Akbulut H. The role of SIRT2 inhibition on the aging process of brain in male rats. Aging Brain. 2023;4:100087.
40. Zhu Y, Prata L, Gerdes EOW, Netto JME, Pirtskhalava T, Giorgadze N, et al. Orally-active, clinically-translatable senolytics restore α-Klotho in mice and humans. EBioMedicine. 2022;77:103912.
41. Sahu MR, Rani L, Subba R, Mondal AC. Cellular senescence in the aging brain: a promising target for neurodegenerative diseases. Mech Ageing Dev. 2022;204:111675.
42. Ramanoël S, Durteste M, Delaux A, de Saint Aubert JB, Arleo A. Future trends in brain aging research: Visuo-cognitive functions at stake during mobility and spatial navigation. Aging Brain. 2022;2:100034.
43. Preininger MK, Kaufer D. Blood-brain barrier dysfunction and astrocyte senescence as reciprocal drivers of neuropathology in aging. Int J Mol Sci. 2022;23(11)
44. Matias I, Diniz LP, Damico IV, Araujo APB, Neves LDS, Vargas G, et al. Loss of Lamin-B1 and defective nuclear morphology are hallmarks of astrocyte senescence in vitro and in the aging human hippocampus. Aging Cell. 2022;21(1):e13521.
45. Liu RM. Aging, cellular senescence, and Alzheimer's disease. Int J Mol Sci. 2022;23(4)
46. Han X, Lei Q, Xie J, Liu H, Li J, Zhang X, et al. Potential regulators of the senescence-associated secretory phenotype during senescence and aging. J Gerontol A Biol Sci Med Sci. 2022;77(11):2207–18.

47. Brown RM, Gruijters SLK, Kotz SA. Prediction in the aging brain: merging cognitive, neurological, and evolutionary perspectives. J Gerontol B Psychol Sci Soc Sci. 2022;77(9):1580–91.
48. Behfar Q, Ramirez Zuniga A, Martino-Adami PV. Aging, senescence, and dementia. J Prev Alzheimers Dis. 2022;9(3):523–31.
49. Balietti M, Conti F. Environmental enrichment and the aging brain: is it time for standardization? Neurosci Biobehav Rev. 2022;139:104728.
50. Aguado J, Wolvetang EJ. Exploring aging interventions in human brain organoids. Aging (Albany NY). 2022;14(4):1592–3.
51. Panagiotou M, Michel S, Meijer JH, Deboer T. The aging brain: sleep, the circadian clock and exercise. Biochem Pharmacol. 2021;191:114563.
52. Iourov IY, Yurov YB, Vorsanova SG, Kutsev SI. Chromosome instability, aging and brain diseases. Cells. 2021;10(5)
53. Guerrero A, De Strooper B, Arancibia-Cárcamo IL. Cellular senescence at the crossroads of inflammation and Alzheimer's disease. Trends Neurosci. 2021;44(9):714–27.
54. Demeneix BA. Environmental influences on brain aging. Aging Brain. 2021;1:100003.
55. Choudhury ME, Miyanishi K, Takeda H, Tanaka J. Microglia and the aging brain: are geriatric microglia linked to poor sleep quality? Int J Mol Sci. 2021;22(15)
56. Blinkouskaya Y, Caçoilo A, Gollamudi T, Jalalian S, Weickenmeier J. Brain aging mechanisms with mechanical manifestations. Mech Ageing Dev. 2021;200:111575.
57. Avila J. Brain aging, epigenetic changes, tau and neurodegeneration. Aging Brain. 2021;1:100004.
58. Andromidas F, Atashpanjeh S, Myers AJ, MacKinnon BE, Shaffer MM, Koob AO. The astrogenic balance in the aging brain. Curr Neuropharmacol. 2021;19(11):1952–65.
59. Almeida OFX. Aging brain is born. Aging Brain. 2021;1:100006.
60. Salas IH, Burgado J, Allen NJ. Glia: victims or villains of the aging brain? Neurobiol Dis. 2020;143:105008.
61. Hill C, Van Gemmert AWA, Fang Q, Hou L, Wang J, Pan Z. Asymmetry in the aging brain: a narrative review of cortical activation patterns and implications for motor function. Laterality. 2020;25(4):413–29.
62. Graves SI, Baker DJ. Implicating endothelial cell senescence to dysfunction in the ageing and diseased brain. Basic Clin Pharmacol Toxicol. 2020;127(2):102–10.
63. Gordleeva S, Kanakov O, Ivanchenko M, Zaikin A, Franceschi C. Brain aging and garbage cleaning: modelling the role of sleep, glymphatic system, and microglia senescence in the propagation of inflammaging. Semin Immunopathol. 2020;42(5):647–65.
64. Benvegnù S. Nucleus-cytoplasm cross-talk in the aging brain. J Neurosci Res. 2020;98(2):247–61.
65. Walton CC, Andersen JK. Unknown fates of (brain) oxidation or UFO: close encounters with neuronal senescence. Free Radic Biol Med. 2019;134:695–701.
66. Schmeer C, Kretz A, Wengerodt D, Stojiljkovic M, Witte OW. Dissecting aging and senescence-current concepts and open lessons. Cells. 2019;8(11)
67. Lupo G, Gaetani S, Cacci E, Biagioni S, Negri R. Molecular signatures of the aging brain: finding the links between genes and phenotypes. Neurotherapeutics. 2019;16(3):543–53.
68. Lerner AJ, Pieper AA. Neurotherapeutics of the aging brain: complexity meets complexity. Neurotherapeutics. 2019;16(3):539–42.
69. Koen JD, Rugg MD. Neural dedifferentiation in the aging brain. Trends Cogn Sci. 2019;23(7):547–59.
70. Hou Y, Dan X, Babbar M, Wei Y, Hasselbalch SG, Croteau DL, et al. Ageing as a risk factor for neurodegenerative disease. Nat Rev Neurol. 2019;15(10):565–81.
71. Cohen J, Torres C. Astrocyte senescence: evidence and significance. Aging Cell. 2019;18(3):e12937.
72. Bah TM, Goodman J, Iliff JJ. Sleep as a therapeutic target in the aging brain. Neurotherapeutics. 2019;16(3):554–68.
73. Angelova DM, Brown DR. Microglia and the aging brain: are senescent microglia the key to neurodegeneration? J Neurochem. 2019;151(6):676–88.

74. Zannas AS. Gene-environment interactions in late life: linking psychosocial stress with brain aging. Curr Neuropharmacol. 2018;16(3):327–33.
75. Shwe T, Pratchayasakul W, Chattipakorn N, Chattipakorn SC. Role of D-galactose-induced brain aging and its potential used for therapeutic interventions. Exp Gerontol. 2018;101:13–36.
76. Santoro A, Spinelli CC, Martucciello S, Nori SL, Capunzo M, Puca AA, et al. Innate immunity and cellular senescence: the good and the bad in the developmental and aged brain. J Leukoc Biol. 2018;103(3):509–24.
77. Li J, Zhou Y, Du G, Qin X, Gao L. Bioinformatic analysis reveals key genes and pathways in aging brain of senescence-accelerated mouse P8 (SAMP8). CNS Neurol Disord Drug Targets. 2018;17(9):712–22.
78. Barter JD, Foster TC. Aging in the brain: new roles of epigenetics in cognitive decline. Neuroscientist. 2018;24(5):516–25.
79. Tse KH, Herrup K. DNA damage in the oligodendrocyte lineage and its role in brain aging. Mech Ageing Dev. 2017;161(Pt A):37–50.
80. Young JS, Hobbs JG, Bailes JE. The impact of traumatic brain injury on the aging brain. Curr Psychiatry Rep. 2016;18(9):81.
81. Tong J, Fitzmaurice PS, Moszczynska A, Mattina K, Ang LC, Boileau I, et al. Do glutathione levels decline in aging human brain? Free Radic Biol Med. 2016;93:110–7.
82. Correia-Melo C, Marques FD, Anderson R, Hewitt G, Hewitt R, Cole J, et al. Mitochondria are required for pro-ageing features of the senescent phenotype. EMBO J. 2016;35(7):724–42.
83. Bigagli E, Luceri C, Scartabelli T, Dolara P, Casamenti F, Pellegrini-Giampietro DE, et al. Long-term neuroglial cocultures as a brain aging model: hallmarks of senescence, microRNA expression profiles, and comparison with in vivo models. J Gerontol A Biol Sci Med Sci. 2016;71(1):50–60.
84. Jellinger KA, Attems J. Challenges of multimorbidity of the aging brain: a critical update. J Neural Transm (Vienna). 2015;122(4):505–21.
85. Chinta SJ, Woods G, Rane A, Demaria M, Campisi J, Andersen JK. Cellular senescence and the aging brain. Exp Gerontol. 2015;68:3–7.
86. Tan FC, Hutchison ER, Eitan E, Mattson MP. Are there roles for brain cell senescence in aging and neurodegenerative disorders? Biogerontology. 2014;15(6):643–60.
87. Stern P, Hines PJ, Travis J. The aging brain. Introduction. Science. 2014;346(6209):566–7.
88. Lauri A, Pompilio G, Capogrossi MC. The mitochondrial genome in aging and senescence. Ageing Res Rev. 2014;18:1–15.
89. Karlawish J, Green RC. Minding the aging brain: are we ready for personalized medicine? Am J Psychiatry. 2014;171(2):137–9.
90. Voytek B, Gazzaley A. Stimulating the aging brain. Ann Neurol. 2013;73(1):1–3.
91. Rossini PM, Ferilli MA, Rossini L, Ferreri F. Clinical neurophysiology of brain plasticity in aging brain. Curr Pharm Des. 2013;19(36):6426–39.
92. Peters A, Folger C. A website entitled "the fine structure of the aging brain". J Comp Neurol. 2013;521(6):1203–6.
93. Dorszewska J. Cell biology of normal brain aging: synaptic plasticity-cell death. Aging Clin Exp Res. 2013;25(1):25–34.
94. Brown EN, Purdon PL. The aging brain and anesthesia. Curr Opin Anaesthesiol. 2013;26(4):414–9.
95. Yang S, Lu W, Zhou DS, Tang Y. Enriched environment and white matter in aging brain. Anat Rec (Hoboken). 2012;295(9):1406–14.
96. Salminen A, Ojala J, Kaarniranta K, Haapasalo A, Hiltunen M, Soininen H. Astrocytes in the aging brain express characteristics of senescence-associated secretory phenotype. Eur J Neurosci. 2011;34(1):3–11.
97. Salat DH. The declining infrastructure of the aging brain. Brain Connect. 2011;1(4):279–93.
98. Martin M, Dotti CG, Ledesma MD. Brain cholesterol in normal and pathological aging. Biochim Biophys Acta. 2010;1801(8):934–44.
99. Yankner BA, Lu T, Loerch P. The aging brain. Annu Rev Pathol. 2008;3:41–66.

100. Rossini PM, Rossi S, Babiloni C, Polich J. Clinical neurophysiology of aging brain: from normal aging to neurodegeneration. Prog Neurobiol. 2007;83(6):375–400.
101. Drachman DA. Aging of the brain, entropy, and Alzheimer disease. Neurology. 2006;67(8):1340–52.
102. Conde JR, Streit WJ. Microglia in the aging brain. J Neuropathol Exp Neurol. 2006;65(3):199–203.
103. Hof PR, Morrison JH. The aging brain: morphomolecular senescence of cortical circuits. Trends Neurosci. 2004;27(10):607–13.
104. Bondy SC, Lahiri DK, Perreau VM, Sharman KZ, Campbell A, Zhou J, et al. Retardation of brain aging by chronic treatment with melatonin. Ann N Y Acad Sci. 2004;1035:197–215.
105. Mattson MP, Chan SL, Duan W. Modification of brain aging and neurodegenerative disorders by genes, diet, and behavior. Physiol Rev. 2002;82(3):637–72.
106. Horner PJ, Gage FH. Regeneration in the adult and aging brain. Arch Neurol. 2002;59(11):1717–20.
107. Cotrina ML, Nedergaard M. Astrocytes in the aging brain. J Neurosci Res. 2002;67(1):1–10.
108. Rockwood K. What metaphor for the aging brain? Neurology. 2007;69(20):1896–7.

Neurodegeneration in the Aging Brain: Role of Cellular Senescence

5

Abida Khan, Saleh I. Alaqel, Mashael N. Alanazi, Muhammad Irfan Siddique, and Mohd Imran

Introduction

Advanced lifespan rates worldwide have boosted age-related neurodegenerative illnesses such as Alzheimer's and Parkinson's, along with dementias, which now strain healthcare resources and economic operations. Brain aging usually produces mounting neuron operational issues and reduced synaptic links, leading to cognitive decline through cellular and molecular changes. The multifactorial nature of neurodegeneration, including genetic susceptibility, environmental factors, oxidative stress, and chronic inflammation, complicates both diagnosis and treatment [1–3]. Present-day research needs to deliver speedy solutions for fighting this brain-related global health emergency because the elderly population is steadily rising. Persons initially gain protection from tissue damage and cancer development through senescence, yet over time, tissue collapse occurs together with persistent inflammation. The tissue homeostasis of aging tissues gets disrupted because senescent cells

A. Khan · M. Imran (✉)
Center for Health Research, Northern Border University, Arar, Saudi Arabia
e-mail: abeda.mohammed@nbu.edu.sa; mohammad.baks@nbu.edu.sa

S. I. Alaqel · M. N. Alanazi
Department of Pharmaceutical Chemistry, College of Pharmacy, Northern Border University, Rafha, Saudi Arabia
e-mail: saleh.alaqel@nbu.edu.sa; Mashael.Alanazi@nbu.edu.sa

M. I. Siddique
Department of Pharmaceutics, Faculty of Pharmacy, Northern Border University, Rafha, Saudi Arabia
e-mail: muhammad.siddique@nbu.edu.sa

G. Gupta (ed.), *Cellular Senescence and Brain Aging*,
https://doi.org/10.1007/978-981-96-8873-9_5

disrupt tissue environments and regeneration pathways while sending inflammatory signals forward. Neural cells, astrocytes, microglia, and oligodendrocyte progenitor cells show senescence-related changes inside the brain that possibly heighten cognitive deficits and functional decline with aging [4–6]. The toxic substance released by aging glial cells causes neural breakdown and cell fatality by enabling prolonged inflammation, oxidative stress, and extracellular matrix modifications. The pro-inflammatory state speeds up brain cell degeneration because it worsens the formation of proteins, synaptic destruction, and blood–brain barrier dysfunction. Multiple human studies and experimental research show that brain tissue senescent cells connect to declining cognition and disease progression, which proves that cellular aging advances neurodegenerative brain degeneration [7, 8]. The study uses experimental research and clinical data assessment to outline the relationship between senescent cell growth and brain inflammation and aging-related degenerative processes. The evaluation explores current and developing therapeutic senescence cell solutions with senomorphics and senolytics to assess their medical potential for age-related neurological conditions. Research understanding about neurodegenerative aspects receives enhancement through this section, which identifies primary challenges with identified knowledge gaps to guide forthcoming developments in cellular senescence therapeutic approaches [9–11].

Hallmarks of Brain Aging and Cellular Senescence

Structural and Functional Changes in the Aging Brain

Lifespan deterioration in the brain occurs through the weakening of both structural elements and functional systems that involve cortical thinning and hippocampal atrophy with synaptic degradation, resulting in cognitive decline and heightened exposure to neurodegenerative disorders. Advanced age reduces the plasticity of neuronal cells and hippocampal neurogenesis to a much greater degree. Old age affects vascular structures, which causes problems with the blood–brain barrier's functioning and produces higher levels of neuroinflammation [12, 13]. Cell metabolism problems occur when aging deteriorates mitochondria function and increases oxidative stress, resulting in protein damage that destroys cellular balance. Modified neurotransmission co-occurs with decreased synaptic connectivity, resulting in reduced information processing speed. Changes that develop with age create neurodegenerative disease-causing environments by generating inflammation together with metabolic breakdown and oxidative damage. Studying how the brain naturally ages allows healthcare providers to identify characteristic Alzheimer's and Parkinson's disease markers because they need to understand the typical age-related changes [14, 15].

Characteristics of Cellular Senescence

Cellular senescence is activated by different stress factors comprising telomere shortening, oxidative damage, and oncogenic signals leading to permanent cell cycle halt. Senescence protects against the cellular proliferation of damaged cells at the start; however, this safety mechanism becomes an aging issue over time. Through SASP, cells acquire the capacity to release pro-inflammatory cytokines with growth factors, proteases, and chemokines. Secreted compounds modify tissue conditions and break down cell-to-cell connectivity, subsequently driving uninterrupted inflammatory responses [16, 17]. The main morphological features of senescence cells encompass lysosomal enrichment, DNA damage foci formation, and enlarged flattened shapes. The molecular studies show that p16INK4a and p21CIP1 expression climbs as cells activate the DNA damage response system through chromatin modifications. Tissues within the brain become damaged by accumulating senescent cells, resulting in poor healing and neuroinflammation in parts that trigger neuronal problems that foster the onset of neurodegenerative diseases [18, 19].

Cellular Senescence in the Central Nervous System

Neurons

Postmitotic neurons do not fall victim to traditional senescence, but researchers have documented senescence-like characteristics in these cells. Physical changes to DNA accumulate while disturbances in mitochondrial operation, calcium regulation, and inflammatory mediator outputs occur during this process. Neurodegeneration with its senescent-like neurons creates synaptic transmission defects and toxic signaling elements that harm glial cells. Alzheimer's disease patients develop cognitive impairments that lead to tau hyperphosphorylation when senescent neurons appear within their brains. Studies show that brain equilibrium deterioration from senescence depends on neuronal physiological modifications and secretory changes, while the cells remain non-arrested during this process [20–22].

Astrocytes

The combination of oxidative stress and inflammation enhances astrocytes' vulnerability to developing senescence. Astrocytic senescence initiation switches off protective mechanisms, which activate harmful secretory activity in the group known as SASP. Pro-inflammatory molecules IL-6 and IL-1β, together with MMPs and chemokines, exit from these cells to activate neuroinflammation that damages neurons. Research shows that senescent astrocytes tend to gather near amyloid plaques in aging and Alzheimer's patients since they speed up disease evolution by disrupting synaptic connections and escalating inflammation at their location [23, 24].

Microglia Cell

Microglia cells that reside in the brain demonstrate functional programming changes during aging, culminating in the senescence state where cells fail to perform phagocytosis while releasing excessive inflammatory signals. During aging, the microglia cells form misshapen structures and increase their production of inflammatory molecules, reducing beta-amyloid clearance and removing cellular waste. These cells experience long-term inflammation, resulting in degenerative events that harm vital nerve cells. The primary role of microglial cells in research models of Alzheimer's and Parkinson's diseases allows them to establish strong connections with disease pathology through their driving action of age-related neuroinflammation [25].

Oligodendrocyte Lineage Cells

Repairing white matter and remyelination depend on oligodendrocyte precursor cells (OPCs) because these glial cells act as proliferative cells. During aging, the cellular characteristics of OPCs transform to show senescence through poor cell multiplication power and elevated p16INK4a expression, which modifies their differentiation routes. Doubling amounts of senescent OPCs lead to white matter damage and maintenance problems for myelin while simultaneously creating cognitive problems. Worsening deterioration of the neural microenvironment occurs because senescent OPC cells produce SASP factors. The condition has acquired significance in age-related processes along with diseases such as multiple sclerosis and Alzheimer's disease because it simultaneously causes demyelination along with white matter deterioration [26, 27].

Biomarkers and Diagnostic Indicators of Senescence

Identifying brain cell senescence remains challenging because different cellular forms of its multiple types cannot be distinguished from other related cell states. Numerous signs have been confirmed through research to help detect senescent cells at both molecular and functional levels. Studies demonstrate that cellular senescence manifests as elevated SA-β-gal levels, and two primary indications include p16INK4a, p21^CIP1, and p53 cell cycle inhibitors together with persistent DNA damage markers γ-H2AX foci and chromatin changes such as senescence-associated heterochromatic foci (SAHF) [28]. The combined presence of IL-6 with IL-1β and MMP enables medical tests to determine these indicators in serum or cerebrospinal fluid analysis. Scientists conduct extensive research to establish evolving imaging procedures combined with transcriptomic methods, which will enhance neuroimaging analysis for live subjects. Correct aging cell identification is a fundamental step to understanding their brain-deteriorating influence because it allows researchers to develop targeted medications that include senolytics and SASP modulators for eliminating senescent cells [29] (Fig. 5.1).

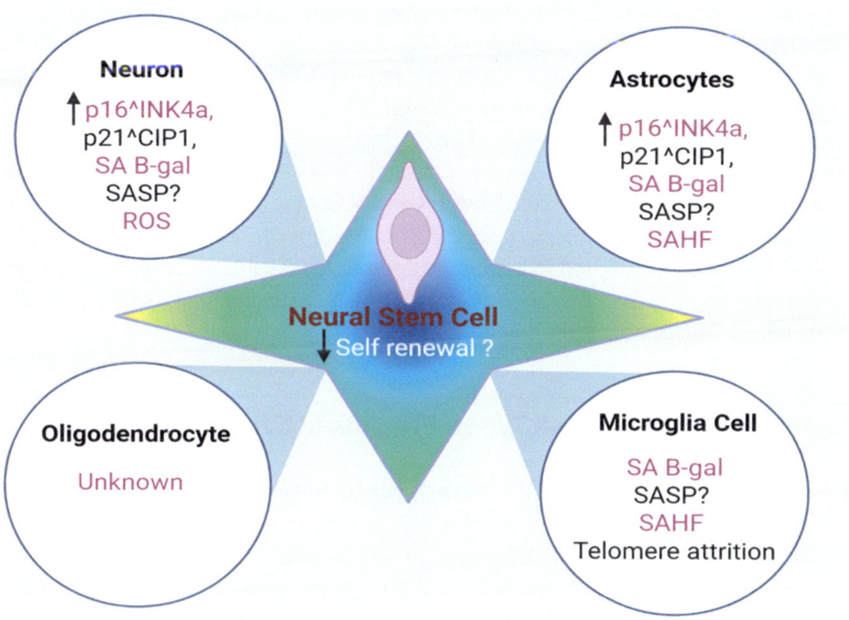

Fig. 5.1 Senescence marker profiles in neural and glial cells. Figure shows how aging molecular markers affect neurons and their related cells, including astrocytes, microglia, and oligodendrocytes, and their potential impact on neural stem cells. The senescent condition is indicated by elevated p16INK4a, p21CIP1, and SA-β-gal expression in neurons alongside astrocytes. Researchers currently lack knowledge about the senescence state of oligodendrocytes despite evidence of microglial SA-β-gal activity alongside telomere attrition and potential SASP and SAHF involvement

Cellular Senescence: Mechanistic Insights

Numerous stressors produce distinct changes in cell appearance, leading to permanent cell proliferation arrest toward senescence while accelerating potential neurodegeneration. Old age brings about three main aging processes that affect the nervous system: DNA damage combined with continuous inflammation while oxidative stress increases simultaneously. Stress-inducing factors activate two tumor suppressor pathways, p53/p21CIP1 and p16INK4a/Rb, to create permanent growth arrest before forming SASP [30]. The hippocampus and cortex, together with substantia nigra face high levels of tissue damage from harmful agents because these brain regions possess complex synaptic structures and high metabolic needs together with limited protection against oxidative damage. Neuroinflammation, synaptic connection damage, and glial support dysfunction emerge as brain-specific functions of SASP through increased IL-6 and IL-1β interleukins with matrix metalloproteinases and reactive oxygen species secretion. The neurodegenerative processes become more severe as cells experience environmental changes that enable cell-to-cell signaling to distribute wisdom [31, 32]. Research laboratories detect senescent cells in neural tissue through SA-β-gal activity markers and elevated proteins p16INK4a and p21CIP1 as well as γ-H2AX foci markers and evidence of SASP-related molecule secretion along with increased p16INK4a protein expression. Research combining brain senescence models with Alzheimer's and Parkinson's disease studies has demonstrated markers as effective tools for understanding temporal and spatial patterns of brain senescence. Studies of these molecular pathways lead to the simultaneous understanding of age-related cognitive decline mechanisms and proper methods for treating senescent cells while managing SASP to protect brain neural functions during aging [33, 34].

Senescence in Brain Cell Types: Preclinical Evidence

Neurons: Resistance and Vulnerability to Senescence

Chronic stress triggers the development of stress-induced cellular features in neurons even though these cells do not usually undergo aging processes because they are postmitotic cells. The activation sequence of p21, p16INK4a, and SASP-like responses is triggered by DNA damage, mitochondrial dysfunction, and oxidative stress conditions. Synaptic impairments, together with inflammatory reactions, increase the risk of developing Alzheimer's disease and Parkinson's disease. Based on preclinical study findings, neural cells develop abnormal cellular states in their aging process that cause glial damage and disrupt neuronal connections [35, 36].

Astrocytes: Drivers of Neuroinflammation

Astrocytes transform into sensitive cells to age-related processes upon exposure to extreme oxidative damage or shortened telomeres and DNA deterioration. During senescence, astrocytes transform into cells that release inflammatory cytokines by producing IL-6, IL-1β, and TNF-α. Neuroinflammation worsens from these factors, which harm the brain's support structures and place blood–brain barrier stability at risk [37]. Preclinical testing environments result in aged astrocytes accumulating until cognitive drops and synaptic changes become prominent, especially in Alzheimer's disease model experiments, according to research. Therapy strategies must concentrate on the cells since they are primary contributors to inflammatory pathway activation in age-related neuronal systems [38].

Microglia: Senescence-Associated Neurotoxicity

Microglia operate as brain immune cells responsible for aging processes when people age due to enduring neuroinflammation. The form of aged microglial cells changes while they produce increasingly large proportions of reactive oxygen molecules and inflammatory cytokines, including IL-1β, TNF-α, and nitric oxide. The modified cellular environment develops toxic conditions that negatively affect nervous system tissues together with synaptic network reduction. The evolution of Parkinson's and Alzheimer's disease progresses faster when microglia experience senescence because it enhances brain inflammation while impairing the elimination of harmful brain proteins. Microglia perform their first function as disease initiators but continue to serve as disease perpetrators of neurodegenerative pathology [39, 40].

Oligodendrocyte Precursor Cells and Myelin Aging

The regenerative abilities of oligodendrocyte precursor cells (OPCs) decrease because of aging, although they can persist with proliferative and differentiating attributes before true senescence appears. The aging cycle transforms OPCs through three damaging effects. It decreases myelin gene expression while producing more oxidative damage and generating defective signaling during differentiation response [41]. The loss of OPC cells results in white matter degeneration through broken remyelination pathways, which produce effects similar to those of normal aging conditions, multiple sclerosis, and Alzheimer's disease. Research on aging rodents indicates that as OPCs move toward senescence, brain function impairment develops because brain functions directly result from myelin structure conditions. Research indicates that aging OPCs could restore their myelin regeneration capacity after resolving their senescence state, which would help protect aging brain neural circuits effectively [42, 43].

Animal Models of Brain Aging and Senescence

Preclinical investigations of brain senescence rely on various animal models for research purposes. Animal models of rodent aging demonstrate effective age-related physiological changes through their natural occurrence of cognitive decline and glial senescence accumulation. Early development of oxidative stress occurs in SAMP8 mice while their brains show neuroinflammatory changes and cognitive impairments [44]. Senescent cell number tracking becomes possible through transgenic models that include p16INK4a or luciferase reporter genes. Observational studies enable scientists to detect how aging affects particular tissue cells and also allow them to investigate disease progression while determining the most beneficial times for treatment application. Brain aging processes are better understood through these combined methods, which also serve as research tools for evaluating unsigned cell treatments [45].

Targeting Senescence in Preclinical Trials: Senolytics and SASP Inhibitors

The administration of senolytics such as dasatinib and quercetin, together with SASP inhibitors consisting of rapamycin and JAK inhibitors, has shown promising outcomes in brain therapy. Administered drugs effectively eliminate aged cells and

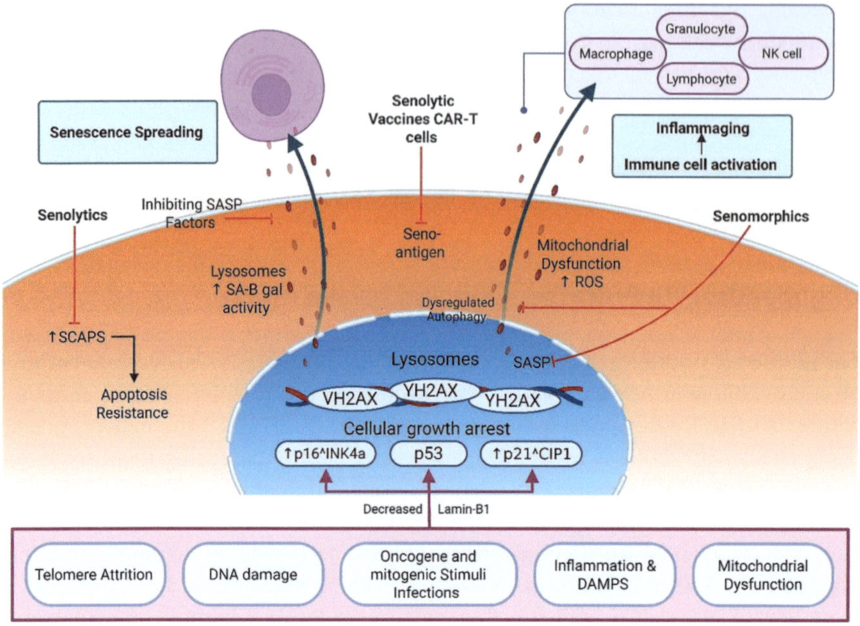

Fig. 5.2 Core molecular mechanisms of cellular senescence. Figure shows cellular senescence's leading molecular causes and characteristics. The process of growth arrest occurs through combinations of four factors: mitochondrial malfunction, telomere attrition, DNA damage, and the elevated expression of p16INK4a, p21CIP1, and p53. Senescent cells produce SASP factors that help activate immune cells and create inflammation while spreading the onset of cellular senescence. Age-related disease treatment options may emerge through therapeutic methods that block SASP signaling and improve immune clearance while restoring tissue balance to target senescent cells

suppress the inflammatory activities of these cells to enhance brain function and neuroinflammation reduction along with synaptic support in brain tissue of aged or disease-compromised rodent brains [46]. Medical models of Alzheimer's disease show improved outcomes after receiving senolytic therapy that both prevents pathological tau development and protects glial cells from deterioration. Two critical barriers exist when working with cell type–specific substances through the blood–brain barrier. The inability to remove or control pro-inflammatory cells makes it difficult to use treatments targeting cellular senescence; however, their preclinical use shows potential benefits for brain aging health treatment [47] (Fig. 5.2).

Clinical Relevance: Senescence in Human Neurodegenerative Disorders

Alzheimer's Disease

Senescent Astrocytes and Microglia

The senescent state affects astrocytes and microglia cells in Alzheimer's patients, leading to structural changes in cells, increased expression of p16INK4a and p21CIP1 genes, and the production of inflammatory SASP factors. Glial cells age from protective operation to inflammatory function, which causes synaptic wreckage and leads to neuronal cell death. Autopsy studies of AD brains reveal senescent glial cells accumulate in the hippocampus and cortex regions of the patient's postmortem, thus confirming their active involvement with Alzheimer's pathology [48, 49].

Tau and Amyloid-β Interaction with SASP

Cell stress occurs through the combination of tau tangles and amyloid-β plaques, while the existing SASP profiles in neighboring senescent cells become enhanced. The SASP cytokines IL-6 and TNF-α sustain an ongoing toxic connection between neuroinflammatory damage through increased tau hyperphosphorylation and amyloid aggregation. The cellular interaction between neurons accelerates neuron degradation, leading to cognitive decline in AD patients. Measuring SASP compounds in cerebrospinal fluid reveals their elevation, and this strengthens our understanding of the disease process, demonstrating the therapeutic importance of targeting senescent pathways for AD management [47, 50].

Parkinson's Disease

Senescence in Dopaminergic Neurons

Dopaminergic neurons in substantia nigra display signs of senescence while a patient has Parkinson's disease because normal aging processes activate markers, including p21 protein expression and cell activity decline. These neurons experience a reduction of dopamine production combined with diminished resistance to toxic substances. Research analysis of PD brains confirms that inflammation sites where neurons contain α-synuclein deposits coexist, creating possible links to neurodegeneration and disease advancement [51].

Role in Mitochondrial Dysfunction

The key disease marker of PD is strongly connected to mitochondrial breakdown and cellular aging processes. Higher ROS production stems from mitochondrial damage in combination with faulty mitophagy processes in aged neurons and glial cells, causing declining metabolism. The excessive dysfunction significantly increases the SASP expression levels and the inflammatory response of neuroglia cells, thus damaging nearby neural tissue. Research on clinical PD specimens and mitochondrial biomarkers demonstrates a need to solve the harmful impact of senescent cells on mitochondria to develop PD treatment [52].

ALS and Other Neurodegenerative Conditions

Clinical Evidence of Senescence Markers

Research of deceased ALS brain tissue and patient cell studies demonstrated that p16INK4a, p53, and SA-β-gal activity markers elevated significantly in glial cells and spinal cord cells. Advanced astroglial cells damage motor neurons by using two destructive pathways that cause inflammation and lower glutamate-breaking ability. SASP factors show increased levels in cerebrospinal fluid tests among individuals diagnosed with ALS, and this indicates that widespread inflammation originates from cellular aging processes [53, 54].

Correlation with Disease Severity and Progression

Greater numbers of senescent cells found in ALS patients lead to unfavorable disease progression as it speeds up degeneration, worsens motor abilities, and reduces life expectancy. The quantity of senescence markers in the brain accelerates neurocellular death and intensifies functional disabilities simultaneously. The results from Huntington's disease and multiple sclerosis examinations match these analyses because neurodegeneration and demyelination patterns cause senescence to appear along with chronic inflammation. The research indicates senescence biomarkers have medical value for numerous neurodegenerative diseases because they may lead to therapeutic options [55, 56].

Therapeutic Targeting of Senescence: Preclinical and Clinical Strategies

Senolytics: Compounds in Animal and Early-Phase Human Trials

The medical breakthrough senolytics operates specifically on pro-survival mechanisms of senescent cells to eliminate those cells from existence. Research shows that the pharmaceutical compounds dasatinib, quercetin navitoclax (ABT-263), and fisetin demonstrate preclinical effectiveness in neurodegeneration, improving brain performance and quality of life [57]. Therapeutic agents use BCL-2/BCL-XL and PI3K/AKT antiapoptotic targets to develop senescent cell apoptosis susceptibility. The beginnings of pharmaceutical readiness to assess how senolytics work in age-related diseases through early-stage clinical research appear in the UNITY trial (UBX0101), while few specific studies target neurodegeneration. Research translation programs concentrate on potential side effects and correct drug delivery practices of these compounds [58].

Senomorphics: SASP Modulation and Anti-Inflammatory Agents

The primary goal at senomorphics is to neutralize harmful cell aging through SASP suppression without harming essential cellular structures. Pro-inflammatory cytokines, chemokine partners, and matrix-degrading enzymes get interrupted by senior cell agents during their operation to stop neuroinflammation and tissue destruction. Scientific research indicates that senomorphics comprising rapamycin and metformin and inhibitors of the JAK class can produce positive effects through mTOR disruption and AMPK regulation [59]. Senomorphics are safer than senolytics since they maintain vital senescent cells while reducing harmful consequences. Current laboratory research shows these compounds have protective effects on neural structures, while human clinical evaluation primarily occurs at basic neurodegenerative disease development stages [60].

Antioxidants and Mitochondria-Targeted Therapies

The aging brain uses oxidative stress functioning with mitochondrial dysfunction to drive cellular senescence as its central mechanism. Cellular senescence damage shows less severity in patients when they receive treatments promoting mitochondrial wellness and ROS removal capabilities. Research shows that the neuron-protective effects of MitoQ SkQ1 and SS-31 antioxidants stem from their ROS-clearing ability and their simultaneous improvements in mitochondrial membrane potential and maintenance of cellular balance [61]. NAD^+ precursors (e.g., nicotinamide riboside) also support mitochondrial function and DNA repair. The studied approaches demonstrate two key results: delaying the progression of aging while enhancing brain function based on animal experiments. Research involving human participants continues, but scientists must uphold their findings by performing clinical neurodegenerative assessments on human subjects [62].

Epigenetic and Gene-Based Interventions

The primary cause of senescence stems from epigenetic dysregulation that manipulates gene expression linked to metabolism, DNA repair, and inflammation. Modern research aims to reprogram senescent traits through the popular use of HDAC inhibitors, BET inhibitors, and DNA methyltransferase inhibitors targeting chromatin modifiers. Academic research shows that the senescence-related pathways could be modified through CRISPR/Cas9 and RNA-based therapeutic approaches targeting p16INK4a or p21CIP1 pathways [63]. Animal studies have demonstrated that therapeutic approaches can both slow down brain-deteriorating processes and activate dormant cells in the brain. Safety concerns, delivery methods, and specificity challenges remain the main barriers to employing sophisticated techniques for treating age-related brain conditions in clinical environments [64].

Current Clinical Trials and Translational Challenges

The field of therapeutic research about senescence primarily concentrates on systemic aging conditions and persistent diseases, but direct treatment of neurodegenerative disorders receives little attention. Present-day clinical research evaluates the effects of UBX1325 and DQ combinations, metformin and rapamycin combination therapies, and mitochondrial modulators on cognitive performance, inflammation, and age-related deterioration [65]. Research extensions in senescence come with fundamental barriers to identifying brain senescence biomarkers, achieving precise blood–brain barrier distribution, and achieving an optimal therapeutic toxicity balance. Using senescent cell research in clinical settings encounters multiple challenges because senescence patterns alter while different senescent cell types exist during disease progression [66].

Future Perspectives and Emerging Directions

Early Detection and Senescence-Specific Biomarkers

It is essential for aging brain research to detect cellular senescence early so that neurodegenerative interventions can be initiated before tissue damage takes place. The combination of p16INK4a and p21CIP1 proteins, SA-β-gal activity, and SASP molecule data provides tools for distinguishing pathological aging from normal brain aging processes. The main ongoing research limitations arise from needing brain-specific markers that avoid body intrusions while also needing algorithms to differentiate heterogeneously composed tissues [67]. Better senescence detection should emerge because scientists create innovative ways to test cerebrospinal fluid (CSF) and build neuroimaging techniques and transcriptomic data evaluation systems. Biomarker detection systems developed for preclinical and clinical settings would make early diagnosis possible alongside clinical treatment monitoring, patient classification, and therapeutic response monitoring [68, 69].

Personalized Therapies for Age-Associated Neurodegeneration

Patients with age-related neurodegenerative diseases can receive personalized treatments that rely on assessments of their exclusive genetic makeup together with molecular and lifestyle attributes. Targeted therapeutic approaches can use individual molecular signatures, including senescence-associated genes, inflammatory profiles, and metabolic indicators, because cellular senescence shows differences between brain regions and people [70]. A medical practice that combines proteomics with pharmacogenomics and genomics delivers the best treatment outcomes and decreases adverse side effects. The application of precision medicine to neurodegeneration remains in the early development stage, and biomarker-assisted research and patient-specific senomorphic and senolytic treatment methods drive its current advancement. Strategies for managing brain aging and cognitive decline must incorporate patient diversity to create sustainable, efficient solutions [71].

Combination Therapies for Targeting Multiple Aging Pathways

Treating aging through monotherapies for a single feature shows insufficient capability for tackling complex neurodegenerative diseases with various underlying causes. Combatting affected aging pathways between neuroinflammation proteostasis failure and mitochondrial dysfunction requires joint treatment methods using both senolytics and anti-inflammatory medications with antioxidants and autophagy-stimulating agents [72]. Ordinary therapeutic medicines fail to match the effectiveness of combination treatments, which merge different therapeutic methods to restore brain tissue and enhance mental processes in research settings. Many challenges remain before we can achieve maximum dose delivery beyond toxicity limits and perform long-term assessments of treatment outcomes. Clinical trial frameworks and combinatorial regimen design require more focus because they will enable the full exploitation of antiaging treatments with multiple targets [73].

Integrating Artificial Intelligence and Big Data for Senescence Mapping

Combining big data analytics with artificial intelligence (AI) technologies almost revolutionizes how researchers study senescence through complex biological networks and multi-omics studies at high speeds. Machine learning algorithms use patient senescence profile classification to determine disease progression together with the ability to identify new senescence indicators. Scientists can construct brain senescence atlases with intervention precision models by merging large datasets from multiple sources, including genomic, transcriptomic, imaging, and clinical documentation information [74]. Its potential requires resolving three main issues: ensuring data uniformity and open algorithm access and validation across heterogeneous populations. Future medical practice will experience a total transformation of age-related neurodegenerative disease diagnosis, prognostication, and treatment preparation due to AI-driven technological advancements [75].

Challenges in Translating Preclinical Findings to Clinical Settings

Preclinical models have clarified cellular senescence's important role in neurodegeneration, yet several clinical barriers exist to their application. An animal-to-human translation of therapy effects remains uncertain because human models differ from animal models in their immunological environment, cell variety, and complex brain anatomy. The brain penetration of medications modulating senescence is restricted, while these drugs typically affect unintended body parts [76]. The path to clinical use becomes longer because ethical concerns combine with regulatory rules alongside the need for solid biomarkers. Adaptive trial designs, enhanced pharmacokinetic profiles, and better human-relevant models that include organoids and iPSC-derived neurons need implementation to address these difficulties. The transformation of senescence medicine into practical therapy for aging brains depends on closing the existing translational barriers [77].

Regenerative Approaches: Senescence and Stem Cell Therapy

Regenerative medicine, particularly using stem cells, represents one approach to managing cellular senescence symptoms in neurodegenerative diseases. Stem cell transplants show the potential to recover brain function by replacing dead nerve cells and adjusting glial cells' aging processes despite environmental restrictions to endogenous restoration [78]. The precision of such methods advances because of new discoveries within neural progenitor cells, induced pluripotent stem cells (iPSCs), and gene-editing technologies using CRISPR-Cas9. Serious barriers to cell transplants remain in the form of immune rejection, potential carcinogenic effects, and cellular aging-induced issues. Researchers believe that therapies that unite stem cell therapy with senolytic drugs will probably become the primary strategy to optimize neuronal niche revitalization while promoting functional outcomes [79].

Policy and Ethical Considerations in Senolytic Drug Development

The medical implementation of senolytic drugs for neurodegeneration treatment creates vital ethical conflicts that deal with protective drug specifications while also addressing problems with availability and continued application solutions. The improper use of drugs, together with the disagreement about longer lives versus enhanced health duration, leads to the main ethical issues and health risks for non-aging cells. Medical progress must be balanced by complete evaluation and assessment of safety risks in monitoring elderly populations in framework regulations. The solution requires equal treatment of high-cost senotherapeutics and a prepared public understanding of aging interventions. Physicians must conduct ethical clinical trials that monitor patients and gain consent to utilize senolytic therapies for neurological diseases associated with aging [80] (Table 5.1).

Table 5.1 This table illustrates the role of cellular senescence in neurodegeneration within the aging brain. It highlights key cellular players such as senescent astrocytes, microglia, and neurons, depicting their contributions to neuroinflammation, oxidative stress, and synaptic dysfunction. The schematic also outlines therapeutic strategies targeting senescence to mitigate neurodegenerative progression

Neurodegeneration in aging brain	Cellular senescence	Mechanism	Preclinical/clinical evidence	Therapeutic implementation	Reference
Alzheimer's disease (AD)	Senescent astrocytes and microglia	SASP-driven neuroinflammation, impaired amyloid clearance	Senescent glial cells identified in AD brains; cognitive recovery seen post-senolytic treatment	Dasatinib + Quercetin, JAK inhibitors (senolytics/senomorphics)	[81]
Parkinson's disease (PD)	Senescent dopaminergic neurons	Mitochondrial dysfunction, α-synuclein accumulation, oxidative stress	Senescence markers (p16, SA-β-gal) upregulated in PD brain tissue and models	Mitochondria-targeting senolytics; metformin as senomorphic agent	[9]
Cognitive aging	Neural stem and progenitor cell senescence	Decreased neurogenesis, stem cell exhaustion	Aging mice show impaired memory linked to senescent NSCs; reversal via senescent cell clearance	FOXO4-DRI peptide, caloric restriction mimetics, senolytics	[82]
Vascular dementia	Endothelial cell senescence	Blood–brain barrier (BBB) breakdown, vascular inflammation	Senescent endothelial cells correlated with cognitive dysfunction and microvascular pathology	NAD+ precursors, sirtuin activators, senolytics targeting vasculature	[83]
Frontotemporal dementia (FTD)	Senescent oligodendrocytes and astrocytes	Myelin degradation, altered glial support, SASP-mediated toxicity	Increased p21/p16 in glial cells in FTD models; inflammation and synapse loss observed	Senescence-targeted gene therapy; anti-inflammatory senomorphics	[83, 84]

Conclusion

The development of neurodegenerative diseases, as well as brain aging, occur primarily through challenging processes, which result in cellular senescence becoming a significant factor. The aging individual's brain becomes inflamed, neurogenesis declines, and neurons function poorly because astrocytes, microglia, and oligodendrocyte progenitors accumulate through SASP. Research performed on animal models has revealed the mechanisms explaining how oxidative stress, mitochondrial damage, DNA damage, and proteostasis imbalances cause neurodegeneration in senescence. Recent exploratory research about senescence biomarkers indicates their use in patient sifting techniques and disease monitoring processes for Alzheimer's and Parkinson's disease, as well as ALS. Research data demonstrates how age-related neurological treatments using senolytics that kill senescent cells and senomorphics that suppress SASP combined with multiple aging pathway-altering methods could become novel therapeutic approaches. The introduction of specific drugs managed by the blood–brain barrier remains one of several requirements that must be addressed for the clinical implementation of these treatments. However, research models show promise in resolving these barriers. The alignment of precision medicine approaches with senescence diagnostic methods creates opportunities for the future development of early disease detection capabilities together with customized treatments and live therapy performance tracking. Research about cellular senescence has transformed it into a valuable marker for brain aging and an effective therapeutic action target. By applying these findings, the therapy of neurodegeneration has the potential to transform itself, which creates new prospects to safeguard both the quality of life and cognitive abilities of older adults.

References

1. Hou Y, Dan X, Babbar M, Wei Y, Hasselbalch SG, Croteau DL, et al. Ageing as a risk factor for neurodegenerative disease. Nat Rev Neurol. 2019;15(10):565–81.
2. Kesidou E, Theotokis P, Damianidou O, Boziki M, Konstantinidou N, Taloumtzis C, et al. CNS ageing in health and neurodegenerative disorders. J Clin Med. 2023;12(6)
3. Shaikh MAJ, Gupta G, Bagiyal P, Gupta S, Singh SK, Pillappan R, et al. Enhancing drug bioavailability for Parkinson's disease: the promise of chitosan delivery mechanisms. Ann Pharm Fr. 2025;83(2):195–210.
4. Klement RJ. Cancer as a global health crisis with deep evolutionary roots. Global Transit. 2024;6:45–65.
5. Lynch SM, Guo G, Gibson DS, Bjourson AJ, Rai TS. Role of senescence and aging in SARS-CoV-2 infection and COVID-19 disease. Cells. 2021;10(12)
6. Sati A, Kyllait P, Gahtori P, Bhat HR, Hussain MS, Gupta G, et al. Behavioral and histopathological insights into phenylthiazolyl-1,3,5-triazines: potential antidepressant candidates in a rat model of depression. J Complement Integr Med. 2025;22:270.
7. Adamu A, Li S, Gao F, Xue G. The role of neuroinflammation in neurodegenerative diseases: current understanding and future therapeutic targets. Front Aging Neurosci. 2024;16:1347987.

8. Kazmi I, Yadav HKS, Al-Abbasi FA, Afzal M, Nadeem MS, Altayb HN, et al. Design of in-situ implant for the brain-targeted drug delivery using cross-linked gellan gum polymer through response surface methodology. Ann Pharm Fr. 2025;83(1):68–80.
9. Zheng L, He S, Wang H, Li J, Liu Y, Liu S. Targeting cellular senescence in aging and age-related diseases: challenges, considerations, and the emerging role of senolytic and senomorphic therapies. Aging Dis. 2024;15(6):2554–94.
10. Riessland M, Orr ME. Translating the biology of aging into new therapeutics for Alzheimer's disease: senolytics. J Prev Alzheimers Dis. 2023;10(4):633–46.
11. Kaur P, Khan H, Grewal AK, Dua K, Singh SK, Gupta G, et al. Exploring therapeutic strategies: the relationship between metabolic disorders and FOXO signalling in Alzheimer's disease. CNS Neurol Disord Drug Targets. 2025;24(3):196–207.
12. Marzola P, Melzer T, Pavesi E, Gil-Mohapel J, Brocardo PS. Exploring the role of neuroplasticity in development, aging, and neurodegeneration. Brain Sci. 2023;13(12)
13. Bashir B, Gulati M, Vishwas S, Gupta G, Dhanasekaran M, Paudel KR, et al. Bridging gap in the treatment of Alzheimer's disease via postbiotics: current practices and future prospects. Ageing Res Rev. 2025;105:102689.
14. Błaszczyk JW. Energy metabolism decline in the aging brain-pathogenesis of neurodegenerative disorders. Meta. 2020;10(11)
15. Vishwas S, Bashir B, Birla D, Khandale N, Chaitanya M, Chellappan DK, et al. Neuroprotective role of phytoconstituents-based nanoemulsion for the treatment of Alzheimer's disease. Curr Top Med Chem. 2024;24(19):1683–98.
16. Huang W, Hickson LJ, Eirin A, Kirkland JL, Lerman LO. Cellular senescence: the good, the bad and the unknown. Nat Rev Nephrol. 2022;18(10):611–27.
17. Thapa R, Ahmad Bhat A, Shahwan M, Ali H, PadmaPriya G, Bansal P, et al. Proteostasis disruption and senescence in Alzheimer's disease pathways to neurodegeneration. Brain Res. 2024;1845:149202.
18. Sikora E, Mosieniak G, Sliwinska MA. Morphological and functional characteristic of senescent cancer cells. Curr Drug Targets. 2016;17(4):377–87.
19. Thapa R, Afzal M, Goyal A, Gupta G, Bhat AA, Almalki WH, et al. Exploring ncRNA-mediated regulation of EGFR signalling in glioblastoma: from mechanisms to therapeutics. Life Sci. 2024;345:122613.
20. von Zglinicki T, Wan T, Miwa S. Senescence in post-mitotic cells: a driver of aging? Antioxid Redox Signal. 2021;34(4):308–23.
21. Samuel VP, Moglad E, Afzal M, Kazmi I, Alzarea SI, Ali H, et al. Exploring Ubiquitin-specific proteases as therapeutic targets in Glioblastoma. Pathol Res Pract. 2024;260:155443.
22. Gadhave DG, Sugandhi VV, Jha SK, Nangare SN, Gupta G, Singh SK, et al. Neurodegenerative disorders: mechanisms of degeneration and therapeutic approaches with their clinical relevance. Ageing Res Rev. 2024;99:102357.
23. Carnero A. MAP 17 and the double-edged sword of ROS. Biochim Biophys Acta. 2012;1826(1):44–52.
24. Bashir B, Alam S, Khandale N, Birla D, Vishwas S, Pandey NK, et al. Opening avenues for treatment of neurodegenerative disease using post-biotics: breakthroughs and bottlenecks in clinical translation. Ageing Res Rev. 2024;95:102236.
25. Harry GJ. Microglia during development and aging. Pharmacol Ther. 2013;139(3):313–26.
26. Nishiyama A, Shimizu T, Sherafat A, Richardson WD. Life-long oligodendrocyte development and plasticity. Semin Cell Dev Biol. 2021;116:25–37.
27. Marangon D, Castro ESJH, Cerrato V, Boda E, Lecca D. Oligodendrocyte progenitors in glial scar: a bet on remyelination. Cells. 2024;13(12)
28. Zhao H, Liu Z, Chen H, Han M, Zhang M, Liu K, et al. Identifying specific functional roles for senescence across cell types. Cell. 2024;187(25):7314–34.e21.
29. Garbarino VR, Palavicini JP, Melendez J, Barthelemy N, He Y, Kautz TF, et al. Evaluation of exploratory fluid biomarker results from a phase 1 senolytic trial in mild Alzheimer's disease. Res Sq. 2024; https://doi.org/10.21203/rs.3.rs-3994894/v1.

30. Ho CY, Dreesen O. Faces of cellular senescence in skin aging. Mech Ageing Dev. 2021;198:111525.
31. Dash UC, Bhol NK, Swain SK, Samal RR, Nayak PK, Raina V, et al. Oxidative stress and inflammation in the pathogenesis of neurological disorders: mechanisms and implications. Acta Pharm Sin B. 2025;15(1):15–34.
32. Lee KH, Cha M, Lee BH. Neuroprotective effect of antioxidants in the brain. Int J Mol Sci. 2020;21(19)
33. Kritsilis M, Rizou SV, Koutsoudaki PN, Evangelou K, Gorgoulis VG, Papadopoulos D. Ageing, cellular senescence and neurodegenerative disease. Int J Mol Sci. 2018;19(10):2937.
34. Gillispie GJ, Sah E, Krishnamurthy S, Ahmidouch MY, Zhang B, Orr ME. Evidence of the cellular senescence stress response in mitotically active brain cells-implications for cancer and neurodegeneration. Life (Basel, Switzerland). 2021;11(2)
35. Herdy JR, Traxler L, Agarwal RK, Karbacher L, Schlachetzki JCM, Boehnke L, et al. Increased post-mitotic senescence in aged human neurons is a pathological feature of Alzheimer's disease. Cell Stem Cell. 2022;29(12):1637–52.e6.
36. de Magalhães JP, Passos JF. Stress, cell senescence and organismal ageing. Mech Ageing Dev. 2018;170:2–9.
37. Han X, Zhang T, Liu H, Mi Y, Gou X. Astrocyte senescence and Alzheimer's disease: a review. Front Aging Neurosci. 2020;12:148.
38. Pal A, Rani I, Pawar A, Picozza M, Rongioletti M, Squitti R. Microglia and astrocytes in Alzheimer's disease in the context of the aberrant copper homeostasis hypothesis. Biomol Ther. 2021;11(11)
39. Brawek B, Skok M, Garaschuk O. Changing functional signatures of microglia along the axis of brain aging. Int J Mol Sci. 2021;22(3)
40. Fornari Laurindo L, Aparecido Dias J, Cressoni Araújo A, Torres Pomini K, Machado Galhardi C, Rucco Penteado Detregiachi C, et al. Immunological dimensions of neuroinflammation and microglial activation: exploring innovative immunomodulatory approaches to mitigate neuroinflammatory progression. Front Immunol. 2023;14:1305933.
41. de la Fuente AG, Queiroz RML, Ghosh T, McMurran CE, Cubillos JF, Bergles DE, et al. Changes in the oligodendrocyte progenitor cell proteome with ageing. Mol Cell Proteomics. 2020;19(8):1281–302.
42. Zou P, Wu C, Liu TC, Duan R, Yang L. Oligodendrocyte progenitor cells in Alzheimer's disease: from physiology to pathology. Transl Neurodegener. 2023;12(1):52.
43. Leenders F, Koole L, Slaets H, Tiane A, van den Hove D, Vanmierlo T. Navigating oligodendrocyte precursor cell aging in brain health. Mech Ageing Dev. 2024;220:111959.
44. Liu RM. Aging, cellular senescence, and Alzheimer's disease. Int J Mol Sci. 2022;23(4)
45. Safwan-Zaiter H, Wagner N, Wagner KD. P16INK4A-more than a senescence marker. Life (Basel, Switzerland). 2022;12(9)
46. Islam MT, Tuday E, Allen S, Kim J, Trott DW, Holland WL, et al. Senolytic drugs, dasatinib and quercetin, attenuate adipose tissue inflammation, and ameliorate metabolic function in old age. Aging Cell. 2023;22(2):e13767.
47. Gaikwad S, Senapati S, Haque MA, Kayed R. Senescence, brain inflammation, and oligomeric tau drive cognitive decline in Alzheimer's disease: evidence from clinical and preclinical studies. Alzheimers Dement. 2024;20(1):709–27.
48. Rim C, You MJ, Nahm M, Kwon MS. Emerging role of senescent microglia in brain aging-related neurodegenerative diseases. Transl Neurodegener. 2024;13(1):10.
49. Shreeya T, Ansari MS, Kumar P, Saifi M, Shati AA, Alfaifi MY, et al. Senescence: a DNA damage response and its role in aging and neurodegenerative diseases. Front Aging. 2023;4:1292053.
50. Zhu J, Wu C, Yang L. Cellular senescence in Alzheimer's disease: from physiology to pathology. Transl Neurodegener. 2024;13(1):55.
51. Riessland M, Kolisnyk B, Kim TW, Cheng J, Ni J, Pearson JA, et al. Loss of SATB1 induces p21-dependent cellular senescence in post-mitotic dopaminergic neurons. Cell Stem Cell. 2019;25(4):514–30.e8.

52. Jomova K, Raptova R, Alomar SY, Alwasel SH, Nepovimova E, Kuca K, et al. Reactive oxygen species, toxicity, oxidative stress, and antioxidants: chronic diseases and aging. Arch Toxicol. 2023;97(10):2499–574.
53. Maximova A, Werry EL, Kassiou M. Senolytics: a novel strategy for neuroprotection in ALS? Int J Mol Sci. 2021;22(21)
54. Jagaraj CJ, Shadfar S, Kashani SA, Saravanabavan S, Farzana F, Atkin JD. Molecular hallmarks of ageing in amyotrophic lateral sclerosis. Cell Mol Life Sci. 2024;81(1):111.
55. Yildiz O, Schroth J, Tree T, Turner MR, Shaw PJ, Henson SM, et al. Senescent-like blood lymphocytes and disease progression in amyotrophic lateral sclerosis. Neurol Neuroimmunol Neuroinflamm. 2023;10(1)
56. Dashtmian AR, Darvishi FB, Arnold WD. Chronological and biological aging in amyotrophic lateral sclerosis and the potential of senolytic therapies. Cells. 2024;13(11)
57. Richardson M, Richardson DR. Pharmacological targeting of senescence with senolytics as a new therapeutic strategy for neurodegeneration. Mol Pharmacol. 2024;105(2):64–74.
58. Qiu Z, Jia J, Zou H, Ao Y, Liu B, Wang Z. Targeting senescent cell clearance: an approach to delay aging and age-associated disorders. Transl Med Aging. 2021;5:1–9.
59. Zhang XX, He SH, Liang X, Li W, Li TF, Li DF. Aging, cell senescence, the pathogenesis and targeted therapies of osteoarthritis. Front Pharmacol. 2021;12:728100.
60. Luís C, Maduro AT, Pereira P, Mendes JJ, Soares R, Ramalho R. Nutritional senolytics and senomorphics: implications to immune cells metabolism and aging - from theory to practice. Front Nutr. 2022;9:958563.
61. Miwa S, Kashyap S, Chini E, von Zglinicki T. Mitochondrial dysfunction in cell senescence and aging. J Clin Invest. 2022;132(13)
62. Kobayashi H, Imanaka S. Mitochondrial DNA damage and its repair mechanisms in aging oocytes. Int J Mol Sci. 2024;25(23)
63. Dasgupta N, Arnold R, Equey A, Gandhi A, Adams PD. The role of the dynamic epigenetic landscape in senescence: orchestrating SASP expression. npj Aging. 2024;10(1):48.
64. Achar A, Myers R, Ghosh C. Drug delivery challenges in brain disorders across the blood-brain barrier: novel methods and future considerations for improved therapy. Biomedicines. 2021;9(12)
65. Si Z, Sun L, Wang X. Evidence and perspectives of cell senescence in neurodegenerative diseases. Biomed Pharmacother = Biomedecine & pharmacotherapie. 2021;137:111327.
66. Tenchov R, Sasso JM, Wang X, Zhou QA. Antiaging strategies and remedies: a landscape of research progress and promise. ACS Chem Neurosci. 2024;15(3):408–46.
67. Zhang L, Pitcher LE, Yousefzadeh MJ, Niedernhofer LJ, Robbins PD, Zhu Y. Cellular senescence: a key therapeutic target in aging and diseases. J Clin Invest. 2022;132(15)
68. Hatami-Fard G, Anastasova-Ivanova S. Advancements in cerebrospinal fluid biosensors: bridging the gap from early diagnosis to the detection of rare diseases. Sensors (Basel, Switzerland). 2024;24(11)
69. Wen Q, Wittens MMJ, Engelborghs S, van Herwijnen MHM, Tsamou M, Roggen E, et al. Beyond CSF and neuroimaging assessment: evaluating plasma miR-145-5p as a potential biomarker for mild cognitive impairment and Alzheimer's disease. ACS Chem Neurosci. 2024;15(5):1042–54.
70. Strianese O, Rizzo F, Ciccarelli M, Galasso G, D'Agostino Y, Salvati A, et al. Precision and personalized medicine: how genomic approach improves the management of cardiovascular and neurodegenerative disease. Genes. 2020;11(7)
71. Sadee W, Wang D, Hartmann K, Toland AE. Pharmacogenomics: driving personalized medicine. Pharmacol Rev. 2023;75(4):789–814.
72. Reddy PH. Mitochondrial medicine for aging and neurodegenerative diseases. NeuroMolecular Med. 2008;10(4):291–315.
73. Margulies S, Hicks R. Combination therapies for traumatic brain injury: prospective considerations. J Neurotrauma. 2009;26(6):925–39.
74. Fu C, Chen Q. The future of pharmaceuticals: artificial intelligence in drug discovery and development. J Pharm Anal. 2025; https://doi.org/10.1016/j.jpha.2025.101248.

75. Kumar R, Waisberg E, Ong J, Paladugu P, Amiri D, Saintyl J, et al. Artificial intelligence-based methodologies for early diagnostic precision and personalized therapeutic strategies in neuro-ophthalmic and neurodegenerative pathologies. Brain Sci. 2024;14(12)
76. Lei S-Y, Qu Y, Yang Y-Q, Liu J-C, Zhang Y-F, Zhou S-Y, et al. Cellular senescence: a novel therapeutic target for central nervous system diseases. Biomed Pharmacother. 2024;179:117311.
77. Korn EL, Freidlin B. Adaptive clinical trials: advantages and disadvantages of various adaptive design elements. J Natl Cancer Inst. 2017;109(6)
78. Deng S, Xie H, Xie B. Cell-based regenerative and rejuvenation strategies for treating neurodegenerative diseases. Stem Cell Res Ther. 2025;16(1):167.
79. Merlin JPJ, Abrahamse H. Optimizing CRISPR/Cas9 precision: mitigating off-target effects for safe integration with photodynamic and stem cell therapies in cancer treatment. Biomed Pharmacother. 2024;180:117516.
80. Christopher C, Kc B, Shrestha S, Blebil AQ, Alex D, Mohamed Ibrahim MI, et al. Medication use problems among older adults at a primary care: a narrative of literature review. Aging Med (Milton). 2022;5(2):126–37.
81. Alshaebi F, Sciortino A, Kayed R. The role of glial cell senescence in Alzheimer's disease. J Neurochem. 2025;169(3):e70051.
82. Fan R, Peng X, Xie L, Dong K, Ma D, Xu W, et al. Importance of Bmal1 in Alzheimer's disease and associated aging-related diseases: mechanisms and interventions. Aging Cell. 2022;21(10):e13704.
83. Han Y, Kim SY. Endothelial senescence in vascular diseases: current understanding and future opportunities in senotherapeutics. Exp Mol Med. 2023;55(1):1–12.
84. de la Fuente AG, Pelucchi S, Mertens J, Di Luca M, Mauceri D, Marcello E. Novel therapeutic approaches to target neurodegeneration. Br J Pharmacol. 2023;180(13):1651–73.

Cellular Senescence in Alzheimer's Disease (AD)

6

Rahamat Unissa Syed, S. Afsar, and Fulvia di Pillo

Introduction

Worldwide, the most common form of dementia exists as AD, which progresses as a neurological disease that affects brain functions through behavioral changes, memory deterioration, and cognitive loss [1]. The two central hallmarks of Alzheimer's disease include Aβ plaques that form outside cells and tau protein aggregates that occur inside cells after hyperphosphorylation. The chemical processes mainly affect the cortex and hippocampus structures, causing brain reduction, the death of neurons, and dysfunction in the synapses [2, 3]. AD causes a significant burden on society because millions suffer from it while enduring medical needs and the absence of treatments that can modify its course. Scientists have so far failed to determine the root cause of AD, even after extensive research, while it contains elements from genetics, environment, and behavior. Researchers need to understand the relationship between pathological alterations and age-related processes to discover new therapeutic approaches and disease promotion mechanisms for AD [4, 5]. DNA damage, shortened telomeres, increased oxidative stress, and

R. U. Syed (✉)
Department of Pharmaceutics, College of Pharmacy, University of Hai'l, Hai'l, Saudi Arabia
e-mail: ru.syed@uoh.edu.sa

S. Afsar
Department of Virology, Sri Venkateswara University, Tirupathi, Andhra Pradesh, India
e-mail: mammoth.afsar1@gmail.com

F. di Pillo
Centre for Medicinal Chemistry (CQMED), Campinas, SP, Brazil
e-mail: fulviadipillo@unicamp.br

G. Gupta (ed.), *Cellular Senescence and Brain Aging*,
https://doi.org/10.1007/978-981-96-8873-9_6

activated oncogenes result in cellular senescence, which becomes an irreversible, permanent cell cycle stop [6]. The basic purpose of senescence within tumors is to protect against cancer development, while the process simultaneously aids embryonic development, tissue healing, and tissue remodeling.

The SASP, along with enhanced senescence-associated β-galactosidase (SA-β-gal) activity and cell shape alterations, defines senescent cell features [7, 8]. Each of these elements—growth factors with matrix-remodeling enzymes and cytokines and chemokines—constitutes the secretory components of SASP, which influence the local microenvironment. Although transient senescent cells benefit short-term needs, they perpetuate various aging-related diseases involving tissue impairment and chronic inflammation. Researchers are studying age-related diseases through studies of these targets due to their accumulation and dysfunctional removal processes during aging. Cells in senescence state cause breaks in tissue structures and persistent inflammation, making scientists explore its effects on degenerative brain conditions and AD [9, 10]. The senescent cells of astrocytes, microglia, and neurons generate proinflammatory cytokines, reactive oxygen species, and matrix-degrading enzymes in aged and AD brains due to SASP signaling. The disease-promoting substances released from such environments accelerate the progression of tau disease, together with amyloid-beta accumulation and synapse depletion. The pace of glial regulation failure and neuronal damage increases due to senescent mitochondrial defects and defective autophagic processes [11, 12]. The decreased immune ability that comes with aging allows senescent cells to remain in the body, which triggers extended neuroinflammation, thus quickening disease progression. Knowledge regarding the relationships between aging cells and pivotal Alzheimer's disease symptoms becomes important for studying neurodegeneration mechanisms and creating new treatments [13, 14]. This paper will analyze cellular senescence's role in Alzheimer's disease development by reviewing current preclinical and clinical literature [15]. First, researchers explore cellular and molecular mechanisms of senescence through SASP and mitochondrial dysfunction, which relate to AD traits that include tau protein accumulation and amyloid-β plaque formation. Postmortem investigations of human brains and results from animal models, which show that neurodegeneration occurs due to senescence cell accumulation, will be discussed in this chapter. The research will evaluate whether senolytics and senomorphics can reduce AD pathology while improving cognitive functions. The study finds ways to develop senescence-based therapeutic approaches for AD while examining knowledge gaps through an extensive review of existing data [16, 17] (Fig. 6.1).

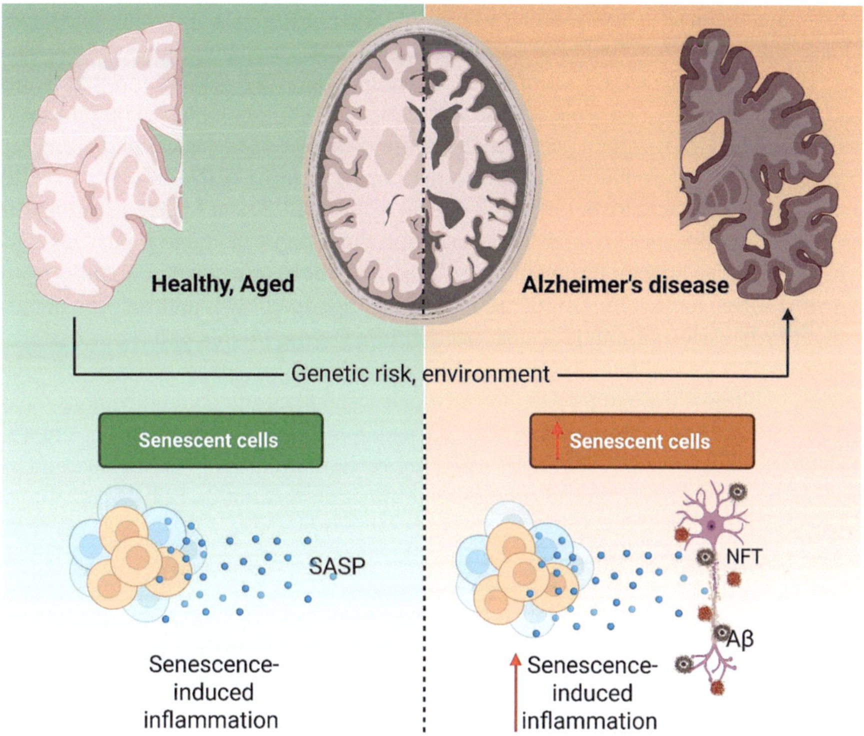

Fig. 6.1 Cellular senescence: AD vs. healthy aging. The comparison of cellular senescence functions appears within this figure between AD patients and healthy elderly individuals. Healthy older brains achieve and sustain low inflammation with homeostasis through SASP factors supported by minimal senescent cell numbers. The accumulation of senescent cells in AD worsens SASP-mediated inflammation that results in neurofibrillary tangle (NFT) development together with amyloid-β (Aβ) plaque formation. Environment and heredity factors worsen the modification process by stimulating the degeneration of nerve cells

Understanding Cellular Senescence

Overview of Cellular Senescence: Types, Biomarkers, SASP, and CNS Involvement

Cellular senescence represents an irreversible cell cycle block that functions as a vital aging mechanism and develops into multiple age-related medical conditions, including AD. Cellular senescence develops due to various types of cellular stress, such as oxidative damage, oncogene activation, and telomere attrition, leading to replicative senescence and DNA damage resulting from stress- or oncogene-induced senescence [18, 19]. Senescent cells' characteristics include elevated

SA-β-galactosidase activity alongside p16INK4a and p21CIP1 cyclin-dependent kinase inhibitor overexpression and prolonged DDR signaling at telomeres. The SASP in cells represents a set of characteristics manifested through growth factor and matrix-degrading enzyme release with pro-inflammatory cytokine outputs (that include IL-6, IL-1β), which generates tissue homeostasis disturbances and promotes both persistent inflammatory responses and impacts nearby cells through paracrine actions [20–22]. Scientists commonly recognize cellular senescence as an active agent in neurodegenerative tissue degradation throughout the central nervous system (CNS). The metabolic dysfunction of senescent neurons, inflammation from SASP in astrocytes and microglia, and synaptic impairment will intensify the pathological progression of amyloid-beta and tau in AD. The brain's aging process creates dysfunctional neural plasticity, quick cognitive decline, and disease-permissive tissue conditions. Research on central nervous system cellular and molecular senescence bases needs understanding for AD pathophysiology studies, and it creates new therapeutic possibilities for eliminating or reprogramming senescent cells to prevent brain deterioration and promote late-function restoration [23–25].

Mechanistic Links Between Senescence and Alzheimer's Disease

Glial Senescence, Genomic Instability, SASP, and Mitochondrial Stress in Alzheimer's Disease

The processes linking cellular senescence result in significant neurodegeneration, which occurs within AD. The deposition of Amyloid-β (Aβ) along with tau disease worsens due to the aging process of glial cells, such as astrocytes and microglia, that expand with age while developing inflammatory properties that disrupt neuronal maintenance and homeostasis. These unhealthy glial cells deteriorate into secretion factories that generate SASP factors alongside pro-inflammatory cytokines and chemokines with proteases to progressively damage the neurons and boost neuroinflammation [26–28]. Neurons that experience telomere shortening and DNA damage display decreased cell regeneration ability and damaged genomic integrity because aging processes make them susceptible to tau hyperphosphorylation and Aβ oligomers. The pathogenic cascade becomes worse because of continuous DNA damage response activation in neurons, which leads to senescence-like cellular changes. The supporting elements of SASP promote protein deterioration and aggregation of tau and Aβ while expanding the cycle of cellular senescence and protein pathology and generating long-term inflammatory responses [29, 30]. The pathologic conditions worsen because senescent cells show reduced metabolic efficiency in mitochondria and increased ROS production, which damages cellular structures while worsening DNA damage and causing difficulties in clearing protein aggregates through autophagy. Neurodegenerative progression spreads because of this oxidative brain environment that leads to deterioration in different brain areas, while driving forward senescence. Research shows cellular senescence functions as a driving force in AD development instead of being an aging consequence because of its interdependent molecular and cellular pathways. The therapeutic benefits of senescence-targeted therapies support their use as a preventive method or disease progression inhibitor [31–33].

Preclinical Evidence of Senescence in AD Models

Rodent Models of AD and Senescence Markers

Research on AD neurodegeneration reveals cellular senescence functions through studies with transgenic mice carrying mutant tau, PSEN1, and APP proteins. These models reproduce all essential AD features, including neural tau pathology, amyloid-β clumping, and cognitive decline signs. A recent study indicates glial cells in these models, consisting mainly of astrocytes and microglia, show high amounts of the senescence markers p16INK4a, p21CIP1, and SA-β-gal. The presence of senescent cells within amyloid plaques creates a link between these plaques and brain dysfunction and neuroinflammation. The preclinical research provides a practical method for assessing treatments that target senescence while supporting the notion that cellular senescence actively contributes to AD pathogenesis beyond being only a result [34–36].

Genetic and Pharmacologic Induction/Reduction of Senescence in Animal Models

The function of cellular senescence during AD progression has become clearer through animal model experiments that either genetically or pharmacologically control senescence. Studies using suicide genes with p16INK4a control mechanisms showed that the deletion of senescent cells leads to decreased neuroinflammation and preserved neuronal health (INK-ATTAC model) [37]. Research studies have tested pharmacologic agents, including the senomorphics rapamycin and the senolytics dasatinib and quercetin, to eliminate or control the existence of senescent cells. The observed treatment methods achieve positive effects by protecting neurons while decreasing tau and amyloid pathology and restoring synaptic function. These models demonstrate that senescence actively drives disease progression in AD while establishing a possibility to treat this condition through targeting senescence [38].

Behavioral and Cognitive Outcomes in Senescence-Targeted Studies

Behavioral tests confirm that mouse models of AD experience improved cognitive function when treatment aims to eliminate cellular senescence pathways. The

Morris water maze, Y-maze, and new objects recognition tasks perform better when treated with senolytic medications and genetic senescent cell elimination procedures [39]. Research indicates an association between senescent cell numbers and cognitive decline because these treatment effects often match decreased neuroinflammation, better synaptic connections, and AD pathology reduction. The research data demonstrates how pharmacological control of senescence presents promising chances for clinical use in age-related cognitive decline while validating the theory for preventing or delaying AD behavioral effects [17].

Limitations and Gaps in Preclinical Research

The study of senescence in AD models through preclinical research encounters major limitations, even though it shows promising outcomes. The general applicability to sporadic AD remains restricted because most rodent models extensively introduce familial AD mutations, while familial AD constitutes only a tiny portion of all human AD cases. The interpretation process remains complex because different studies present varying maps of senescence markers and their timing patterns throughout models and brain segmentations [40]. Medicine fields have not developed defined ways or universal biomarkers for detecting in vivo cell aging conditions. Researchers have not sufficiently established the long-term safety profile and effectiveness of older animal test subjects receiving senolytic therapies. Research has not identified the effects that environmental contexts, along with genetic backgrounds and sexual conditions, have on Alzheimer's disease development. Research methods and sophisticated models need development because they are essential for verifying senescence treatment as a future therapeutic avenue in Alzheimer's disease pathology [41, 42].

Clinical Observations and Translational Insights

Evidence of Senescent Cells in Postmortem AD Brains

Pathological analysis of AD brains has shown increased cell numbers of astrocytes, microglia, and oligodendrocyte progenitor cells. Cells with senescence markers show β-galactosidase activity and p16INK4a and p21CIP1/WAF1, as well as other traditional markers. Elevated tau and amyloid-beta pathology areas frequently contain senescent glial cells, whose presence suggests possible involvement in illness development. Neuroinflammation and neuronal dysfunction develop worse because of these cells' inflammatory secretory actions. Research into treatments targeting cellular senescence represents a promising solution for neurodegenerative diseases since AD pathology directly corresponds to cellular senescence at a molecular and geographic level [17, 43].

Senescence-Related Biomarkers in Human AD Patients

Research has shown that AD patients contain senescence-associated biomarkers in their cerebrospinal fluid and peripheral blood, which support monitoring and identifying the disorder early on. The extent of neurodegeneration correlates with elevated p16INK4a molecules and inflammatory cytokines (e.g., IL-6, IL-8) together with SASP factors. Researchers have proposed that exosomal markers, including miRNAs, together with DNA methylation changes related to cellular aging, have the potential for utilization for diagnostic purposes [44]. These markers establish a connection between peripheral health signs and central AD neuropathological changes, which represent the combined effect of systemic and age-related brain deterioration. The medical worth of these biomarkers for tracking AD therapy, clinical patient categorization, and early disease detection can improve through standardization and validation efforts [45].

Correlation Between Senescence and Cognitive Decline

Rising evidence through clinical research indicates that progressive cognitive decline in AD directly correlates with increasing senescent cell numbers. The persistent inflammatory condition formed by microglia and senescent astrocytes disrupts neuroplasticity and synaptic function. Cognitive impairment becomes stronger because of elevated senescence markers detected in patient-derived test samples, particularly within memory and executive function domains. The disease progression appears to increase senescence cell abundance, resulting in reduced neuronal integrity. Evidence supports the idea that targeting senescence through therapeutic measures can fight against neurodegeneration and cognitive aging, improving the therapeutic value of anti-senescence treatments in AD management [46, 47].

Emerging Clinical Studies Targeting Senescence in AD

Recent clinical research evaluates potential senotherapeutics that target the elimination of senescent cells in cases of AD. The safety, along with effectiveness, of dasatinib and quercetin medications for treating moderate cognitive impairment or early-stage AD is being studied in current initial-stage clinical assessments. The aim of these studies is to measure cognitive developments, senescence changes, and improvements in inflammatory markers. Preliminary research regarding aging has shown evidence of potential neuroprotective outcomes. The current experimental trials provide early indications for translating senescence-focused strategies into potential Alzheimer's disease treatment methods. More substantial results from therapeutic trials will become certain after additional trials with extended follow-up periods using larger patient populations [38, 48] (Fig. 6.2).

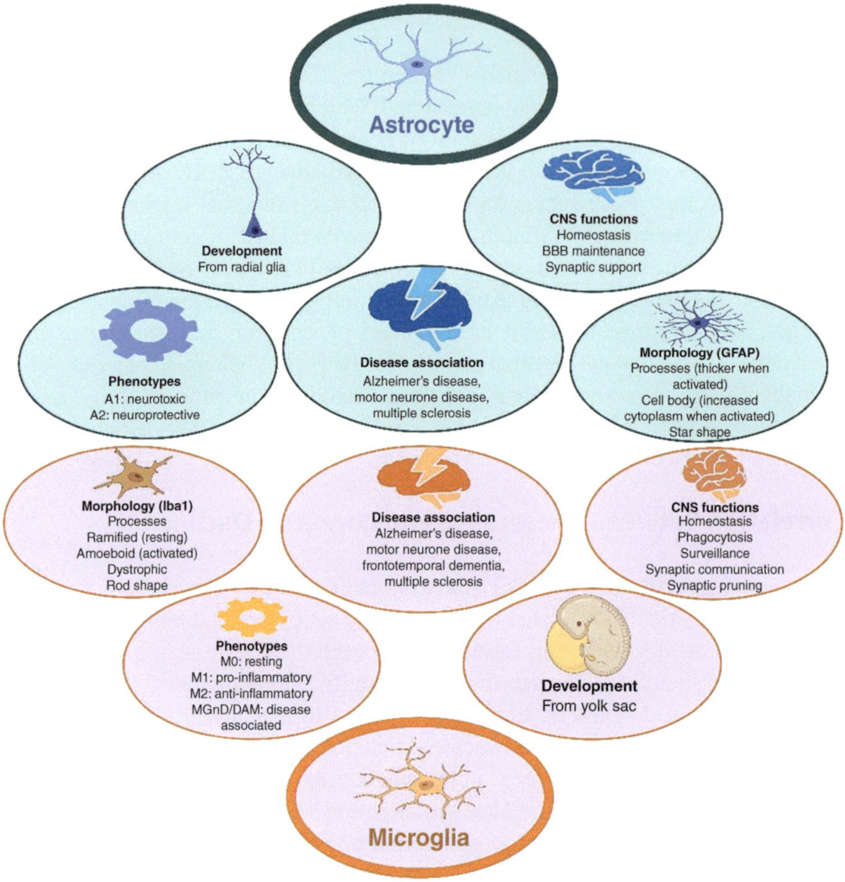

Fig. 6.2 Astrocyte and microglia phenotypes in CNS health and disease. This figure describes the contrasting processes of astrocyte and microglia development, cellular phenotypes, overall structure, central nervous system functioning and their associations with diseases. These cells stem from radial glia, thereby producing a star shape that helps preserve synaptic function and maintain blood–brain barrier integrity and achieve homeostasis throughout the central nervous system. The yolk sac serves as the origin of microglial cells, which take either a ramified or an amoeboid form and exhibit Iba1-positive properties. Microglia execute the duties of surveillance together with phagocytosis and synaptic pruning functions. The different characteristics of these cells emerge through cellular origins, while their research connection extends to multiple sclerosis along with Alzheimer's disease and motor neuron disease. The development of frontotemporal dementia is related to microglia cells

Therapeutic Targeting of Senescence in AD

Senolytics and Senomorphics

Medical researchers call the substances "senolytics" that target survival pathways, BCL-2, BCL-xL, and PI3K/AKT to trigger apoptosis in mature cells. The treatment known as senomorphics transforms senescent conditions without generating cell mortality. The substances operate by both mTOR and JAK/STAT and NF-κB signaling pathways to block the release of SASP [49]. Through chronic SASP release, the buildup of senescent glial and neuronal cells in Alzheimer's disease aggravates neuroinflammation and neurodegeneration. Using therapeutic strategies with senolytics or senomorphics helps target senescence to reduce harmful effects while lowering the pathogenic load and maintaining cognitive abilities. The mechanistic nature of these medications seems suitable for early-stage AD treatment since senescent patient cells help initiate AD by causing tau pathology and synaptic dysfunction before extensive disease spread occurs [50, 51].

Preclinical Trials of Senolytic Agents in AD Models

The preclinical research shows that senolytic drugs minimize Alzheimer's disease in animal models. Recent research eliminates senescent astrocytes, neurons, and microglia by using several drugs such as fisetin, navitoclax, dasatinib, and quercetin (D+Q) in transgenic Alzheimer's disease mouse models. The combination of treatments enhanced synaptic connectivity alongside spatial memory and caused a reduction of neuroinflammation, as well as minimized tau hyperphosphorylation alongside the amyloid-beta burden. Intermittent administration of drugs produced limited side effects, preserving healthy cell numbers in the study. The experimental findings support the notion that senescent cells cause AD development, making senolytic treatment an excellent research direction for potential therapeutic applications [38, 52].

Challenges in Brain-Targeted Delivery of Senotherapeutics

The principal challenge to reach effective brain delivery of senotherapeutic drugs lies in the blood–brain barrier that restricts entry of multiple small molecules and biologic agents. Senolytic distribution follows a nonspecific pattern to cause tissue damage as it affects healthy regions, requiring temporary cell aging for proper growth. Scientists build contemporary drug delivery methods using liposomes, nanoparticle carriers, and receptor-mediated delivery systems to overcome the blood–brain barrier and exclusively increase delivery rates to senescent cells. Research on prodrugs and senolytic penetration into CNS tissue is growing in interest because of their low systemic toxicity properties. Successful therapeutic use of senotherapeutics in AD requires developing technological improvements in translation and safety measures [53, 54].

Potential Combination Therapies (e.g., Anti-Amyloid + Senolytic)

The treatment of Alzheimer's disease with synergy can be achieved through the simultaneous usage of senolytic drugs together with established Alzheimer's medications that fight amyloid beta or tau proteins. Senolytics work on the cell environment through their ability to eliminate senescent cells, which enables damage to tissues and neuroinflammation via SASP. However, conventional treatments focus primarily on clearing extracellular aggregates [55]. Preclinical evidence suggests that reducing the population of senescent cells enhances how well treatment methods impact amyloid by helping decrease glial activity and reestablishing immune function. The combined approach, which tackles aging cell factors before aggregating plaque and tangles, could create stronger positive therapeutic outcomes. The potential benefits of combined neuroprotection therapies for AD treatment require additional future research on average doses and safety profiles, timing of administration, and dosing schedules [56].

Future Perspectives and Emerging Directions

Advancing Senescence Biomarker Discovery in AD

The main obstacle in AD research involves discovering specific cellular senescence markers that demonstrate precision and accuracy. The current tissue nonspecific markers, senescence-associated β-galactosidase, p16INK4a, and p21CIP1, often show overlapping results with normal aging markers. Biomarker identification needs multi-method approaches that unite transcriptomic data with proteomic data and metabolomic data obtained from preclinical models and AD patient-derived samples [57]. Examining senescent cell populations within the disease-affected brain tissue can be achieved through single-cell approaches and spatial transcriptomic methods. Fluid-based biomarkers extracted from blood and cerebrospinal fluid show potential for treatment observation and early disease recognition. Future research must validate prospective candidates among chronic human populations while establishing diagnostic methods that separate senescence from other diseases affecting aging patients [58].

Integration of AI and Omics for Predictive Senescence Profiling

Through a convergence of artificial intelligence (AI) and omics technologies, there will be a revolutionary turning point in senescence research for Alzheimer's disease because of extensive and predictable profiling of senescent cell phenotypes. Research obstacles remain because of the complex nature and diverse types of omics data, although they are plentiful in quantity. Machine learning algorithms process multimodal data that combines transcriptomics, proteomics, epigenomics, and genomics to reveal hidden patterns and subtype senescent cells and predict clinical progression [59]. Early intervention strategies can benefit from predictive models to recognize people whose neurodegeneration appears prematurely to reach the senescence stage. Multiple challenges persist regarding standardizing data formats and disclosing algorithms while acquiring lengthy data series that fulfil high technical requirements. AI-driven senescence profiling in AD requires an open-source platform and standardized database development through cooperative initiatives to reach its full potential [59, 60].

Personalized Senescence-Targeted Interventions

Senescence-modulation treatment of Alzheimer's disease shows potential under the concept of personalized medicine. Different patient populations show unique participation of cell types along with unique molecular causes and systemic effects of senescence. Organizing patients into groups based on the analysis of biomarkers alongside omics data plus imaging methods enables efficient use of senomorphics and senolytics with minimized possible adverse effects. Brain-penetrant prodrugs and nanoparticle-based carriers can be applied in precision techniques to act on and eliminate or modify particular senescent cell populations [61, 62]. The healthcare outcome can improve when neuroprotective substances are combined with anti-inflammatory treatments and medications targeting amyloid. For treatment optimization, healthcare professionals must account for individual patient conditions, including personal genetic influences, sex-linked factors, and other health complications. Neurologists, pharmacologists, and systems biologists must work transdisciplinarily to turn aging biology into patient-centered clinical practice by establishing customized therapy design [63, 64].

Roadmap for Clinical Translation: Regulatory and Ethical Challenges

The development of senescence-targeted treatments for Alzheimer's disease remains blocked because of regulatory uncertainties, together with ethical issues. Potential adverse effects of senotherapeutics at both their molecular level and their off-target site, along with their distinct methods of functioning, create safety risks that are especially worrisome for elderly individuals with cognitive limitations. New treatment methods should include fundamental regulations defining trial methodology, biomarker assessment, and risk versus benefit assessments in contemporary regulatory standards [65]. The ethical concerns include fair accessibility and abuse risk of age-modulation drugs, together with informed consent challenges that afflict disadvantaged population groups. Solving the social impact issues and fighting stigma demands leadership from the public and important stakeholder groups. Multidisciplinary consortia, real-world data registries, and adaptive clinical trial platforms need to be established because they help with deep examinations and speed up approval periods. A complete strategic path combining scientific progress with ethical oversight and regulatory administration must exist to convert new AD treatments [66] (Table 6.1).

Table 6.1 The aspects of cellular senescence in Alzheimer's disease, highlighting major triggers, biomarkers, and the role of SASP in disease progression. It also outlines current senotherapeutic strategies and their evidence, emphasizing translational challenges such as blood–brain barrier penetration, delivery methods, and the need for precision medicine approaches

Aspect	Mechanism/marker	Pathological role in AD	Preclinical/clinical evidence	Therapeutic strategies	Reference
Triggers of senescence	Telomere shortening, oxidative stress, DNA damage, oncogene activation	Induces irreversible cell cycle arrest in glial and neuronal cells	Observed in aged AD brains and transgenic models	Target upstream stressors (e.g., antioxidants, DNA repair enhancers)	[67, 68]
Senescence biomarkers	SA-β-gal, p16^INK4a^, p21^CIP1/WAF1^, persistent DDR	Reflect senescent phenotype in astrocytes, microglia, and neurons	Elevated in AD patient samples (brain, CSF, blood); colocalized with plaques	Develop noninvasive fluid-based biomarkers for diagnosis and monitoring	[69, 70]
SASP factors	IL-6, IL-1β, MMPs, chemokines	Amplify neuroinflammation, disrupt synaptic function, promote Aβ and tau pathology	Found in both AD models and human postmortem tissues	Senomorphics to suppress SASP; combination with anti-inflammatory agents	[71–73]
Senotherapeutics	Senolytics (e.g., dasatinib + quercetin, navitoclax), senomorphics (e.g., rapamycin)	Reduce senescent cell burden, mitigate inflammation and pathology	Improved cognition, reduced Aβ/tau in AD mouse models	Clinical trials in early AD underway; personalized dosing under exploration	[55, 74]
Delivery and translation	BBB penetration, off-target toxicity, patient heterogeneity	Limits efficacy and safety in CNS; complicates clinical application	Variable responses in models; need for patient stratification	Nanocarriers, CNS-permeable agents, precision medicine approaches	[75, 76]

Conclusion

Research demonstrates that AD pathogenesis significantly depends on cellular senescence because studies have documented its role in both starting and progressing brain tissue deterioration. The elderly brain and brain with disease foster accumulation of senescent cells among astrocytes, microglia, and oligodendrocyte progenitor cells. Neurotoxic and inflammatory chemicals called SASP emerge from these affected cells to advance cognitive deterioration. When SASP factors remain active chronically, the pathology intensifies to affect both amyloid-beta and tau proteins, thus worsening inflammation while damaging synapses in a destructive symptomatic cycle. New therapeutic approaches have become possible because scientists can redesign treatments by recognizing cell aging as a pathological process that can be modified. Research in senescence biology has generated two new diagnostic biomarkers by discovering both senescence-specific gene expression patterns and different secretory mediators. The discovery of biomarkers leads to better early-stage disease identification and more effective disease assessment. Research into senotherapeutic drugs has brought optimism to preclinical AD models through the combination of neuroinflammation reduction and better cognitive outcomes achieved by senolytics, which target senescent cells and senomorphics that minimize SASP. The potential application of cellular senescence investigations in neurodegenerative diseases can be observed through ongoing clinical research at their early stages. Independently using senescence-targeted therapy is not adequate for treating the complex nature of AD. These therapeutic approaches must amalgamate with wider treatment plans that examine all elements that cause damage, including mitochondrial dysfunction alongside oxidative stress, poor neurogenesis, and amyloid, as well as tau pathology. Significant effective therapeutic benefits may require combination approaches to succeed. Knowledge about cellular senescence in AD development has brought significant progress toward biomarker innovation and innovative treatment solutions. Research investigating the senescence-pathogenic process interactions will be essential to creating successful multimodal prevention and treatment solutions for Alzheimer's disease.

References

1. Nadeem MS, Khan JA, Al-Abbasi FA, Alqurashi MM, Bawadood AS, Alzarea SI, et al. Barbigerone against lipopolysaccharide-induced memory deficit in rodents via alteration of inflammatory and oxidative stress pathway: in vivo and molecular dynamics simulations study. CNS Neurol Disord Drug Targets. 2025;24:554.
2. Karantzoulis S, Galvin JE. Distinguishing Alzheimer's disease from other major forms of dementia. Expert Rev Neurother. 2011;11(11):1579–91.
3. Kaur P, Khan H, Grewal AK, Dua K, Singh SK, Gupta G, et al. Exploring therapeutic strategies: the relationship between metabolic disorders and FOXO signalling in Alzheimer's disease. CNS Neurol Disord Drug Targets. 2025;24(3):196–207.
4. Wainberg ML, Scorza P, Shultz JM, Helpman L, Mootz JJ, Johnson KA, et al. Challenges and opportunities in global mental health: a research-to-practice perspective. Curr Psychiatry Rep. 2017;19(5):28.

5. Bashir B, Gulati M, Vishwas S, Gupta G, Dhanasekaran M, Paudel KR, et al. Bridging gap in the treatment of Alzheimer's disease via postbiotics: current practices and future prospects. Ageing Res Rev. 2025;105:102689.
6. Vishwas S, Bashir B, Birla D, Khandale N, Chaitanya M, Chellappan DK, et al. Neuroprotective role of phytoconstituents-based nanoemulsion for the treatment of Alzheimer's disease. Curr Top Med Chem. 2024;24(19):1683–98.
7. Chen JH, Hales CN, Ozanne SE. DNA damage, cellular senescence and organismal ageing: causal or correlative? Nucleic Acids Res. 2007;35(22):7417–28.
8. Thapa R, Moglad E, Afzal M, Gupta G, Bhat AA, Hassan Almalki W, et al. The role of sirtuin 1 in ageing and neurodegenerative disease: a molecular perspective. Ageing Res Rev. 2024;102:102545.
9. Alavimanesh S, Nayerain Jazi N, Choubani M, Saeidi F, Afkhami H, Yarahmadi A, et al. Cellular senescence in the tumor with a bone niche microenvironment: friend or foe? Clin Exp Med. 2025;25(1):44.
10. Ohtani N. The roles and mechanisms of senescence-associated secretory phenotype (SASP): can it be controlled by senolysis? Inflamm Regen. 2022;42(1):11.
11. Deng Q, Wu C, Parker E, Liu TC, Duan R, Yang L. Microglia and astrocytes in Alzheimer's disease: significance and summary of recent advances. Aging Dis. 2024;15(4):1537–64.
12. Thapa R, Ahmad Bhat A, Shahwan M, Ali H, PadmaPriya G, Bansal P, et al. Proteostasis disruption and senescence in Alzheimer's disease pathways to neurodegeneration. Brain Res. 2024;1845:149202.
13. Nadeem MS, Khan JA, Kazmi I, Moglad E, Afzal M, Alzarea SI, et al. Synthesis, DFT, ADMET and molecular docking studies of thiazole derived thiazolidinone-based chalcone derivatives: Alzheimer's disease current therapies. Future Med Chem. 2024; https://doi.org/10.1080/17568919.2024.2421158.
14. Gadhave DG, Sugandhi VV, Jha SK, Nangare SN, Gupta G, Singh SK, et al. Neurodegenerative disorders: mechanisms of degeneration and therapeutic approaches with their clinical relevance. Ageing Res Rev. 2024;99:102357.
15. Cao W, Zheng H. Peripheral immune system in aging and Alzheimer's disease. Mol Neurodegener. 2018;13(1):51.
16. Musi N, Valentine JM, Sickora KR, Baeuerle E, Thompson CS, Shen Q, et al. Tau protein aggregation is associated with cellular senescence in the brain. Aging Cell. 2018;17(6):e12840.
17. Hudson HR, Sun X, Orr ME. Senescent brain cell types in Alzheimer's disease: pathological mechanisms and therapeutic opportunities. Neurotherapeutics. 2025;22(3):e00519.
18. Saito Y, Yamamoto S, Chikenji TS. Role of cellular senescence in inflammation and regeneration. Inflamm Regen. 2024;44(1):28.
19. Chakraborty S, Vishwas S, Harish V, Gupta G, Paudel KR, Dhanasekaran M, et al. Exploring nanoparticular platform in delivery of repurposed drug for Alzheimer's disease: current approaches and future perspectives. Expert Opin Drug Deliv. 2024;21(12):1771–92.
20. Kumari R, Jat P. Mechanisms of cellular senescence: cell cycle arrest and senescence associated secretory phenotype. Front Cell Dev Biol. 2021;9:645593.
21. O'Sullivan EA, Wallis R, Mossa F, Bishop CL. The paradox of senescent-marker positive cancer cells: challenges and opportunities. npj Aging. 2024;10(1):41.
22. Ashique S, Sirohi E, Kumar S, Rihan M, Mishra N, Bhatt S, et al. Aducanumab in Alzheimer's disease: a critical update. Curr Med Chem. 2024;31(31):5004–26.
23. Lei S-Y, Qu Y, Yang Y-Q, Liu J-C, Zhang Y-F, Zhou S-Y, et al. Cellular senescence: a novel therapeutic target for central nervous system diseases. Biomed Pharmacother. 2024;179:117311.
24. Si Z, Sun L, Wang X. Evidence and perspectives of cell senescence in neurodegenerative diseases. Biomed Pharmacother. 2021;137:111327.
25. Alharbi KS, Almalki WH, Alzarea SI, Kazmi I, Al-Abbasi FA, Afzal O, et al. Anaesthesia-induced changes in genomic expression leading to neurodegeneration. CNS Neurol Disord Drug Targets. 2024;23(4):411–9.

26. Gaikwad S, Senapati S, Haque MA, Kayed R. Senescence, brain inflammation, and oligomeric tau drive cognitive decline in Alzheimer's disease: evidence from clinical and preclinical studies. Alzheimers Dement. 2024;20(1):709–27.
27. Alshaebi F, Sciortino A, Kayed R. The role of glial cell senescence in Alzheimer's disease. J Neurochem. 2025;169(3):e70051.
28. Acharya M, Singh N, Gupta G, Tambuwala MM, Aljabali AAA, Chellappan DK, et al. Vitamin D, Calbindin, and calcium signaling: unraveling the Alzheimer's connection. Cell Signal. 2024;116:111043.
29. Zhang P, Dilley C, Mattson MP. DNA damage responses in neural cells: focus on the telomere. Neuroscience. 2007;145(4):1439–48.
30. Singh N, Vishwas S, Kaur A, Kaur H, Kakoty V, Khursheed R, et al. Harnessing role of sesamol and its nanoformulations against neurodegenerative diseases. Biomed Pharmacother. 2023;167:115512.
31. Baechle JJ, Chen N, Makhijani P, Winer S, Furman D, Winer DA. Chronic inflammation and the hallmarks of aging. Mol Metab. 2023;74:101755.
32. Zhang W, Xiao D, Mao Q, Xia H. Role of neuroinflammation in neurodegeneration development. Signal Transduct Target Ther. 2023;8(1):267.
33. Alharbi KS, Javed Shaikh MA, Imam SS, Alshehri S, Ghoneim MM, Almalki WH, et al. Role of flavonoids in management of various biological targets in Alzheimer's disease: evidence from preclinical to clinical studies. Curr Med Chem. 2023;30(18):2061–74.
34. Elder GA, Gama Sosa MA, De Gasperi R. Transgenic mouse models of Alzheimer's disease. Mt Sinai J Med. 2010;77(1):69–81.
35. Sanchez-Varo R, Mejias-Ortega M, Fernandez-Valenzuela JJ, Nuñez-Diaz C, Caceres-Palomo L, Vegas-Gomez L, et al. Transgenic mouse models of Alzheimer's disease: an integrative analysis. Int J Mol Sci. 2022;23(10)
36. Pandey SN, Singh G, Semwal BC, Gupta G, Alharbi KS, Almalki WH, et al. Therapeutic approaches of nutraceuticals in the prevention of Alzheimer's disease. J Food Biochem. 2022;46(12):e14426.
37. Yan J, Chen S, Yi Z, Zhao R, Zhu J, Ding S, et al. The role of p21 in cellular senescence and aging-related diseases. Mol Cells. 2024;47(11):100113.
38. Gonzales MM, Garbarino VR, Marques Zilli E, Petersen RC, Kirkland JL, Tchkonia T, et al. Senolytic therapy to modulate the progression of Alzheimer's disease (SToMP-AD): a pilot clinical trial. J Prev Alzheimers Dis. 2022;9(1):22–9.
39. Puzzo D, Lee L, Palmeri A, Calabrese G, Arancio O. Behavioral assays with mouse models of Alzheimer's disease: practical considerations and guidelines. Biochem Pharmacol. 2014;88(4):450–67.
40. Hossain MK, Kim H-R, Chae HJ. Aging phenotype in AD brain organoids: track to success and challenges. Ageing Res Rev. 2024;96:102256.
41. Muthamil S, Kim HY, Jang HJ, Lyu JH, Shin UC, Go Y, et al. Biomarkers of cellular senescence and aging: current state-of-the-art, challenges and future perspectives. Adv Biol. 2024;8(9):e2400079.
42. Olivieri F, Prattichizzo F, Grillari J, Balistreri CR. Cellular senescence and inflammaging in age-related diseases. Mediat Inflamm. 2018;2018:9076485.
43. Tamatta R, Pai V, Jaiswal C, Singh I, Singh AK. Neuroinflammaging and the immune landscape: the role of autophagy and senescence in aging brain. Biogerontology. 2025;26(2):52.
44. Huang Z, Xu P, Hess DC, Zhang Q. Cellular senescence as a key contributor to secondary neurodegeneration in traumatic brain injury and stroke. Transl Neurodegener. 2024;13(1):61.
45. Ausó E, Gómez-Vicente V, Esquiva G. Biomarkers for Alzheimer's disease early diagnosis. J Pers Med. 2020;10(3)
46. Ghimire A, Rehman SA, Subhani A, Khan MA, Rahman Z, Iqubal MK, et al. Mechanism of microglia-mediated neuroinflammation, associated cognitive dysfunction, and therapeutic updates in Alzheimer's disease. hLife. 2025;3(2):64–81.

47. Li N, Deng M, Hu G, Li N, Yuan H, Zhou Y. New insights into microglial mechanisms of memory impairment in Alzheimer's disease. Biomol Ther. 2022;12(11)
48. Riessland M, Orr ME. Translating the biology of aging into new therapeutics for Alzheimer's disease: senolytics. J Prev Alzheimers Dis. 2023;10(4):633–46.
49. Robbins PD, Jurk D, Khosla S, Kirkland JL, LeBrasseur NK, Miller JD, et al. Senolytic drugs: reducing senescent cell viability to extend health span. Annu Rev Pharmacol Toxicol. 2021;61:779–803.
50. Miller SJ, Campbell CE, Jimenez-Corea HA, Wu GH, Logan R. Neuroglial senescence, α-synucleinopathy, and the therapeutic potential of senolytics in Parkinson's disease. Front Neurosci. 2022;16:824191.
51. Wang Y, Kuca K, You L, Nepovimova E, Heger Z, Valko M, et al. The role of cellular senescence in neurodegenerative diseases. Arch Toxicol. 2024;98(8):2393–408.
52. Melo dos Santos LS, Trombetta-Lima M, Eggen BJL, Demaria M. Cellular senescence in brain aging and neurodegeneration. Ageing Res Rev. 2024;93:102141.
53. Khalil R, Diab-Assaf M, Lemaitre JM. Emerging therapeutic approaches to target the dark side of senescent cells: new hopes to treat aging as a disease and to delay age-related pathologies. Cells. 2023;12(6)
54. Niazi SK. Non-invasive drug delivery across the blood-brain barrier: a prospective analysis. Pharmaceutics. 2023;15(11)
55. Orr M, Gonzales MM, Garbarino VR, Kautz T, Palavicini JP, Lopez-Cruzan M, et al. Senolytic therapy to modulate the progression of Alzheimer's disease (SToMP-AD) - outcomes from the first clinical trial of senolytic therapy for Alzheimer's disease. Res Sq. 2023; https://doi.org/10.21203/rs.3.rs-2809973/v1.
56. Boccardi V, Orr ME, Polidori MC, Ruggiero C, Mecocci P. Focus on senescence: clinical significance and practical applications. J Intern Med. 2024;295(5):599–619.
57. Kohli J, Wang B, Brandenburg SM, Basisty N, Evangelou K, Varela-Eirin M, et al. Algorithmic assessment of cellular senescence in experimental and clinical specimens. Nat Protoc. 2021;16(5):2471–98.
58. Ma S, Ji Z, Zhang B, Geng L, Cai Y, Nie C, et al. Spatial transcriptomic landscape unveils immunoglobin-associated senescence as a hallmark of aging. Cell. 2024;187(24):7025–44.e34.
59. Fabrizio C, Termine A, Caltagirone C, Sancesario G. Artificial intelligence for Alzheimer's disease: promise or challenge? Diagnostics (Basel, Switzerland). 2021;11(8)
60. Song S, Li T, Lin W, Liu R, Zhang Y. Application of artificial intelligence in Alzheimer's disease: a bibliometric analysis. Front Neurosci. 2025;19:1511350.
61. Shafqat A, Khan S, Omer MH, Niaz M, Albalkhi I, AlKattan K, et al. Cellular senescence in brain aging and cognitive decline. Front Aging Neurosci. 2023;15:1281581.
62. Bhatti JS, Khullar N, Mishra J, Kaur S, Sehrawat A, Sharma E, et al. Stem cells in the treatment of Alzheimer's disease – promises and pitfalls. Biochim Biophys Acta (BBA) - Mol Basis Dis. 2023;1869(6):166712.
63. Arafah A, Khatoon S, Rasool I, Khan A, Rather MA, Abujabal KA, et al. The future of precision medicine in the cure of Alzheimer's disease. Biomedicines. 2023;11(2)
64. Singh B, Day CM, Abdella S, Garg S. Alzheimer's disease current therapies, novel drug delivery systems and future directions for better disease management. J Control Release. 2024;367:402–24.
65. Li Z, Zhang Z, Ren Y, Wang Y, Fang J, Yue H, et al. Aging and age-related diseases: from mechanisms to therapeutic strategies. Biogerontology. 2021;22(2):165–87.
66. Corbière M, Samson E, Villotti P, Pelletier JF. Strategies to fight stigma toward people with mental disorders: perspectives from different stakeholders. TheScientificWorldJOURNAL. 2012;2012:516358.
67. Liu B, Peng Z, Zhang H, Zhang N, Liu Z, Xia Z, et al. Regulation of cellular senescence in tumor progression and therapeutic targeting: mechanisms and pathways. Mol Cancer. 2025;24(1):106.

68. Amaya-Montoya M, Pérez-Londoño A, Guatibonza-García V, Vargas-Villanueva A, Mendivil CO. Cellular senescence as a therapeutic target for age-related diseases: a review. Adv Ther. 2020;37(4):1407–24.
69. Kritsilis M, Rizou SV, Koutsoudaki PN, Evangelou K, Gorgoulis VG, Papadopoulos D. Ageing, cellular senescence and neurodegenerative disease. Int J Mol Sci. 2018;19(10)
70. Zhu J, Wu C, Yang L. Cellular senescence in Alzheimer's disease: from physiology to pathology. Transl Neurodegener. 2024;13(1):55.
71. Herbstein F, Sapochnik M, Attorresi A, Pollak C, Senin S, Gonilski-Pacin D, et al. The SASP factor IL-6 sustains cell-autonomous senescent cells via a cGAS-STING-NFκB intracrine senescent noncanonical pathway. Aging Cell. 2024;23(10):e14258.
72. Lu Y, Jarrahi A, Moore N, Bartoli M, Brann DW, Baban B, et al. Inflammaging, cellular senescence, and cognitive aging after traumatic brain injury. Neurobiol Dis. 2023;180:106090.
73. Wei L, Yang X, Wang J, Wang Z, Wang Q, Ding Y, et al. H3K18 lactylation of senescent microglia potentiates brain aging and Alzheimer's disease through the NFκB signaling pathway. J Neuroinflammation. 2023;20(1):208.
74. Islam MT, Tuday E, Allen S, Kim J, Trott DW, Holland WL, et al. Senolytic drugs, dasatinib and quercetin, attenuate adipose tissue inflammation, and ameliorate metabolic function in old age. Aging Cell. 2023;22(2):e13767.
75. Chacko AM, Li C, Pryma DA, Brem S, Coukos G, Muzykantov V. Targeted delivery of antibody-based therapeutic and imaging agents to CNS tumors: crossing the blood-brain barrier divide. Expert Opin Drug Deliv. 2013;10(7):907–26.
76. Sun B, Li R, Ji N, Liu H, Wang H, Chen C, et al. Brain-targeting drug delivery systems: the state of the art in treatment of glioblastoma. Mater Today Bio. 2025;30:101443.

7 Cellular Senescence in Parkinson's Disease (PD)

Muhammad Afzal

Introduction

The neurodegenerative brain disorder Parkinson's disease causes lasting damage to the nervous system over time by creating four basic motor symptoms: tremors, rigidity, bradykinesia, and postural instability [1, 2]. The motor symptoms arise because dopamine-producing neurons degenerate in the midbrain area controlling motor functions at substrate nigra pars compacta. PD patients acquire several non-motor symptoms that join motor symptoms to reveal how this degenerative disease affects the entire body [3]. Several studies have investigated PD heredity patterns and environmental triggers, but scientists still have basic knowledge regarding neuron degeneration in PD patients [4, 5]. The main factor in PD vulnerability determination is aging because aging dramatically accelerates the formation of this neurological condition. Advances in molecular understanding are essential to create effective treatments that specifically target cellular senescence in aging people

M. Afzal (✉)
Department of Pharmaceutical Sciences, Pharmacy Program, Batterjee Medical College, Jeddah, Saudi Arabia
e-mail: mohmmad.afzal@bmc.edu.sa

G. Gupta (ed.), *Cellular Senescence and Brain Aging*,
https://doi.org/10.1007/978-981-96-8873-9_7

because PD incidence continues to rise globally [6]. At first, researchers treated senescence as tumor prevention. Still, it is the origin of aging alongside linked diseases because it produces rising senescent cell numbers as tissues age. The key elements defining senescent cells comprise two cell cycle–inhibiting proteins, p16INK4a and p21CIP1, persistent DDR signals, SA-β-gal-positive activity, and the senescence-associated secretory phenotype (SASP) [7, 8]. The SASP contains pro-inflammatory cytokines, chemokines, growth factors, and proteases that modify tissue circumstances while affecting neighboring cell health [9, 10]. Likewise, tissue regeneration benefits from activated senescent cell processes, yet prolonged uncontrolled senescence activities generate health problems leading to Parkinson's disease and other neurodegenerative disorders. Senescent cells in the central nervous system composed of astrocytes, microglia, and neurons manifest SASP properties that produce neurotoxicity with pro-inflammatory effects and lead to sustained neuroinflammation, leading to neuronal dysfunction [11, 12]. PD causes glial cells and neurons to develop senescence-like features stemming from the combination of oxidative stress and mitochondrial dysfunction and α-synuclein aggregation, creating a destructive inflammatory cycle [13]. Developing high levels of senescent cells causes neurogenesis issues that deteriorate the normal stability of neuronal microenvironment conditions [14, 15]. The research link between senescence and neurodegeneration enables investigations of senescent cell elimination through senolytics to use and regulatory SASP activity through senomorphics. A complete understanding of the effects of senescence on PD progression is the fundamental requirement for developing treatment methods and biomarkers for Parkinson's disease [16]. A combination of preclinical and clinical research in the text demonstrates how senescent cells induce damage to dopaminergic neurons, neuroinflammation, and disease progression [17, 18]. Data analysis explains important molecular pathways responsible for senescence alongside detecting specific markers in PD experimental research of human samples and describes strategies to lower senescent cell numbers [19, 20]. The research project will investigate the effects of SASP on PD neuroimmune environments through a specialized study of their interacting mechanisms. The objective of this chapter involves creating an integrated understanding of cellular senescence properties as both a disease-initiating agent and therapeutic potential within Parkinson's disease investigation [21, 22] (Fig. 7.1).

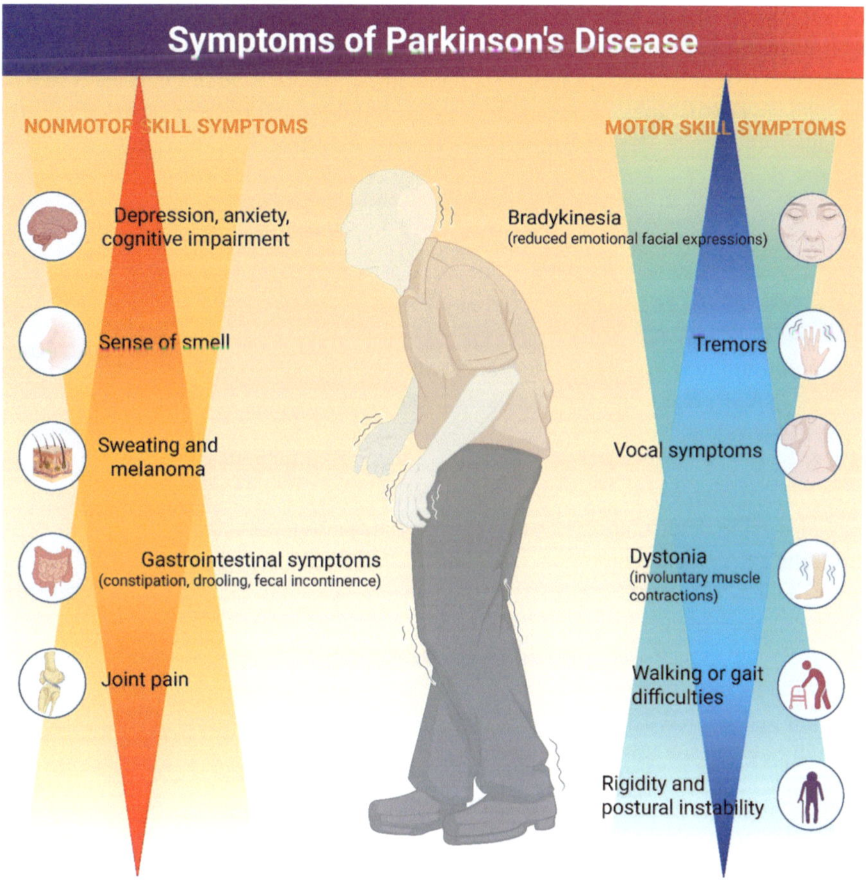

Fig. 7.1 EGFR/IL-2β-mediated STAT3 signaling. The Parkinson's disease symptoms can be seen where motor and nonmotor symptoms are represented separately. Some Parkinson's disease symptoms include depression, anxiety, melanoma, sweating, cognitive decline, joint pain, and loss of smell, which represent nonmotor symptoms. Bradykinesia joins vocal abnormalities with tremors and dystonia to create stiffness that leads to postural instability and gait issues, which are motor symptoms. The clinical manifestation of all symptoms is depicted through a central figure that displays trembling hands and back hunching

Molecular Mechanisms of Cellular Senescence

DNA Damage Response and Senescence

DNA damage persistency is essential in aging cells because this condition activates the DNA damage response (DDR) pathway [23]. When senescent cells experience double-strand breaks, ATM/ATR protein kinases activate, which results in both p53 phosphorylation and p21 induction that are sufficient to prevent new cell development. Brain aging, together with Parkinson's disease, intensifies DNA damage throughout glial cells and neuronal cells because of increased oxidative stress with defective mitochondria [24, 25]. The continued activation of DDR leads to two simultaneous outcomes: mobile cell growth inhibition and inflammation-triggering events in the tissue. The multiple DNA damages in dopaminergic neurons lead to neuron death but generate pro-inflammatory cellular states within glial cells that participate in ongoing Parkinson's disease progression [26, 27].

Senescence-Associated Secretory Phenotype (SASP)

Multiple immune response molecules create senescence-associated secretory phenotype (SASP) from senescent cells by producing IL-6 and IL-1β and chemokines growth factors and proteases [6, 28]. SASP induces the strengthening of cellular senescence but causes harmful effects to neighboring cells by sending signals through cell membrane channels. The manifestation of Parkinson's disease pathology alongside deteriorating chronic neuroinflammation results from PD glial cells exhibiting SASP characteristics [29]. NF-κB and C/EBPβ transcription factors regulate the genes needed for SASP component synthesis, but DDR signaling stability preserves their gene expression patterns [30, 31]. The combination of higher neuronal stress and synaptic abnormalities generated by SASP likely decreases your ability to regenerate neurons and thus accelerates neurodegenerative processes. When treating age-related diseases, SASP represents a primary therapeutic target since it both initiates tissue remodeling and degeneration patterns [32].

Mitochondrial Dysfunction and Oxidative Stress

The primary indicator of aging appears through mitochondrial dysfunction, although this condition is essential for Parkinson's disease development. Damaged mitochondria generate high amounts of ROS that lead to the destruction of DNA together with proteins and lipids [33, 34]. The DNA damage signaling mechanisms activated by oxidative stress activate p53 and p21 pathways to cause cellular aging [35]. PD develops cell death in dopaminergic neurons due to PINK1 and Parkin gene mutations that impair the degeneration process of damaged mitochondria. The active neurons are most susceptible to ROS-mediated cellular aging when metabolically active. The activation of SASP happens because of mitochondrial dysfunction that connects metabolic dysregulation to progressive neurodegeneration in PD [36].

Epigenetic Modifications and Cell Cycle Arrest

Cellular senescence establishment demands three different epigenetic modifications, consisting of DNA methylation changes and histone modifications, such as H3K9me3 and H4K20me3, together with structural features of chromatin. DNA modifications suppress cell growth by intensifying cell cycle inhibitors p16INK4a and p21CIP1 transcription while blocking additional cellular proliferation [37]. The epigenetic structures within aging neurons and glial cells in Parkinson's disease become affected to generate deregulated gene networks tied to aging mechanisms. SAHF structures achieve gene silencing functions as a stabilizing system for cellular senescence. Various environmental stressors and oxidative damage activate epigenetic pattern modifications that sustain cellular senescence until PD-related neurodegeneration becomes evident [38].

Senescence in the Brain: Focus on Parkinson's Disease

Senescent Cell Types in PD (Astrocytes, Microglia, Neurons)

Parkinson's Disease drives the cellular aging process for astrocytes, microglia, and neurons across brain regions. As astrocytes become senescent, their p16INK4a and p21CIP1 expressions increase, releasing inflammatory elements that inflict damage on the brain tissue [39, 40]. Microglial aging leads cells to lose their phagocytic ability, yet continue producing sustained IL-6 and TNF-α cytokines that cause neuroinflammatory damage. Neurons develop features similar to cellular senescence because they are postmitotic cells, so they accumulate DNA damage and impaired mitochondria and initiate SASP-like secretion events that advance degenerative damage [41, 42]. Studies of the aging brain and PD models reveal high numbers of damaged senescent cells that disrupt neural homeostasis and accelerate neurodegeneration. This supports the requirement for strategic PD senescence-related damage investigations into various brain cell types [43, 44].

Role of α-Synuclein Aggregation in Inducing Senescence

Neuronal and glial cell populations undergo cellular senescence because of the high concentration of α-synuclein protein that defines the PD hallmark. URL Exception (ubicmental modify) sets up oxidative pressure and damages mitochondria, triggering DNA harm replies that begin the senescence program through p53/p21 and p16INK4a [45, 46]. As astrocytes and microglia cells incorporate α-synuclein inclusions, they develop both inflammatory responses and cellular senescence and generate exaggerated SASP production [47, 48]. A-synuclein causes damage to lysosomes that leads to autophagy system malfunction, which functions as the central mechanism for controlling senescence development [49, 50]. Neurodegeneration advances through the toxic cascade released by senescent cells that stimulate α-synuclein aggregate spread between cells. Expert brain research demonstrates that α-synuclein accumulation functions as a pathology biomarker in Parkinson's disease while simultaneously accelerating cellular aging of Parkinson's disease structures [51, 52].

Neuroinflammation and SASP in PD Progression

Neuroinflammation in Parkinson's disease patients persists mainly due to SASP. Aged cells of astrocytes and microglia release SASP factors IL-1β, IL-6, TNF-α, and MMPs through their secretions, resulting in blood–brain barrier destruction, synaptic dysfunction, and peripheral immune cell attraction [53, 54]. Multiple chemical compounds released by these cells create an environment of inflammation that enables quick dopaminergic neuron death alongside PD progression. Prolonged SASP signal exchange triggers cell senescence and intensifies neurodegenerative responses between cells [55, 56]. Studies show that PD disease qualities diminish with increased time. SASP factors remain active in preclinical, experimental models and human biological specimens, validating SASP factors as potential targets for PD treatment development [57, 58].

Dopaminergic Neuron Vulnerability and Senescence

The high-level dopamine metabolism in substantia nigra pars compacta dopaminergic neurons exposes them to severe aging effects and linked oxidative stress conditions. The activation of both p53/p21 and mitochondrial unfolded protein response pathways occurs through neurons that experience DNA damage to their mitochondria and the generation of reactive oxygen species (ROS) [59, 60]. Terminal dopaminergic neurons exhibit characteristics of senescence, which become apparent through γH2AX and SA-β-gal observations, together with diminished neural plasticity and stronger immune system responses. PD-related factors and genetic components affecting LRRK2 and PARKIN make dopaminergic neurons essential for PD development through aging processes that lead to their demise [49, 61].

Preclinical Evidence of Cellular Senescence in PD

Animal Models of PD Showing Senescence Markers

Preclinical studies using MPTP-treated mice, 6-OHDA-lesioned rats, and α-synuclein transgenic models demonstrated that vulnerable brain regions cause increased expressions of senescence markers according to the experiments conducted by researchers [62, 63]. Studies confirm that both p16INK4a and p21CIP1 gene quantities increase simultaneously with SA-β-gal accumulation in dopaminergic neurons of the substantia nigra [44]. The disease severity in these models intensifies because aged astrocytes and microglia produce SASP as part of their neuroinflammatory response. These models confirm the associative link between behavioral symptoms and senescence marker elevation, which assists in PD pathogenesis study and makes them valuable tools for senescence-related research [64].

In Vitro Studies Using Human and Rodent Neuronal Cultures

PD-related stressors, including α-synuclein aggregates and MPP+ and rotenone together with hydrogen peroxide, trigger cellular senescence-like effects in human iPSC-derived neurons as well as SH-SY5Y cells and rodent primary cultures [65, 66]. Sa-beta-gal enzyme activity increases along with mitochondrial dysfunction and DNA damage as p53 and p21 proteins exhibit elevated amounts in the cells. The combination of astrocytes and microglia activates the inflammatory signals derived from SASP through oxidative stress-induced conditions [67, 68]. Recent research proves that PD causes neuronal vulnerability through cellular senescence processes while creating test environments to investigate molecular mechanisms and develop anti-senescence therapeutic approaches [69, 70].

Genetic and Pharmacologic Manipulations of Senescence Pathways

Scientists have generated substantial data through genetics and pharmacologic approaches aimed at PD models that follow cellular senescence pathways. According to scientific research, targeted p53 and p16INK4a and mTOR gene regulation element inhibition or knockout leads to decreased neurocellular mortality and restored motor function [71]. Normal activation of these genes leads to simultaneous PD disease progression and acceleration of cellular aging processes. Lab tests confirmed that Navitoclax, dasatinib, and quercetin serve as selective cell targets that eliminate senescent cells and decrease neuroinflammatory responses in animal subjects. New research involving senescence pathway investigations demonstrates this strategy could function as a treatment for controlling PD pathogenesis while preserving neuronal viability [72].

Biomarkers of Senescence in Experimental Models

Assessing cellular senescence in PD models requires two markers: SA-β-gal and p16INK4a, p21CIP1, γH2AX, and the SASP inflammatory components IL-6 and IL-1β [73, 74]. Medical research demonstrates these markers progressively accumulate within both substantia nigra and striatum segments of animals exposed to toxins or having transgenic conditions while also connecting to severe tissue damage [75, 76]. The increasing molecular techniques have enabled scientists to discover fresh biomarkers for senescence as part of their research on transcriptomic profiling and proteomics methods. Further research needs humanized experimental models to affirm the validity of these biomarkers because they present value as diagnostic indicators and treatment assessment measures [77, 78].

Clinical Insights and Biomarker Discovery

Evidence of Senescence Markers in Human PD Brains

Autopsy investigations of PD brains reveal elevated expression of senescence marker proteins p16INK4a, p21CIP1/WAF1, and γ-H2AX in dopaminergic neurons along with glial cells inside the substantia nigra region [79, 80]. Research evidence reveals that senescence activation causes neurodegenerative damage. The existence of phenotypic senescence in cells manifests through elevated SA-β-gal activity combined with increased levels of IL-6 and IL-8 proinflammatory cytokines in cellular secretion [81, 82]. The co-occurrence of α-synuclein aggregates with senescent markers increases neurological dysfunction to a greater extent. Research findings through molecular evaluation confirm PD brain senescence presence and prove that this cellular process leads to progressive neuronal cell death to produce opportunities for therapeutic development targeting PD senescence pathways [83, 84].

Circulating Biomarkers: Potential Diagnostic Tools

The potential for PD biomarkers becomes more promising because scientists measure increased levels of senescence-associated molecules in patient samples from PD peripheral blood and cerebrospinal fluid. Analysis showed elevated IL-6, MCP-1, and MMP-3 SASP factors in blood serum and cerebrospinal fluid, thus proving senescence burden throughout the system [85, 86]. The aging processes of neurons can be monitored without invasive procedures through extracellular vesicles, which contain senescence-linked microRNAs and mitochondrial DNA fragments [87]. Numerous medical studies show that measuring p16INK4a levels in leukocytes yields results related to PD disease progression while correlating with symptom intensification. The addition of longitudinal evidence processing for these biomarkers would create diagnostic tools for early PD detection and monitoring of PD advancement and medical responses throughout disease management [88].

Clinical Correlations: Aging, Disease Severity, and Senescence

The clinical features of PD emerge from the combination of healthy aging patterns and cellular aging that occurs prematurely in patients. Research evidence demonstrates that PD disease feature development tracks directly with senescence marker expression levels along with an increase in motor impairment and cognitive decline [89, 90]. Procurement of PD treatment becomes more challenging with advancing patient age because increased SASP factor levels create a worsening inflammatory condition that causes neurodegeneration [91]. The rapidness of disease progression in patients leads to a faster accumulation of senescence burden because senescence controls disease changes during its course. Age-associated conditions aside, the research confirms that senescence works independently to drive the development of PD disease pathology. The clinical assessment approach strengthens through senescence tests yielding data beneficial for precise diagnosis while building potential treatment alternatives [91].

Senescence in Non-Motor Symptoms of PD

Emerging evidence implicates cellular senescence in the development of non-motor symptoms such as depression, sleep disturbances, autonomic dysfunction, and cognitive decline in PD. The release of inflammatory mediators from aged glial cells inflicts damage to both the hippocampal region and the limbic system, which controls memory function and mood behavior according to scientific research acceptance [92]. Studies demonstrate that increased SASP factors in cerebrospinal fluid create both executive function deterioration and higher apathy due to their spread from motor regions. PD patients exhibit commonly occurring autonomic symptoms because senescent tissue affects both the gut and cardiovascular systems. Research evidence shows that senescence affects every part of the body, which warrants comprehensive analysis of its role throughout the PD disease process [93].

Therapeutic Strategies Targeting Cellular Senescence

Senolytic and Senomorphic Drugs: Preclinical Successes

Senomorphics modify harmful cell secretions in bodies without damaging senescent cells, but senolytics function completely by removing these cells from the body. Middle-stage PD research demonstrates that dasatinib, in combination with navitoclax and quercetin, reduces brain inflammation by targeting aging brain cells that protect cells carrying dopamine and prevent neuron damage. According to animal model research, the use of rapamycin and metformin as senomorphics led to cellular homeostasis recovery and superior motor function. Therapeutic approaches reduce disease progression and decrease α-synuclein aggregation by modifying senescence environments. More research will lead to the development of optimal brain-targeted treatment delivery methods and dosage protocols [94, 95].

Anti-Inflammatory Approaches and SASP Modulation

The senescence-associated secretory phenotype (SASP) of PD brains releases neuroinflammatory pro-inflammatory cytokines, chemokines, and proteases originating from senescent cells. Inhibition of NF-κB in combination with JAK/STAT and mTOR creates fresh methods to handle SASP inflammatory needs, according to research. Research on animals demonstrates that the treatment of neurotoxicity by SASP is effective when using rapamycin, JAK inhibitors, and glucocorticoids [96]. Cells near neural cells gain protection during SASP reduction, which lowers inflammatory events and offers preservation against PD disease progression. SASP modulation provides an approach to combine neuroprotective therapy treatments when managing age-related neurodegeneration [97, 98].

Targeting Mitochondrial Function and ROS

The combination of elevated ROS levels and damaged mitochondria starts the cellular aging process, affecting PD patients. The blocking of cellular senescence through therapy becomes possible when mitochondrial dynamics are restored, and oxidative stress is reduced. Protective results occur in PD models after administration of coenzyme Q10 along with MitoQ and SS-31, which enhance mitochondrial bioenergetic functions while reducing ROS. Oxidative DNA damage decreases with cellular senescence markers through the antioxidant substance N-acetylcysteine (NAC) [99, 100]. The dual advantage of specialized treatments for mitochondrial health consists of decreased cellular aging and preserved dopaminergic neuron functions, which suggests the potential for modifying PD disease development [101, 102].

Challenges in Translation to Clinical Trials

Available preclinical evidence shows various obstacles that prevent the successful translation of senescence-targeted therapies into clinical trials for PD treatment. The absence of cell-targeted functions within senolytics creates safety risks to healthy cells and the necessity to study the extended-term effects of senomorphic medications. Scientific teams face two major barriers: the physiological difficulty of delivering drugs across the blood–brain barrier and insufficient development of senescence-tracking biomarkers for human systems [103, 104]. Evaluation of

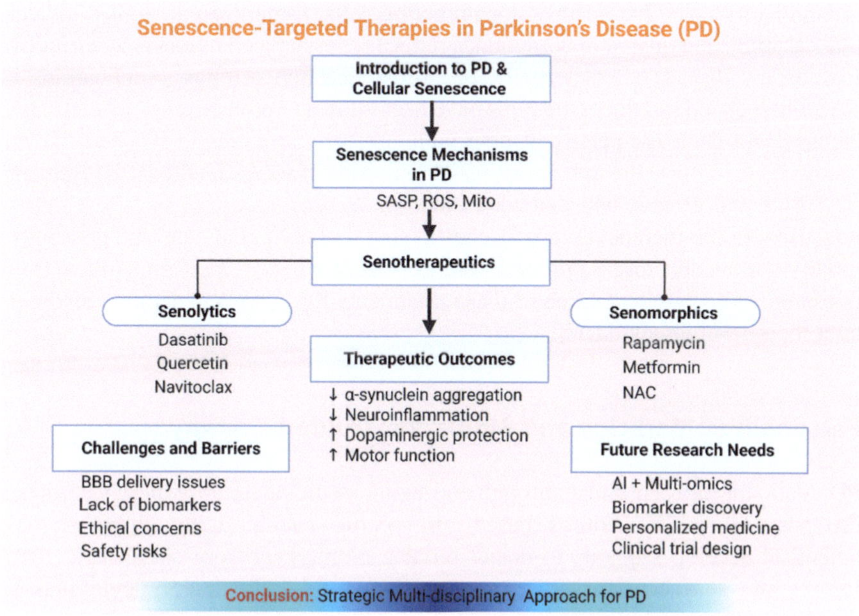

Fig. 7.2 Apatinib's multimodal mechanisms in NSCLC. This figure describes senescence-targeted treatments for Parkinson's disease (PD). The research begins with a summary of PD and cellular senescence before examining the senescence processes that involve mitochondrial failure and ROS alongside SASP. The article focuses on senomorphics, including rapamycin metformin and NAC together with senolytics, which consist of dasatinib quercetin and navitoclax as potential therapeutic approaches to enhance dopaminergic protection alongside improving motor performance while decreasing α-synuclein aggregation and neuroinflammation. The paper addresses BBB delivery challenges and ethical aspects but stresses the necessity to explore customized medications and biomarker development further

treatments becomes complicated because aging-related coexisting conditions pose design complications that affect both the interpretation of outcomes and experimental approaches. The combination of ethical opposition to aging process manipulation and insufficient research on human safety causes the process to move slowly. The development of precise medical strategies with new disease research methodologies serves as a requirement to eliminate barriers to neurodegenerative treatment advancement [105] (Fig. 7.2).

Future Perspectives and Translational Challenges

Potential of Senescence-Targeted Therapies in PD

The mixture of senolytics with senomorphics presents a new possibility to control Parkinson's disease development by providing targeted treatment against inflammatory cells that destroy dopaminergic cells. Experimental studies indicate that dasatinib and quercetin compounds reduce senescent cell numbers and improve Parkinson's disease characteristics in animal research settings. Applications of SASP-associated treatment methods have been shown to protect neuronal cells from damage and decrease persistent neuroinflammatory activities in the body [106]. Medical use of these therapies requires effective brain access methods, assessments of detrimental effects, and extended duration security requirements. Researchers now use specific therapeutic approaches to spot and treat glial cells and neurons to modify Parkinson's disease progression. Researchers need to perform clinical trials to establish the practical value of these treatments for neurodegenerative disorders after their development [107].

Personalized Medicine and Biomarker-Guided Therapy

PD treatment shows promise through anti-aging medicine because these techniques use personalized procedures based on specific patient aging characteristics. Scientists develop methods to detect reliable biomarkers from circulating SASP factors and senescence-associated microRNAs and inflammatory cytokines to enable health providers to measure patient senescent cell amounts and treatment responses of anti-senescence therapies [108]. Using individual patient genomic and transcriptomic information allows doctors to improve treatment approaches, thus enabling them to detect illnesses early while monitoring disease evolution and forecasting therapeutic outcomes. Biomarker-based individual treatments will help patients obtain better outcomes from senolytic and anti-inflammatory medicines as they face decreased PD pathology diversity. Future research must confirm biological indicators in prolonged clinical examinations to develop operational criteria enabling their clinical application for enhancing personalized medicine's effects on PD cellular senescence control [109, 110].

Integrating Multi-Omics and AI for Senescence Mapping

Modern investigations of cellular senescence in Parkinson's disease (PD) demand a revolutionary approach that unites artificial intelligence (AI) with multiple omics techniques incorporating transcriptomics, proteomics, metabolomics, and genomic research. The integration of these technologies enables scientists to locate new molecular markers, pathways, and regulatory networks that trigger senescence in particular cells and brain regions. AI-driven algorithms evaluate high-dimensional information for forecasting disease trajectories, identifying therapeutic targets, and creating personalized treatment strategies [111, 112]. Machine learning methods should be trained to establish the connection between senescence phenotypes and their relationship to clinical data or treatment results. Systemic biological frameworks that combine various omics layers help researchers better understand the interrelated patterns between neurodegeneration and aging and inflammation. This novel approach to neurodegenerative disease treatment based on aging research will gain momentum through these technological developments in PD studies that shorten the search for effective therapeutic approaches and specific disease markers [113].

Gaps in Current Knowledge and Roadmap for Future Research

Research activity related to cellular senescence in Parkinson's disease keeps expanding despite the need for additional scientific discoveries. Research into PD pathophysiology still lacks evidence about when senescence starts, which cell types are affected, and how senescence functions for PD progression. Human research that thoroughly investigates PD clinical characteristics and disease evolution metrics has not appropriately linked these to senescence biomarkers [114]. Scientific progress toward developing brain-accessing senolytics encounters technical obstacles preventing the production of minimally adverse versions. To eliminate current research limitations, researchers need to work on several simultaneous steps including extensive studies with various PD patient populations, and develop reliable methods for clinical marker detection and treatment approaches that combine neuroinflammation intervention with senescence targets. The development of standard methods for identifying senescent development within living organisms needs to remain priority number one. The movement of senescence discoveries toward beneficial PD treatments depends on a strategic approach that unites multidisciplinary groups and combines research at both patient-specific and group levels, as well as preclinical and clinical stages [115, 116] (Table 7.1).

Table 7.1 The table shows that PD cellular senescence primarily relies on DNA damage, mitochondrial failure, SASP signaling, epigenetic modifications, and α-synuclein accumulation. The different routes through therapeutic targets and indicators lead to neuronal loss and inflammation while advancing the illness. The potential treatments available include antioxidants, SASP inhibitors, epigenetic modulators, and senolytics

Cellular senescence	Mechanism	Biomarker	Therapeutic target	Implications in PD	Reference
DNA damage-induced senescence	Persistent DNA damage activates ATM/ATR → p53/p21 axis → cell cycle arrest	γH2AX, p53, $p21^{CIP1}$, SA-β-gal	DDR pathway components (ATM/ATR, p53)	This leads to dopaminergic neuron death and glial inflammation	[117, 118]
Mitochondrial dysfunction	PINK1/Parkin mutations impair mitophagy → ROS overproduction → senescence	Mitochondrial ROS, $p16^{INK4a}$, SA-β-gal	Antioxidants (CoQ10, MitoQ, NAC), mitochondrial protectants	ROS-induced damage accelerates dopaminergic neuron aging	[119]
SASP-driven senescence	NF-κB/C/EBPβ activation → IL-6, IL-1β, MMPs, TNF-α secretion	IL-6, IL-1β, TNF-α, MCP-1, MMP-3	SASP inhibitors (NF-κB, JAK/STAT, mTOR)	Promotes chronic neuroinflammation and BBB disruption	[120, 121]
Epigenetic-mediated senescence	Altered DNA methylation, H3K9me3/H4K20me3 → $p16^{INK4a}$/$p21^{CIP1}$ upregulation	SAHF, histone marks, $p16^{INK4a}$	Epigenetic modulators (HDAC inhibitors)	Gene silencing supports irreversible cell cycle arrest	[122]
α-Synuclein-induced senescence	α-Syn aggregation → lysosomal/autophagy dysfunction → SASP activation	α-Synuclein + SA-β-gal, γH2AX	Autophagy enhancers, senolytics (dasatinib, quercetin)	Triggers senescence in neurons and glia, worsening PD pathology	[123, 124]

Conclusion

The progression of Parkinson's disease (PD) follows cellular senescence highly through several mechanisms that relate brain degeneration to aging processes. Primary data analysis from this chapter confirms that PD progression develops through neuroinflammatory conditions from cells such as astrocytes and microglia and neurons with features of senescence, which emerge due to senescence-associated secretory phenotype (SASP). Studies show that biological stresses of aging, including mitochondrial failure, DNA damage, and oxidative stress create strong factors that lead to premature CNS tissue aging, which increases the risk to dopaminergic nigral neurons. Prior research utilizing preclinical models demonstrates that PD pathology combines with cellular senescence through two measurable effects associated with human PD symptoms. Scientific research on cellular senescence discovered therapeutic methods after scientists developed senomorphic drugs to protect cell functions and senolytic drugs that removed senescent cells effectively. First-stage preclinical PD tests indicate that these potential therapeutic approaches help lower neurodegeneration and restore centralized control systems. Research into brain-specific senescence markers will help develop monitoring tools for therapy assessment and diagnostic and classifying methods for patients. The complexity of aging phenotypes and challenges regarding brain-dependent drugs designed for specific cell types and long-term safety questions slows down therapeutic application development based on current research discoveries. Research must prioritize better developing therapeutic approaches for senescence while identifying senescent cell chronology patterns through advanced imaging and omics methods and establishing clinical trials to prove therapeutic effectiveness. Research demonstrates that focusing on cellular senescence presents an innovative method to combat Parkinson's disease origins related to aging processes. The technique offers prospects to transform therapeutic paradigms by switching from symptom management to disease modification and neural protection practices.

References

1. Ma Y, Erb ML, Moore DJ. Aging, cellular senescence and Parkinson's disease. J Parkinsons Dis. 2025;15(2):239–54.
2. Yang W, et al. Aging-related biomarkers for the diagnosis of Parkinson's disease based on bioinformatics analysis and machine learning. Aging (Albany NY). 2024;16(17):12191–208.
3. Jankovic J, Tan EK. Parkinson's disease: etiopathogenesis and treatment. J Neurol Neurosurg Psychiatry. 2020;91(8):795–808.
4. Wang Y, et al. The role of cellular senescence in neurodegenerative diseases. Arch Toxicol. 2024;98(8):2393–408.
5. Saurat N, et al. Genome-wide CRISPR screen identifies neddylation as a regulator of neuronal aging and AD neurodegeneration. Cell Stem Cell. 2024;31(8):1162–1174.e8.
6. Coleman C, Martin I. Unraveling Parkinson's disease neurodegeneration: does aging hold the clues? J Parkinsons Dis. 2022;12(8):2321–38.
7. Kilfeather P, et al. Single-cell spatial transcriptomic and translatomic profiling of dopaminergic neurons in health, aging, and disease. Cell Rep. 2024;43(3):113784.

8. Kasper J, et al. Resting-state changes in aging and Parkinson's disease are shaped by underlying neurotransmission: a normative modeling study. Biol Psychiatry Cogn Neurosci Neuroimaging. 2024;9(10):986–97.
9. Naylor RM, Baker DJ, van Deursen JM. Senescent cells: a novel therapeutic target for aging and age-related diseases. Clin Pharmacol Ther. 2013;93(1):105–16.
10. McHugh D, Gil J. Senescence and aging: causes, consequences, and therapeutic avenues. J Cell Biol. 2018;217(1):65–77.
11. Barmpa K, et al. Modeling early phenotypes of Parkinson's disease by age-induced midbrain-striatum assembloids. Commun Biol. 2024;7(1):1561.
12. Yazar V, et al. DNA methylation signature of aging: potential impact on the pathogenesis of Parkinson's disease. J Parkinsons Dis. 2023;13(2):145–64.
13. Boccardi V, et al. Focus on senescence: clinical significance and practical applications. J Intern Med. 2024;295(5):599–619.
14. Wolf J, et al. Liquid-biopsy proteomics combined with AI identifies cellular drivers of eye aging and disease in vivo. Cell. 2023;186(22):4868–4884.e12.
15. Min X, Zhou W, Fan X. Risk factors for Parkinson's disease in aging population. Asian J Surg. 2023;46(10):4413–5.
16. Melo dos Santos LS, et al. Cellular senescence in brain aging and neurodegeneration. Ageing Res Rev. 2024;93:102141.
17. Karekal A, et al. Elevated Gaussian-modeled beta power in the cortex characterizes aging, but not Parkinson's disease. J Neurophysiol. 2023;129(5):1086–93.
18. Chhetri JK, et al. New horizons in Parkinson's disease in older populations. Age Ageing. 2023;52(10)
19. Abokyi S, Ghartey-Kwansah G, Tse DY. TFEB is a central regulator of the aging process and age-related diseases. Ageing Res Rev. 2023;89:101985.
20. Wang Z, et al. Dietary restriction against Parkinson's disease: what we know so far. Nutrients. 2022;14(19)
21. Gaikwad S, et al. Senescence, brain inflammation, and oligomeric tau drive cognitive decline in Alzheimer's disease: evidence from clinical and preclinical studies. Alzheimers Dement. 2024;20(1):709–27.
22. Guo J, et al. Aging and aging-related diseases: from molecular mechanisms to interventions and treatments. Signal Transduct Target Ther. 2022;7(1):391.
23. Rani L, Sahu MR, Mondal AC. Age-related mitochondrial dysfunction in Parkinson's disease: new insights into the disease pathology. Neuroscience. 2022;499:152–69.
24. Kouli A, Williams-Gray CH. Age-related adaptive immune changes in Parkinson's disease. J Parkinsons Dis. 2022;12(s1):S93–s104.
25. Heavener KS, Bradshaw EM. The aging immune system in Alzheimer's and Parkinson's diseases. Semin Immunopathol. 2022;44(5):649–57.
26. Wu X, et al. DNA damage response (DDR): a link between cellular senescence and human cytomegalovirus. Virol J. 2023;20(1):250.
27. Clarke TL, Mostoslavsky R. DNA repair as a shared hallmark in cancer and ageing. Mol Oncol. 2022;16(18):3352–79.
28. Ebina J, Ebihara S, Kano O. Similarities, differences and overlaps between frailty and Parkinson's disease. Geriatr Gerontol Int. 2022;22(4):259–70.
29. Chambers CR, et al. Overcoming the senescence-associated secretory phenotype (SASP): a complex mechanism of resistance in the treatment of cancer. Mol Oncol. 2021;15(12):3242–55.
30. Carelli S, Rey F, Cereda C. SNCA-AS1 in aging and Parkinson's disease. Aging (Albany NY). 2022;14(8):3335–6.
31. Asghar M, et al. Mitochondrial biogenesis, telomere length and cellular senescence in Parkinson's disease and Lewy body dementia. Sci Rep. 2022;12(1):17578.
32. Chien Y, et al. Control of the senescence-associated secretory phenotype by NF-κB promotes senescence and enhances chemosensitivity. Genes Dev. 2011;25(20):2125–36.
33. Qi S, et al. Prevalence of Parkinson's disease: a community-based study in China. Mov Disord. 2021;36(12):2940–4.

34. Hur EM, Lee BD. LRRK2 at the crossroad of aging and Parkinson's disease. Genes (Basel). 2021;12(4)
35. Misrani A, Tabassum S, Yang L. Mitochondrial dysfunction and oxidative stress in Alzheimer's disease. Front Aging Neurosci. 2021;13:617588.
36. Pickrell AM, Youle RJ. The roles of PINK1, parkin, and mitochondrial fidelity in Parkinson's disease. Neuron. 2015;85(2):257–73.
37. Crouch J, et al. Epigenetic regulation of cellular senescence. Cells. 2022;11:4.
38. Song H, et al. Epigenetic modification in Parkinson's disease. Front Cell Dev Biol. 2023;11:1123621.
39. Gonzalez-Latapi P, et al. Parkinson's disease and COVID-19: do we need to be more patient? Mov Disord. 2021;36(2):277.
40. Marchetti B, et al. Parkinson's disease, aging and adult neurogenesis: Wnt/β-catenin signalling as the key to unlock the mystery of endogenous brain repair. Aging Cell. 2020;19(3):e13101.
41. Jin H, et al. Association of inflammatory factors and aging in Parkinson's disease. Neurosci Lett. 2020;736:135259.
42. Bohnen NI. Vulnerabilities of aging and biological effects of physical activity provide new clues for interventions in Parkinson's disease. J Gerontol A Biol Sci Med Sci. 2020;75(4):687–9.
43. Liu RM. Aging, cellular senescence, and Alzheimer's disease. Int J Mol Sci. 2022;23(4)
44. Miller SJ, et al. Neuroglial senescence, α-synucleinopathy, and the therapeutic potential of senolytics in Parkinson's disease. Front Neurosci. 2022;16:824191.
45. Barata-Antunes S, et al. Impact of aging on the 6-OHDA-induced rat model of Parkinson's disease. Int J Mol Sci. 2020;21(10)
46. Vila M. Neuromelanin, aging, and neuronal vulnerability in Parkinson's disease. Mov Disord. 2019;34(10):1440–51.
47. Hong B, et al. Glial senescence enhances α-synuclein pathology owing to its insufficient clearance caused by autophagy dysfunction. Cell Death Discov. 2024;10(1):50.
48. Du T, et al. Nuclear alpha-synuclein accelerates cell senescence and neurodegeneration. Immun Ageing. 2024;21(1):47.
49. Trist BG, Hare DJ, Double KL. Oxidative stress in the aging substantia nigra and the etiology of Parkinson's disease. Aging Cell. 2019;18(6):e13031.
50. Riessland M, et al. Loss of SATB1 induces p21-dependent cellular senescence in post-mitotic dopaminergic neurons. Cell Stem Cell. 2019;25(4):514–530.e8.
51. Wong YC, Krainc D. α-Synuclein toxicity in neurodegeneration: mechanism and therapeutic strategies. Nat Med. 2017;23(2):1–13.
52. Dai L, et al. Lysosomal dysfunction in α-synuclein pathology: molecular mechanisms and therapeutic strategies. Cell Mol Life Sci. 2024;81(1):382.
53. Nandi A, et al. Role of catalase in oxidative stress- and age-associated degenerative diseases. Oxidative Med Cell Longev. 2019;2019:9613090.
54. Chandra G, et al. Reinforcing mitochondrial functions in aging brain: an insight into Parkinson's disease therapeutics. J Chem Neuroanat. 2019;95:29–42.
55. Williams-Gray CH, et al. Abnormalities of age-related T cell senescence in Parkinson's disease. J Neuroinflammation. 2018;15(1):166.
56. Pan Y, Nicolazzo JA. Impact of aging, Alzheimer's disease and Parkinson's disease on the blood-brain barrier transport of therapeutics. Adv Drug Deliv Rev. 2018;135:62–74.
57. Wang Q, Liu Y, Zhou J. Neuroinflammation in Parkinson's disease and its potential as therapeutic target. Transl Neurodegener. 2015;4:19.
58. Han T, et al. Microglial response to aging and neuroinflammation in the development of neurodegenerative diseases. Neural Regen Res. 2024;19(6):1241–8.
59. Mattson MP, Arumugam TV. Hallmarks of brain aging: adaptive and pathological modification by metabolic states. Cell Metab. 2018;27(6):1176–99.
60. Dorsey ER, et al. The emerging evidence of the Parkinson pandemic. J Parkinsons Dis. 2018;8(s1):S3–s8.

61. Ni A, Ernst C. Evidence that substantia nigra pars compacta dopaminergic neurons are selectively vulnerable to oxidative stress because they are highly metabolically active. Front Cell Neurosci. 2022;16:826193.
62. Chesnokova AY, Ekimova IV, Pastukhov YF. Parkinson's disease and aging. Adv Gerontol. 2018;31(5):668–78.
63. Calabrese V, et al. Aging and Parkinson's disease: Inflammaging, neuroinflammation and biological remodeling as key factors in pathogenesis. Free Radic Biol Med. 2018;115:80–91.
64. Han X, et al. Astrocyte senescence and Alzheimer's disease: a review. Front Aging Neurosci. 2020;12:148.
65. Zucca FA, et al. Interactions of iron, dopamine and neuromelanin pathways in brain aging and Parkinson's disease. Prog Neurobiol. 2017;155:96–119.
66. Lenka A, Padmakumar C, Pal PK. Treatment of older Parkinson's disease. Int Rev Neurobiol. 2017;132:381–405.
67. Collier TJ, Kanaan NM, Kordower JH. Aging and Parkinson's disease: different sides of the same coin? Mov Disord. 2017;32(7):983–90.
68. Błaszczyk JW. Nigrostriatal interaction in the aging brain: new therapeutic target for Parkinson's disease. Acta Neurobiol Exp (Wars). 2017;77(1):106–12.
69. Dias V, Junn E, Mouradian MM. The role of oxidative stress in Parkinson's disease. J Parkinsons Dis. 2013;3(4):461–91.
70. Zhou M, et al. Targeting ferroptosis in Parkinson's disease: mechanisms and emerging therapeutic strategies. Int J Mol Sci. 2024;25(23)
71. Schmitt CA, et al. A senescence program controlled by p53 and p16INK4a contributes to the outcome of cancer therapy. Cell. 2002;109(3):335–46.
72. Maurer S, et al. Senolytic therapy combining Dasatinib and Quercetin restores the chondrogenic phenotype of human osteoarthritic chondrocytes by the release of pro-anabolic mediators. Aging Cell. 2025;24(1):e14361.
73. Wüllner U, et al. DNA methylation in Parkinson's disease. J Neurochem. 2016;139(Suppl 1):108–20.
74. Wang HY, et al. Mechanism of anti-aging therapy on Parkinson's disease. Zhongguo Zhong Yao Za Zhi. 2016;41(14):2680–6.
75. Mattis J, Sehgal A. Circadian rhythms, sleep, and disorders of aging. Trends Endocrinol Metab. 2016;27(4):192–203.
76. Heemels MT. Neurodegenerative diseases. Nature. 2016;539(7628):179.
77. Kudlova N, De Sanctis JB, Hajduch M. Cellular senescence: molecular targets, biomarkers, and senolytic drugs. Int J Mol Sci. 2022;23(8)
78. Carreno G, Guiho R, Martinez-Barbera JP. Cell senescence in neuropathology: a focus on neurodegeneration and tumours. Neuropathol Appl Neurobiol. 2021;47(3):359–78.
79. Rodriguez M, et al. Parkinson's disease as a result of aging. Aging Cell. 2015;14(3):293–308.
80. Bobela W, Aebischer P, Schneider BL. Alpha-Synuclein as a mediator in the interplay between aging and Parkinson's disease. Biomol Ther. 2015;5(4):2675–700.
81. Abdullah R, et al. Parkinson's disease and age: the obvious but largely unexplored link. Exp Gerontol. 2015;68:33–8.
82. Reeve A, Simcox E, Turnbull D. Ageing and Parkinson's disease: why is advancing age the biggest risk factor? Ageing Res Rev. 2014;14(100):19–30.
83. Vazquez-Villaseñor I, et al. Expression of p16 and p21 in the frontal association cortex of ALS/MND brains suggests neuronal cell cycle dysregulation and astrocyte senescence in early stages of the disease. Neuropathol Appl Neurobiol. 2020;46(2):171–85.
84. Alshaebi F, Sciortino A, Kayed R. The role of glial cell senescence in Alzheimer's disease. J Neurochem. 2025;169(3):e70051.
85. Hipkiss AR. Aging risk factors and Parkinson's disease: contrasting roles of common dietary constituents. Neurobiol Aging. 2014;35(6):1469–72.
86. Butterfield DA, Dalle-Donne I. Redox proteomics. Antioxid Redox Signal. 2012;17(11):1487–9.

87. Basisty N, et al. A proteomic atlas of senescence-associated secretomes for aging biomarker development. PLoS Biol. 2020;18(1):e3000599.
88. Liu Y, et al. Expression of p16(INK4a) in peripheral blood T-cells is a biomarker of human aging. Aging Cell. 2009;8(4):439–48.
89. Olanow CW, Kordower JH. Modeling Parkinson's disease. Ann Neurol. 2009;66(4):432–6.
90. Archibald NK, et al. The retina in Parkinson's disease. Brain. 2009;132(Pt 5):1128–45.
91. Azam S, et al. The ageing brain: molecular and cellular basis of neurodegeneration. Front Cell Dev Biol. 2021;9:683459.
92. Hussein A, et al. Non-motor symptoms of Parkinson's disease: the neurobiology of early psychiatric and cognitive dysfunction. Neuroscientist. 2023;29(1):97–116.
93. Stewart T, et al. Cerebrospinal fluid α-synuclein predicts cognitive decline in Parkinson disease progression in the DATATOP cohort. Am J Pathol. 2014;184(4):966–75.
94. Raffaele M, Vinciguerra M. The costs and benefits of senotherapeutics for human health. Lancet Healthy Longev. 2022;3(1):e67–77.
95. Al-Naggar IMA, Kuchel GA, Xu M. Senolytics: targeting senescent cells for age-associated diseases. Curr Mol Biol Rep. 2020;6(4):161–72.
96. Playfer JR. Ageing and Parkinson's disease. Pract Neurol. 2007;7(1):4–5.
97. Cuollo L, et al. The senescence-associated secretory phenotype (SASP) in the challenging future of cancer therapy and age-related diseases. Biology (Basel). 2020;9(12)
98. Maus M, et al. Iron accumulation drives fibrosis, senescence and the senescence-associated secretory phenotype. Nat Metab. 2023;5(12):2111–30.
99. Pekkonen E. Mismatch negativity in aging and in Alzheimer's and Parkinson's diseases. Audiol Neurootol. 2000;5(3–4):216–24.
100. Playfer JR. Parkinson's disease. Postgrad Med J. 1997;73(859):257–64.
101. De Gaetano A, et al. Mitophagy and oxidative stress: the role of aging. Antioxidants (Basel). 2021;10(5)
102. Pahal S, et al. Mitochondria in aging and age-associated diseases. Mitochondrion. 2025;82:102022.
103. Kirkland JL, Tchkonia T. Senolytic drugs: from discovery to translation. J Intern Med. 2020;288(5):518–36.
104. Kakoty V, et al. Senolytics: opening avenues in drug discovery to find novel therapeutics for Parkinson's disease. Drug Discov Today. 2023;28(6):103582.
105. Reynolds CF 3rd, et al. Mental health care for older adults: recent advances and new directions in clinical practice and research. World Psychiatry. 2022;21(3):336–63.
106. Richardson M, Richardson DR. Pharmacological targeting of senescence with senolytics as a new therapeutic strategy for neurodegeneration. Mol Pharmacol. 2024;105(2):64–74.
107. Buttery PC, Barker RA. Gene and cell-based therapies for Parkinson's disease: where are we? Neurotherapeutics. 2020;17(4):1539–62.
108. Bi J, et al. Drug delivery for age-related bone diseases: from therapeutic targets to common and emerging therapeutic strategies. Saudi Pharm J. 2024;32(12):102209.
109. Molla G, Bitew M. Revolutionizing personalized medicine: synergy with multi-omics data generation, main hurdles, and future perspectives. Biomedicines. 2024;12(12)
110. Khan A, et al. Genomic medicine and personalized treatment: a narrative review. Ann Med Surg (Lond). 2025;87(3):1406–14.
111. Chen C, et al. Applications of multi-omics analysis in human diseases. MedComm (2020). 2023;4(4):e315.
112. Redenšek S, Dolžan V, Kunej T. From genomics to omics landscapes of Parkinson's disease: revealing the molecular mechanisms. OMICS. 2018;22(1):1–16.
113. Theodorakis N, et al. Integrating machine learning with multi-omics technologies in geroscience: towards personalized medicine. J Pers Med. 2024;14(9)
114. Kritsilis M, et al. Ageing, cellular senescence and neurodegenerative disease. Int J Mol Sci. 2018;19(10)

115. Gomez LS, Jurk D. Unlocking the potential of senolytic compounds: advancements, opportunities, and challenges in ageing-related research. Subcell Biochem. 2024;107:91–116.
116. Robbins PD, et al. Senolytic drugs: reducing senescent cell viability to extend health span. Annu Rev Pharmacol Toxicol. 2021;61:779–803.
117. Mijit M, et al. Role of p53 in the regulation of cellular senescence. Biomol Ther. 2020;10(3)
118. Delia D, et al. DNA damage-induced cell-cycle phase regulation of p53 and p21waf1 in normal and ATM-defective cells. Oncogene. 2003;22(49):7866–9.
119. Narendra DP, et al. PINK1 is selectively stabilized on impaired mitochondria to activate Parkin. PLoS Biol. 2010;8(1):e1000298.
120. Coppé JP, et al. The senescence-associated secretory phenotype: the dark side of tumor suppression. Annu Rev Pathol. 2010;5:99–118.
121. Wang WJ, Cai GY, Chen XM. Cellular senescence, senescence-associated secretory phenotype, and chronic kidney disease. Oncotarget. 2017;8(38):64520–33.
122. Zheng Y, et al. H3K9me-enhanced DNA hypermethylation of the p16INK4a gene: an epigenetic signature for spontaneous transformation of rat mesenchymal stem cells. Stem Cells Dev. 2013;22(2):256–67.
123. Klucken J, et al. Alpha-synuclein aggregation involves a bafilomycin A 1-sensitive autophagy pathway. Autophagy. 2012;8(5):754–66.
124. Miller SJ, et al. Senolytic and senomorphic secondary metabolites as therapeutic agents in Drosophila melanogaster models of Parkinson's disease. Front Neurol. 2023;14:1271941.

Cellular Senescence in Huntington's Disease (HD)

8

Waleed Hassan Almalki

Introduction

Huntington's disease (HD) is a fatal autosomal-dominant neurodegenerative disorder caused by the expansion of CAG trinucleotide repeats in exon 1 of the HTT gene, leading to the production of mutant huntingtin (mHTT) protein with an expanded polyglutamine tract. This mutant protein predominantly affects the striatum and cortex, disrupting cellular processes and resulting in progressive neuronal degeneration. The striatum and cortex are the primary areas where this altered protein interferes with different cellular processes and causes the neuron to deteriorate gradually [1, 2]. The clinical presentation of HD includes a blend of mental health including depression, irritability, as well as psychosis, reduced cognitive faculties that are characterized by executive dysfunction, and memory issues, with motor problems also present in the form of chorea, dystonia, and impaired coordination [3]. These symptoms tend to arise sometime in middle age and subsequently, inexorably exacerbate until a severe impairment and an early death occur. While the genetic factor of HD is known, little is known about the molecular mechanisms of neuronal vulnerability, as well as the development of the disease [4, 5]. It is now well appreciated that cellular senescence is critical to the pathogenesis of neurodegenerative diseases, a previously defined state of irreversible loss of cell cycle activity due to stress [6]. One of the most characteristic features of senescent cells is that they display a persistent DNA damage response (DDR), activation of the cyclin-dependent kinase inhibitors, for example, p16INK4a and p21CIP1, signals of mitochondrial dysfunction and altered expression of other pro-apopt. Despite the fact that it has become generally accepted that actively dividing glial cells are linked to senescence in the CNS [7]. Recent studies show that nondividing neurons can also have the features of the cellular senescence such as metabolic disruption, increased genomic instability, and

W. H. Almalki (✉)
Department of Pharmacology and Toxicology, College of Pharmacy,
Umm Al-Qura University, Makkah, Saudi Arabia

G. Gupta (ed.), *Cellular Senescence and Brain Aging*,
https://doi.org/10.1007/978-981-96-8873-9_8

SASP-activity, especially under prolonged stress [8, 9]. Studies have shown that mHTT protein causes oxidative stress, DNA damage, and mitochondrial dysfunction in the HD, the factors which cause cellular senescence [10]. Senescence in astrocytes and microglia, promoted by SASP, aggravates synaptic dysfunction, depletes neuronal supportive roles, and accelerates neurodegeneration through induction of a maladaptive inflammatory state [11]. Features typical of cellular senescence show in lab models and patient-derived tissues; this confirms the relationship with the mechanisms underlying HD. Admitting senescence as a controllable mechanism of neurodegeneration is a compelling reason for studying senescence in HD [12, 13]. Remarkably, in multiple other models of neurodegenerative disease, inhibiting senescence decreased neuroinflammation and improved neuronal health. These have senolytics that kill senescent cells and senomorphics, which control the SASP without causing death of the cell [14]. However, there are still limits for senescence-targeted therapies in HD: the variety of senescence phenotypes of neurodegenerative tissues, the need to rely on biomarkers, and potentially unwanted side effects [15]. Elucidation of spatial and temporal dimensions of senescence, as well as the relationship between mHTT toxicity, will be a cornerstone of successful treatment of HD [16]. Our chapter indicate how cellular senescence represents a critical junction of various mechanisms and present promising therapeutic directions that require further investigation in neurodegenerative diseases [17, 18] (Fig. 8.1).

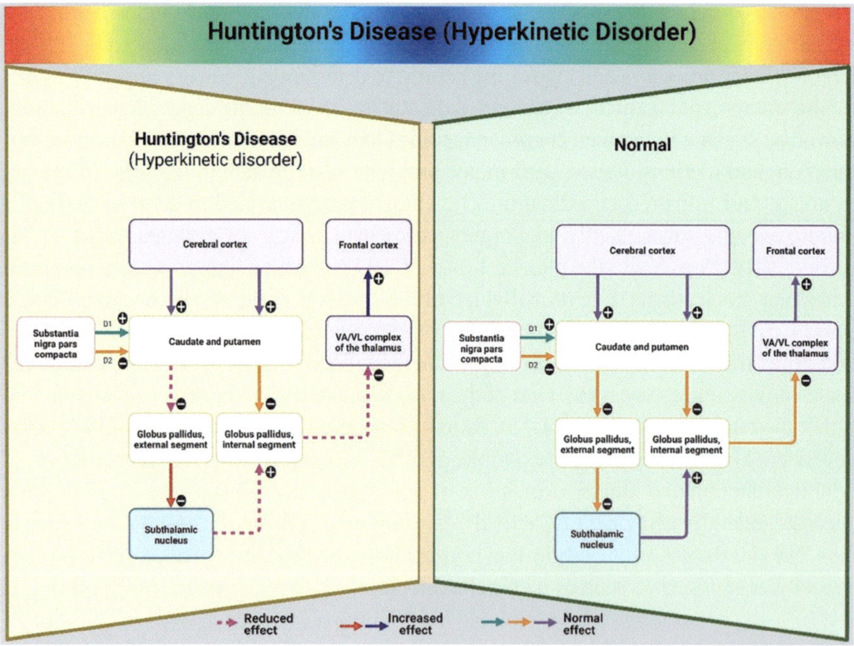

Fig. 8.1 Altered basal ganglia pathways in Huntington's disease. The figure describes the basal ganglia pathways in general brains compared to those in subjects with Huntington's disease, and causes a hyperkinetic nature of the disorder. It demonstrates the altered signaling pathways that indicate a reduction of inhibitory signals from caudate-putamen to the pars externa globus pallidus, thus increasing the activity in the thalamus and frontal cortex. Different-colored arrows represent changes in signaling, as Huntington's pathology changes motor control by degenerating striatal neurons of the indirect pathway

Mechanisms of Cellular Senescence

Cellular senescence arises from a combination of causes, such as oxidative stress and DDR activation, which cause persistent DNA damage and disruption of normal cellular homeostasis. Senescence is promoted by events in malfunctions in mitochondria dysfunction and increased levels of reactive oxygen species (ROS), both of which worsen cell damage. Upon senescence-associated secretory phenotype, senescent cells release matrix-degrading enzymes and pro-inflammatory cytokines, which play key roles in tissue deterioration. The overexpression of both p16INK4a and p21CIP1 caused by epigenetic changes triggers permanent cell cycle arrest, fixation of cellular senescence, and enhancement of inflammatory and degenerative processes [19, 20]. HD is an autosomal dominant disorder characterized by the faulty expansion of CAG trinucleotide repeats in the HTT gene, leading to an extended polyglutamine tract in the huntingtin protein. Striatal neurons are the principal victims of HD caused by protein misfolding, aggregation, and neurotoxicity, leading to escalating motor, cognitive, and psychiatric deficits [21]. A knock-in mouse model of HD was used by Alcalá-Vida et al. to interrogate the chromatin landscape of HD striatum. The study revealed two early actions in the remodeling of the structure of chromatin, which are a local perturbation of 3D chromatin configuration at the Htt locus, associated with CAG repeat expansion and accelerated epigenetic neuronal and glial gene expression changes with age [22]. Neural progenitor cells (NPCs), which are multipotent and are capable of differentiating into neurons and glial cells, have an important role in development. HD due to mHTT inhibits neurogenesis by reducing the viability of NPCs, their differentiation, and proliferation. This illuminates the role of NPC dysfunction in HD in the context of its contribution to early developmental defects and persistent neuronal loss [23]. According to Zhou et al., aging leads to neurodegeneration by destabilising the neuronal microenvironment, especially in terms of neural progenitor cells and microglia. The study highlights aging as a major factor and a critical treatment target in neurodegeneration, demonstrating how senescence of glial cells has an adverse effect on the CNS function and drives disease states such as Huntington's disease [24]. As a nuclear-targeted mitogen, HDGF is associated with angiogenesis, tissue regeneration, and cell proliferation. Although the exact role of HDGF is unclear, abnormal expression of HDGF may contribute to both neuroinflammation and neuronal survival decrease in HD. Further investigation is required to determine whether HDGF is pathogenic or neuroprotective [25, 26]. Voelkl et al. showed that hepatoma-derived growth factor (HDGF) bestows neuroprotection in a model of HD. In neurons that are sensitive to Huntington's, there is reduced HDGF expression; in R6/2 animals that are missing HDGF, there were higher degrees of motor deficits and reduced lifespan. Giving HDGF to human and mouse cellular HD models enhanced neuronal health and reduced accumulation of mutant huntingtin aggregates [27]. Being an essential coenzyme involved in mitochondrial processes, DNA repair and other general cell functions, nicotinamide adenine dinucleotide (NAD^+) is inestimable. There is a negative influence of reduced NAD^+ levels typical of Huntington's disease on neuronal survival and energy metabolism. The protective influence of NAD^+ augmentation enhances its candidacy as a promising therapeutic target for controlling HD [28]. Azam et al. suggest that neurodegeneration that happens during aging is the primary risk factor for Alzheimer's, Parkinson's, Huntington's, and frontotemporal dementia, perpetuated by molecular mechanisms. The findings of this study identified cellular senescence, mitochondrial breakdown, and genomic instability as critical for aging. The study also suggested pharmacological interventions like mitophagy inducers, senescence inhibitors, and NAD^+ precursors [29].

Evidence of Senescence in Huntington's Disease

In Vitro Studies (Cellular Models)

The cellular senescence mechanisms of Huntington's disease are explained quite well in vitro analyses using neurons and glial cells derived from HD iPSC. In these animals, elevated senescence markers p16INK4a and SA-β-gal indicate the toxic effect of a mHTT. mHTT induced DNA damage and oxidative stress result in abnormally reentered cell cycle, thereby causing the progression of neurodegeneration and dysfunctional cellular activity. These experimental platforms provide a useful framework for further understanding senescence in HD and further uncovering promising treatment strategies [30]. Scientists have discovered that the state of HD causes mislocation and aggregation of TDP-43, a DNA/RNA-binding protein essential for RNA metabolism. Misplacement of TDP-43 reveals RNA processing and further damages neuronal function, increasing the risk of neurotoxicity. Mutant together with TDP-43 abnormalities may exacerbate neurodegenerative actions in HD [31]. Bai et al. showed that when the pathogenic TDP-43 is mislocalized into the cytoplasm, it exacerbates motor symptoms in HD mice and hastens the production of toxic exon1 HTT protein by binding with intron1 of HTT pre-mRNA into neurotoxicity through RNA processing [32]. In individuals with HD, amyloid-beta (Aβ), a peptide associated with Alzheimer's disease, was detected in the brain, worsening neurodegenerative processes. Although Aβ contributes to oxidative stress, disrupted mitochondria, and synaptic injury, its interaction with mutant huntingtin induces greater cellular damage and a more rapid course of HD disease [33]. As Pantiya et al. have observed that there is a strong relationship between mitochondrial defects, cognitive decline, and the progression of Alzheimer's, Parkinson's, and Huntington's diseases. Such changes may be detected in laboratory and real-world (in vitro and in vivo) settings for these medical situations. In addition, their findings present promising approaches for the treatment of neurodegenerative diseases based on targeting the mitochondria in the brain and, in some cases, increasing functionality and delaying the advancement of the disease [34].

In Vivo Studies (Animal Models)

In vivo studies with HD animal models, such as R6/2 and YAC128, have demonstrated stunning descriptions of senescence involving increases in DNA damage and the stopping of the cell cycle in neurons and glial cells. Activated microglia and astrocytes in these animals show SASP, which promotes significant neuroinflammation. Increased neurodegeneration in HD is therefore triggered by such age-related changes and inflammation, which further illuminates the role of gliocyte senescence in the disease process [35, 36]. Cellular absorption of carnitine, necessary for the proper functioning of mitochondrial fatty acid β-oxidation, is regulated by OCTN2 (organic cation/carnitine transporter 2). Neuronal damage in Huntington's disease is caused by mitochondrial malfunction and disrupted energy metabolism caused by defective OCTN2 function. Modulation of the behavior of OCTN2 may be another means of metabolic balance and possibly of treating HD [37]. Cristo et al. reported with the help of cell and fly models that the recently developed THP-based carnitine-lowering agents (TCLs), which target OCTN2 while having few changes to production of carnitine, could restore metabolic flexibility in HD models by shifting reliance on energy from fatty acids. This resulted in protein aggregates reduction, an enhanced motor function, and longer lifespan [38]. The protein BPIFB4 is linked to lifespan, and the regulation of the immune system demonstrates neuroprotective properties in neurodegenerative diseases. Overexpression of BPIFB4 can enhance the survival rate of neurons, reduce inflammatory damage, and lessen Huntington's disease-specific oxidative stress. A new investigation reveals that manipulating BPIFB4 may help control HD by disrupting neuroimmune and anti-senescent processes [39]. Pardo et al. have recently reported that LAV-BPIFB4, delivered by adeno-associated viral infection, not only alleviated motor manifestations but also stabilized body mass and reduced mutant huntingtin levels in R6/2 HD mice with a CXCR4-dependent protection from neurodegeneration. The outcomes suggest that LAV-BPIFB4 has potential as a therapeutic intervention for HD, specifically targeting immune and cellular health [38, 39]. Clofazimine, an anti-inflammatory and anti-leprosy drug, seems promising in HD mouse models because it can decrease neuroinflammation, protein clumping, and oxidative stress. Clofazimine may have a potential use as therapy for HD since it can change cellular pathways linked to mHTT toxicity and provide neuroprotective effects [40]. As in the work of Li et al., it was discovered that clofazimine combats polyQ toxicity through activation of PPARγ, which promotes mitochondrial biogenesis. By reducing the toxicity of zebra fish and nerve-specific *C. elegans* models and by repairing the mitochondria in their Htt-Q94-expressing cells, the drug showed promise as a therapy for polyQ disorders [40] (Fig. 8.2).

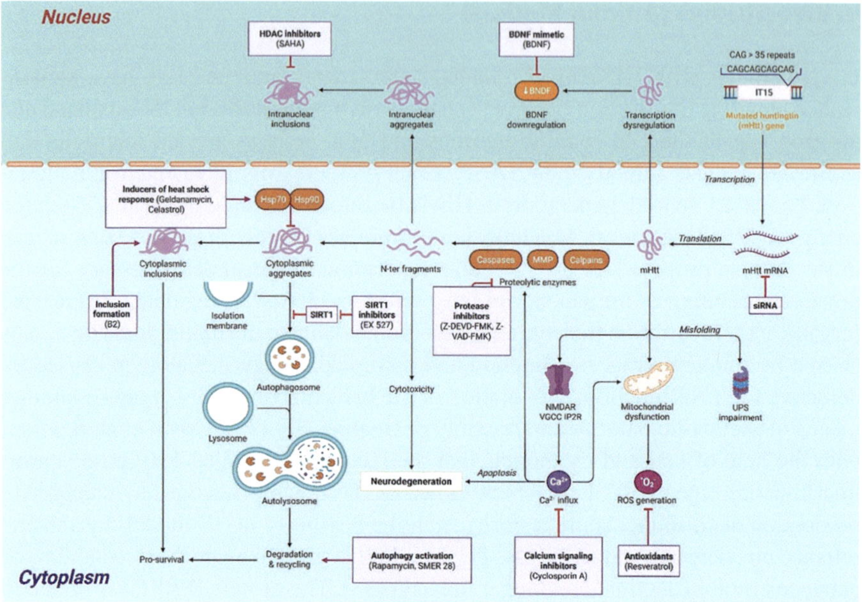

Fig. 8.2 Molecular mechanisms of mHTT toxicity in Huntington's disease. The figure displays the molecular basis of mHTT for Huntington's disease: how it damages protein structure, alters gene regulation and proteolysis, and causes toxic aggregates, inclusion formation, mitochondrial failure, oxidative damage, and cell death. Strategies include inhibitors such as siRNA, protease, HDAC, SIRT1, antioxidants, calcium signaling inhibitors, and autophagy stimulators that may help reduce toxicity, achieve protein balance, and protect neurons against loss in Huntington's disease

Clinical Observations and Biomarkers

The increased levels of senescence markers, including SA-β-gal and p16INK4a, detected in neurons and glial cells in postmortem HD brains emphasize the pathophysiological importance of cellular senescence in HD. Recent research has shown that peripheral markers such as β-gal and p16 are promising markers to follow when monitoring senescence in individuals who have Huntington's disease. In addition, such biomarkers may facilitate the creation of noninvasive techniques for the analysis of the advancement of diseases and treatment effectiveness [41, 42]. In patients with Huntington's disease and some neurological disorders, the expression of N-terminal tau (NTA-tau), a short form of tau protein, rises in these patients. NTA-tau promotes destabilization of microtubules, synaptic dysfunctions, and neuroinflammation, jointly causing neuronal malfunction in HD. Given that NTA-tau might exacerbate HD, it is an interesting option for therapeutic interventions [43, 44].

Alpaugh et al. explored possible relationships between tau levels and the extent of Huntington's disease from blood extracts collected from the patients. It was determined that peripheral blood mononuclear cells, platelets, and plasma had elevated levels of both total tau (t-tau) and N-terminal tau (NTA-tau). Surprisingly, there was a strong correlation between the level of disease, particularly cognitive impairment, and the platelet levels of tau [43]. The various bacteria inhabiting the gut, the gut microbiota, also act on the principles of the gut–brain axis to impact brain function. Disruptions in microbial composition lead to HD, using neurodegeneration, neuroinflammation, and altered metabolism as mechanisms. This finding implies that management of the intestinal microbiota may represent a favourable therapeutic option in HD [45]. Khoshnan et al. studied the effects of the gut microbiota on Huntington's disease mouse model, saying that the mutant huntingtin protein has numerous effects on many organs and causes neurological and behavioral disorders. Dysbiosis in the gut microbiota is associated with HTT aggregation, changes in behavior, and shorter life expectancy in HD, underlining its significance in HD and its prospect for treatment [46].

Senescence in Neuronal vs. Nonneuronal Cells in HD

The fact that senescence is found in both neuronal and nonneuronal cells contributes to a faster progression of the disease in HD. Senescence and astrocyte and microglial senescence play a role in neuroinflammation and neuronal damage. This disruption in the barrier integrity in the blood–brain barrier (BBB) endothelial cells resulting from senescence forms an important pathophysiological process for both initiation and aggravation of neuroinflammation. Also, systemic senescence amplifies inflammation via senescent peripheral immune cells, and thus aids in the escalation of disease. Based on the demonstration that nonneuronal cells contribute to HD through senescence, these results indicate that they are one of the main targets to be addressed [47, 48]. As derived from human embryonic kidneys, HEK293 cells are capable of expressing human genes like mHTT and are therefore used regularly in the study of HD. They provide a platform to understand the cellular mechanism and toxicity of mHTT and to discover new avenues for HD treatments [49]. Using HEK293 and Neuro 2a cell models of HD, Jain et al. showed that RNA aptamers with high binding affinity to the mutant huntingtin (EGFP-74Q) are capable of blocking its aggregation. Evidence indicates that aptamers might be of help to avoid protein aggregation in HD because they are related to increased chaperone activity, lessened oxidative stress, better mitochondrial function, and more viable cells [50].

HD especially affects Purkinje cells in the cerebellum of our brain, which are large inhibitory neurons. The disintegration of Purkinje cells is a factor that leads to motor complications that define HD. The central role of the Purkinje cells in HD, as well as the opportunity for treating it, is reinforced because the cells are directly associated with the motor symptoms provoked by their dysfunction [51]. Cvetanovic et al. explored glial cell function in polyglutamine disorders and highlighted their central position in their path. Although associated with spinocerebellar ataxias (SCA) and striatal medium spiny neuron loss with motor dysfunction in HD, an increasing number of studies indicate that dysfunction of glial cells is, of course, an important contributor to these disorders. Researchers should suggest more targeted treatments for disorders by describing the characteristics of glial phenotypes associated with different disorders [52]. Transient receptor potential (TRP) channels play a role in calcium balance, cell-to-cell communication, and sensory response. Dysregulation of TRPV1 and TRPM8 channels contributes significantly to triggering neurodegeneration and destroying neuronal function in HD by causing excitotoxicity, changing the calcium homeostasis, and inducing neuroinflammation [53]. As reported by Rather et al., the transient receptor potential (TRP) channels play a role in the maintenance of calcium homeostasis, both in neuronal as well as nonneuronal cells, and, therefore, are relevant to neurodegenerative disorders. Alzheimer's, Parkinson's, Huntington's, and ALS turn worse if the function of TRP channels is impaired. The research on TRP processes can open up new treatment approaches aimed at such neurological conditions [53]. The process of both development and repair of neural tissues depends on the functions of neural stem cells (NSCs). The compromised proliferative and differentiation capacity in NSCs in Huntington's disease leads to neuronal loss and neurodegeneration. Enhancing the NSC survival and function has the potential to enhance neuroprotection and tissue regeneration in HD-specific tissues, with further investigations needed in therapy [54, 55]. Ghosh et al. studied ReNcell VM neural stem cells (NSCs) with varying polyglutamines (polyQ) in exon 1 huntingtin (HTT) fragments to create a human neuronal model of HD. Their observations indicated that neurogenesis had polyQ length-dependent intranuclear inclusions upon exposure to longer polyQ tracts. Although pathogenic lines displayed indications of mitochondrial dysfunction, neuronal survival was not influenced [56].

Crosstalk Between Senescence and HD Pathogenesis

The associations between HD development and cellular senescence are increasingly important for the understanding of Huntington's disease progression. Protein mHTT induces cellular senescence in neurons and glial cells, due to oxidative stress, DNA damage, and dysfunctional mitochondrial function. Senescence is characterized by a terminal cell cycle block and activation of the key pathways, such as p53/p21CIP1 and p16INK4a. The SASP represents a mechanism that accompanies the release of matrix-degrading enzymes and pro-inflammatory cytokines, which lead to HD's increasing neurotoxicity and synaptic dysfunction [57, 58]. Also, senescent cells that exhibit defective autophagy and dynamics of the mitochondria increase cellular stress, hinder the elimination of aggregates of proteins, and promote neurodegeneration. When mHTT impacts the autophagic-lysosomal system, toxic cellular waste accumulates and disrupts mitochondrial usage, therefore increasing oxidative stress and driving senescence pathogenesis forward. Analysis of these processes, when considered together, indicates that senescence acts synergistically in HD, and the inclusion of senescent brain cells plays a notable role in the development of the disease. One approach to attenuating neurodegenerative features in HD might be through interventional in these pathways [59, 60].

Therapeutic Approaches Targeting Senescence in HD

Senolytics

Navitoclax and the combination of dasatinib and quercetin (D + Q) and fisetin are senolytic drugs that can recognize and eliminate the senescent cells associated with neurodegenerative diseases. Preclinical studies demonstrate the efficiency of such treatments. Animal trials reveal that these substances are used as cognitive enhancers, neuronal protectors, and attenuate neuroinflammation that arises due to SASP [61]. Navitoclax, which blocks BCL-2 and BCL-xL, is selective in its targeting of senescent cells through apoptosis, while D + Q and fisetin are generally anti-immunogenic, anti-inflammatory, and antioxidant. Still, there are major problems that persist, including insufficient brain access, danger of damaging healthy cells, as well as insufficient selectivity for senescent cells, particularly in neurologic tissues. Strategies to improve the translational application of delivery systems and biomarker-guided targeted approaches are essential in neurodegenerative disorders [62].

Senomorphics

Senomorphics drugs such as rapamycin, metformin, and JAK inhibitors provide a promising way to reduce the harmful impact of cellular senescence without causing cytotoxicity. Efficient control of the SASP through these substances reduces neurotoxicity and chronic inflammatory response typical to neurodegenerative diseases. With the aim of targeting NF-κB-driven inflammation, JAK inhibitors suppress SASP-related cytokine signaling. In contrast, metformin slows mTOR to minimize pro-inflammatory secretome production, and rapamycin modulates AMPK activity. Experimental data from neurodegenerative models show that senomorphic approaches save neurons, decrease glial reactivity, and fine-tune functional recovery. By encouraging repair rather than elimination of cells, these interventions signal a milder approach to treating age-related neurodegeneration [63, 64].

Gene Editing and RNA Therapeutics

Since gene editing and RNA-based therapies already target suppressing the synthesis of mHTT, the underlying cause of HD, they present promising options for disease modification. Using CRISPR/Cas9-based genome editing, the capacity to silence or remove CAG expansions in the HTT gene is used to constrain the production of hazardous mHTT. Transmission of ASOs against mHTT mRNA for mRNA degradation with RNase H-dependent mechanisms provides an additional option and helps to reduce the development of disease-related events [65]. Recent discoveries show that ameliorating mHTT burden has the potential to diminish senescence-associated markers such as oxidative stress, DNA damage signaling, and SASP factor upregulation in cultures of neurons and glial groups. Preclinical data indicate that knocking down mHTT is associated with decreased senescence markers and improved neurodegeneration, which means senescence management is related to targeting mHTT. Such findings will support the combination of RNA therapies and gene-editing as promising, precise methods, which interrupt the harmful linkage between mHTT toxicity and cellular senescence in HD [66].

Lifestyle and Pharmacological Interventions

Interventions such as increased NAD^+, physical activity, and calorie intake all reduce cellular senescence by tuning critical metabolic and stress response pathways. Caloric restriction increases the function of mitochondria and genomic stability via actions of the SIRT1 and AMPK pathways. Exercise blocks the SASP factor production due to its effects of reducing pro–pro-inflammatory cytokines and oxidative stresses. With supplementation through the application of nicotinamide riboside, increasing NAD+ levels can restore SIRT-dependent DNA repair and mitophagy. Combined, these treatments oppose the character of senescence and are of value to preclinical models of aging and age-related diseases [67, 68]. Overrepetition of the CAG repeat in the gene that derives the HTT leads to the development of the mHTT, which creates a shorter form of the huntingtin protein; the latter interferes with the cellular mechanisms that have to do with protein folding, intracellular movement, and mitochondrial health [25]. According to Bhat et al., research shows that accumulation of m-HTT causes oxidative stress, DNA damage, lower autophagy, and increased inflammation processes that promote cellular senescence. The senescent cell excretion of pro-inflammatory cytokines intensifies the neuronal damage and the progression of the disease. When targeting the treatment to animal models, the symptoms of HD are decreased, as well as the lifespan, and potential new therapies enabling to improve of the quality of life are figured [57]. An ATP-binding cassette transporter, P-glycoprotein 3 (Pgp-3), plays an essential role in cellular detoxification, where harmful metabolites are expelled. Poor functioning of Pgp-3 in HD may inhibit the elimination of toxic compounds, resulting in neuronal injury. Mucking with Pgp-3 may cause the emergence of new methods for slowing the development of HD [69]. Machiela et al. explored mitochondrial dynamics in HD-related C. elegans models. With aging, they discovered that mitochondrial fragmentation came through the expansion of polyglutamine tracts. When DRP-1 levels were reduced, it became evident that there were negative consequences for the situation. The introduction of pgp-3, F25B5.6, and alh-12 as therapeutic targets corrected problems with mitochondria, with promising strategies for resolving HD [70].

A major component of the unfolded protein response (UPR) reaction to ER stress is PERK (PKR-like ER kinase). When cellular stress stimulates PERK, it causes neurodegeneration in Huntington's disease patients. The detrimental effects of mutant huntingtin aggravate neuronal dysfunction, the development of HD, and dysregulated PERK signaling [71]. From the work of Rozpędek-Kamińska et al., an overactive ER stress triggered by misfolded proteins emphasises the critical role of PERK-associated UPR in the pathogenesis of neurodegenerative diseases. The UPR promotes cell survival under minimal stress on the ER but extends stress to apoptosis and leads to neurodegeneration. Mechanistic studies on small-molecule inhibitors used to control PERK signaling activity may prove useful for the treatment of Alzheimer's, Parkinson's, Huntington's, and prion diseases [72]. Huntington's disease interferes with GABAergic signaling that is essential for inhibitory neurotransmission. Stria-affected GABAergic neurons, particularly in the striatum, cause increased excitotoxicity and reduced motor coordination. This dysfunction indicates the possible benefits of therapies for HD by exacerbating its characteristic symptoms, such as problems with motor coordination, mental function, and emotional well-being [73]. Traa et al. reported the outcomes of the lesser mitochondrial fragmentation with the help of Neur-67Q, the neural *C. elegans* model of polyglutamine toxicity. They discovered that disruption of drp-1 caused improved mobility, extended lifespan, and reversal of mitochondrial abnormalities. Twenty-four RNA interference (RNAi) clones were identified, 11 new genomic targets that promoted mobility and longevity, suggesting possible treatments for polyglutamine conditions [74]. FKBP5 plays a co-chaperone with heat shock proteins to regulate the glucocorticoid receptor's function and control the organism's stress reaction. FKBP5 is involved in cellular stress pathways and neuroinflammation in HD. Dysregulation of FKBP5 enhances the toxicity of mutant huntingtin, provoking HD pathophysiology and possibly disease onset and progression [75]. According to Bailus et al.'s findings, human HD isogenic stem cells, as well as HD mouse models, have much lower levels of FKPB5. Constituting FKBP5 either genetically or pharmacologically without affecting mTOR, reduces mHTT levels, increases autophagy, and increases mHTT clearance. From these results, it seems that focusing on FKBP5 can be promising in the treatment of Huntington's disease [76].

Challenges and Knowledge Gaps

Although cellular senescence in HD is now considered to be a causative factor in the majority of cases, the development of therapeutic approaches is limited due to important challenges. With an intricate cellular background in the HD brain, finding and characterizing cell type-specific senescence remains both technically and conceptually difficult. The failure of classic senescence markers, including SA-β-gal, p16INK4a, and γH2AX, to discriminate between the two cell types (neuronal and glial cells) confuses understanding of their particular relationships with disease [77]. Second, the mechanisms and paths that human senescence takes and the pace of senescence progression in humans are not congruent with animal models of HD, such as transgenic and knock-in mice, making them not so useful in mechanistic and drug screening pursuits. These models have limited effectiveness in accurately modeling the constant senescent changes that are observed in patients, which are marked by inconsistent mHTT expression and lack elements of human neuroinflammation and aging [78]. Finally, to date, no clinical trials have explicitly identified the targeting of senescent cells in HD patients, which is indicative of the fact that the introduction of senescence-directed therapies in the clinic (like senolytics, senomorphics), has only just begun. Such apparent gaps are, moreover, further amplified by the absence of validated, noninvasive treatment response diagnostic tools and uncertainty as to when to initiate therapy and the possible risks, especially in the case of the central nervous system [79]. In order to progress the senescence-targeted therapies toward HD and other neurodegenerative diseases, it is essential to extend specific methods of senescence detection and prepare more elaborate preclinical models. Eliminating these information gaps is essential to exploit the promise of senescence regulation for treating HD and other neurodegenerative diseases [48] (Table 8.1).

Table 8.1 This table gives an overview of the cellular senescence pathways that play a role in HD, as well as senescence indicators, specific HD factors, and impaired cell types. It describes the adverse effects of the most important molecular processes that participate in senescence, such as DNA-induced damage, malfunctioning telomeres, neuroinflammation, and defective autophagy. Furthermore, the table demonstrates results from both animal experiments as well as patient models of HD, showcasing the multifarious manner in which senescence influences disease development through cell-specific and non-cell-specific mechanisms of action

Mechanism of senescence	Cellular model or tissue	HD-associated trigger	Evidence for senescence (markers/ phenotypes)	Molecular pathways involved	Impact on HD pathology	Reference
Oxidative stress-induced senescence	Primary cortical neurons; striatal-derived ST14A cells	Mutant huntingtin aggregation, mitochondrial dysfunction	↑ SA-β-gal, ↑ p16INK4a, nuclear envelope distortion	p53-p21 pathway, ROS accumulation	Synaptic dysfunction, neuronal loss	[80, 81]
DNA damage response (DDR)-mediated senescence	iPSC-derived astrocytes from HD patients	mHTT-induced genotoxic stress	↑ γH2AX foci, 53BP1 localization, ↑ p21	ATM/ATR-Chk2, p53 signaling	Impaired astrocytic support for neurons	[82, 83]
SASP	Striatal astrocytes in HD mouse models	Chronic neuroinflammation	↑ IL-6, IL-1β, MMP3 secretion	NF-κB activation, JAK/STAT pathway	Amplified neuroinflammation, BBB disruption	[84, 85]
Telomere shortening and dysfunction	Peripheral blood mononuclear cells (PBMCs) from HD patients	Replicative senescence, chronic stress	Shortened telomeres, ↑ TIFs, ↑ p16INK4a	Shelterin complex disruption, TRF2 loss	Biomarker of disease progression	[86, 87]
Epigenetic reprogramming-associated senescence	Cortical neurons, brain organoids	Histone deacetylation, DNA methylation from mHTT	SAHF formation, reduced Lamin B1	PRC2, HDAC1/2 dysregulation	Altered gene expression, chromatin instability	[88, 89]

(continued)

Table 8.1 (continued)

Mechanism of senescence	Cellular model or tissue	HD-associated trigger	Evidence for senescence (markers/ phenotypes)	Molecular pathways involved	Impact on HD pathology	Reference
Impaired autophagy/ mitophagy leading to senescence	Fibroblasts and striatal cells from HD patients	mHTT aggregates block autophagosome clearance	↑ p62, LC3-II accumulation, enlarged lysosomes	mTOR, AMPK, TFEB pathway dysfunction	Cytoplasmic stress, cellular senescence	[90, 91]
Paracrine senescence via SASP	Coculture of senescent astrocytes and healthy neurons	SASP from HD astrocytes	Neuronal SA-β-gal activity, DNA damage	IL-6/IL-8-JAK/ STAT signaling	Non-cell-autonomous neuronal dysfunction	[17]

Conclusion and Future Perspectives

Cellular senescence is identified as an important factor in determining the fundamental processes of HD. The accumulation of mHTT causes senescence through multiple pathways like oxidative stress, impairment in mitochondria, ongoing DNA damage, and activation of the p53/p21^{CIP1} and p16^{INK4a} signaling. Consequentially, both neurons and glial cells also progress to be characterized by a SASP, which will only serve to multiply the neuroinflammatory and neurodegenerative processes. The further rise in this cycle aids in the quick deterioration of symptomatology as well as the disease's progression. Since both neuronal and glial senescence are critical in the exacerbation of HD, the mounting evidence has definitively defined cellular senescence as the chief factor of the disease, not merely a result of neurodegeneration. The direct target that senolytic and senomorphic therapies have on senescent cells has tremendous capabilities as they may contribute to minimising inflammation and returning cellular homeostasis. However, operationalizing these findings into the clinical arena comes with its challenges. The development of customized therapeutics becomes complicated with the wide range of expression that senescent phenotypes exhibit, especially in different cell types in the central nervous system. The pressing requirement of biomarkers that can locally identify senescence and determine treatment effects stands out. Good strategies in the future will include improving early diagnostic tools, enhancing the safety and distribution of senotherapeutics, and fine-tuning the therapies that target the senescence process in a given cell type. Determining the best therapeutic intervals will be based on a longitudinal study on senescence biomarker computations and standard clinical measurements. A combination of senescence-targeting strategies and traditional neuroprotective therapies might change the course of HD, highlighting an immediate need to conduct advanced clinical trials and further investigate senescence in neurodegenerative disease treatment.

References

1. Gatto EM, Rojas NG, Persi G, Etcheverry JL, Cesarini ME, Perandones C. Huntington disease: advances in the understanding of its mechanisms. Clin Park Relat Disord. 2020;3:100056.
2. Bashir B, Vishwas S, Gupta G, Paudel KR, Dureja H, Kumar P, et al. Does drug repurposing bridge the gaps in management of Parkinson's disease? Unravelling the facts and fallacies. Ageing Res Rev. 2025;105:102693.
3. Bashir B, Gulati M, Vishwas S, Gupta G, Dhanasekaran M, Paudel KR, et al. Bridging gap in the treatment of Alzheimer's disease via postbiotics: current practices and future prospects. Ageing Res Rev. 2025;105:102689.
4. Ross CA, Pantelyat A, Kogan J, Brandt J. Determinants of functional disability in Huntington's disease: role of cognitive and motor dysfunction. Mov Disord. 2014;29(11):1351–8.
5. Tiwari C, Khan H, Grewal AK, Dhankhar S, Chauhan S, Dua K, et al. Opiorphin: an endogenous human peptide with intriguing application in diverse range of pathologies. Inflammopharmacology. 2024;32(5):3037–56.

6. Thapa R, Moglad E, Afzal M, Gupta G, Bhat AA, Hassan Almalki W, et al. The role of sirtuin 1 in ageing and neurodegenerative disease: a molecular perspective. Ageing Res Rev. 2024;102:102545.
7. Thapa R, Ahmad Bhat A, Shahwan M, Ali H, PadmaPriya G, Bansal P, et al. Proteostasis disruption and senescence in Alzheimer's disease pathways to neurodegeneration. Brain Res. 2024;1845:149202.
8. Martínez-Cué C, Rueda N. Cellular senescence in neurodegenerative diseases. Front Cell Neurosci. 2020;14:16.
9. Gonzalez-Meljem JM, Martinez-Barbera JP. Implications of cellular senescence in paediatric pituitary tumours. EBioMedicine. 2024;99:104905.
10. Gadhave DG, Sugandhi VV, Jha SK, Nangare SN, Gupta G, Singh SK, et al. Neurodegenerative disorders: mechanisms of degeneration and therapeutic approaches with their clinical relevance. Ageing Res Rev. 2024;99:102357.
11. Alharbi KS, Almalki WH, Alzarea SI, Kazmi I, Al-Abbasi FA, Afzal O, et al. Anaesthesia-induced changes in genomic expression leading to neurodegeneration. CNS Neurol Disord Drug Targets. 2024;23(4):411–9.
12. Zheng J, Winderickx J, Franssens V, Liu B. A mitochondria-associated oxidative stress perspective on Huntington's disease. Front Mol Neurosci. 2018;11:329.
13. Jha NK, Chen WC, Kumar S, Dubey R, Tsai LW, Kar R, et al. Molecular mechanisms of developmental pathways in neurological disorders: a pharmacological and therapeutic review. Open Biol. 2022;12(3):210289.
14. Bhardwaj S, Kesari KK, Rachamalla M, Mani S, Ashraf GM, Jha SK, et al. CRISPR/Cas9 gene editing: new hope for Alzheimer's disease therapeutics. J Adv Res. 2022;40:207–21.
15. Liu X, Sharma RK, Mishra A, Chinnaboina GK, Gupta G, Singh M. Role of aqueous extract of the wood ear mushroom, Auricularia polytricha (Agaricomycetes), in avoidance of haloperidol-induced catalepsy via oxidative stress in rats. Int J Med Mushrooms. 2019;21(4):323–30.
16. Jha NK, Sharma A, Jha SK, Ojha S, Chellappan DK, Gupta G, et al. Alzheimer's disease-like perturbations in HIV-mediated neuronal dysfunctions: understanding mechanisms and developing therapeutic strategies. Open Biol. 2020;10(12):200286.
17. Zhu J, Wu C, Yang L. Cellular senescence in Alzheimer's disease: from physiology to pathology. Transl Neurodegener. 2024;13(1):55.
18. Zheng L, He S, Wang H, Li J, Liu Y, Liu S. Targeting cellular senescence in aging and age-related diseases: challenges, considerations, and the emerging role of senolytic and senomorphic therapies. Aging Dis. 2024;15(6):2554–94.
19. Chen JH, Hales CN, Ozanne SE. DNA damage, cellular senescence and organismal ageing: causal or correlative? Nucleic Acids Res. 2007;35(22):7417–28.
20. Kumari R, Jat P. Mechanisms of cellular senescence: cell cycle arrest and senescence associated secretory phenotype. Front Cell Dev Biol. 2021;9:645593.
21. Gijs M, Jorna N, Datson N, Beekman C, Dansokho C, Weiss A, et al. High levels of mutant huntingtin protein in tear fluid from Huntington's disease gene expansion carriers. J Mov Disord. 2024;17(2):181–8.
22. Alcalá-Vida R, Seguin J, Lotz C, Molitor AM, Irastorza-Azcarate I, Awada A, et al. Age-related and disease locus-specific mechanisms contribute to early remodelling of chromatin structure in Huntington's disease mice. Nat Commun. 2021;12(1):364.
23. Piao X, Li D, Liu H, Guo Q, Yu Y. Advances in gene and cellular therapeutic approaches for Huntington's disease. Protein Cell. 2025;16:307.
24. Zhou D, Lin Y, Han Z, Zhang Z, Lin L, Lin S, et al. The impact of aging on neurological diseases in the elderly: molecular mechanisms and therapeutic perspectives. Aging Dis. 2024; https://doi.org/10.14336/AD.2024.1085.
25. Tong H, Yang T, Xu S, Li X, Liu L, Zhou G, et al. Huntington's disease: complex pathogenesis and therapeutic strategies. Int J Mol Sci. 2024;25(7)
26. Gallitzendoerfer R, Abouzied MM, Hartmann D, Dobrowolski R, Gieselmann V, Franken S. Hepatoma-derived growth factor (HDGF) is dispensable for normal mouse development. Dev Dyn. 2008;237(7):1875–85.

27. Voelkl K, Gutiérrez-Ángel S, Keeling S, Koyuncu S, da Silva Padilha M, Feigenbutz D, et al. Neuroprotective effects of hepatoma-derived growth factor in models of Huntington's disease. Life Sci Alliance. 2023;6(11):e202302018.
28. Pehar M, Harlan BA, Killoy KM, Vargas MR. Nicotinamide adenine dinucleotide metabolism and neurodegeneration. Antioxid Redox Signal. 2018;28(18):1652–68.
29. Azam S, Haque ME, Balakrishnan R, Kim IS, Choi DK. The ageing brain: molecular and cellular basis of neurodegeneration. Front Cell Dev Biol. 2021;9:683459.
30. Kim D, Yoo SH, Yeon GB, Oh SS, Shin WH, Kang HC, et al. Senescent astrocytes derived from human pluripotent stem cells reveal age-related changes and implications for neurodegeneration. Aging Dis. 2024;16(3):1709–31.
31. Hanson KA, Kim SH, Tibbetts RS. RNA-binding proteins in neurodegenerative disease: TDP-43 and beyond. Wiley Interdiscip Rev RNA. 2012;3(2):265–85.
32. Bai D, Deng F, Jia Q, Ou K, Wang X, Hou J, et al. Pathogenic TDP-43 accelerates the generation of toxic exon1 HTT in Huntington's disease knock-in mice. Aging Cell. 2024;23(12):e14325.
33. Sultana R, Perluigi M, Butterfield DA. Oxidatively modified proteins in Alzheimer's disease (AD), mild cognitive impairment and animal models of AD: role of Abeta in pathogenesis. Acta Neuropathol. 2009;118(1):131–50.
34. Pantiya P, Thonusin C, Chattipakorn N, Chattipakorn SC. Mitochondrial abnormalities in neurodegenerative models and possible interventions: focus on Alzheimer's disease, Parkinson's disease, Huntington's disease. Mitochondrion. 2020;55:14–47.
35. Currais A, Farrokhi C, Dargusch R, Armando A, Quehenberger O, Schubert D, et al. Fisetin reduces the impact of aging on behavior and physiology in the rapidly aging SAMP8 mouse. J Gerontol A Biol Sci Med Sci. 2018;73(3):299–307.
36. Tabrizi SJ, Workman J, Hart PE, Mangiarini L, Mahal A, Bates G, et al. Mitochondrial dysfunction and free radical damage in the Huntington R6/2 transgenic mouse. Ann Neurol. 2000;47(1):80–6.
37. Virmani A, Binienda Z. Role of carnitine esters in brain neuropathology. Mol Asp Med. 2004;25(5):533–49.
38. Di Cristo F, Calarco A, Digilio FA, Sinicropi MS, Rosano C, Galderisi U, et al. The discovery of highly potent THP derivatives as OCTN2 inhibitors: from structure-based virtual screening to in vivo biological activity. Int J Mol Sci. 2020;21(19)
39. Di Pardo A, Ciaglia E, Cattaneo M, Maciag A, Montella F, Lopardo V, et al. The longevity-associated variant of BPIFB4 improves a CXCR4-mediated striatum-microglia crosstalk preventing disease progression in a mouse model of Huntington's disease. Cell Death Dis. 2020;11(7):546.
40. Li X, Hernandez I, Koyuncu S, Kis B, Häggblad M, Lidemalm L, et al. The anti-leprosy drug clofazimine reduces polyQ toxicity through activation of PPARγ. EBioMedicine. 2024;103:105124.
41. Safwan-Zaiter H, Wagner N, Wagner KD. P16INK4A-more than a senescence marker. Life (Basel, Switzerland). 2022;12(9)
42. Han X, Zhang T, Liu H, Mi Y, Gou X. Astrocyte senescence and Alzheimer's disease: a review. Front Aging Neurosci. 2020;12:148.
43. Alpaugh M, Lantero-Rodriguez J, Benedet AL, Manseau U, Boutin M, Maiuri M, et al. Tau levels in platelets isolated from Huntington's disease patients serve as a biomarker of disease severity. J Neurol. 2025;272(3):254.
44. Masnata M, Salem S, de Rus Jacquet A, Anwer M, Cicchetti F. Targeting tau to treat clinical features of Huntington's disease. Front Neurol. 2020;11:580732.
45. Kandpal M, Indari O, Baral B, Jakhmola S, Tiwari D, Bhandari V, et al. Dysbiosis of gut microbiota from the perspective of the gut-brain axis: role in the provocation of neurological disorders. Mcta. 2022;12(11)
46. Khoshnan A. Gut microbiota as a modifier of Huntington's disease pathogenesis. J Huntington's Dis. 2024;13(2):133–47.
47. Behfar Q, Ramirez Zuniga A, Martino-Adami PV. Aging, senescence, and dementia. J Prev Alzheimers Dis. 2022;9(3):523–31.

48. Baker DJ, Petersen RC. Cellular senescence in brain aging and neurodegenerative diseases: evidence and perspectives. J Clin Invest. 2018;128(4):1208–16.
49. Sari Y. Huntington's disease: from mutant huntingtin protein to neurotrophic factor therapy. Int J Biomed Sci. 2011;7(2):89–100.
50. Jain S, Roy I. Aptamer reduces aggregation of mutant huntingtin and rescues proteostasis network in non-neuronal and neuronal cells. ACS Chem Neurosci. 2023;14(12):2385–95.
51. Nishiyama H, Nishiyama N, Zemelman BV. Loss of Purkinje cells in the developing cerebellum strengthens the cerebellothalamic synapses. bioRxiv [Preprint]. 2023; https://doi.org/10.1101/2023.11.01.564864.
52. Cvetanovic M, Gray M. Contribution of glial cells to polyglutamine diseases: observations from patients and mouse models. Neurotherapeutics. 2023;20(1):48–66.
53. Rather MA, Khan A, Wang L, Jahan S, Rehman MU, Makeen HA, et al. TRP channels: role in neurodegenerative diseases and therapeutic targets. Heliyon. 2023;9(6):e16910.
54. Cecerska-Heryć E, Pękała M, Serwin N, Gliźniewicz M, Grygorcewicz B, Michalczyk A, et al. The use of stem cells as a potential treatment method for selected neurodegenerative diseases: review. Cell Mol Neurobiol. 2023;43(6):2643–73.
55. Kaminska A, Radoszkiewicz K, Rybkowska P, Wedzinska A, Sarnowska A. Interaction of neural stem cells (NSCs) and mesenchymal stem cells (MSCs) as a promising approach in brain study and nerve regeneration. Cells. 2022;11(9)
56. Ghosh R, Wood-Kaczmar A, Dobson L, Smith EJ, Sirinathsinghji EC, Kriston-Vizi J, et al. Expression of mutant exon 1 huntingtin fragments in human neural stem cells and neurons causes inclusion formation and mitochondrial dysfunction. FASEB J. 2020;34(6):8139–54.
57. Bhat AA, Moglad E, Afzal M, Thapa R, Almalki WH, Kazmi I, et al. Therapeutic approaches targeting aging and cellular senescence in Huntington's disease. CNS Neurosci Ther. 2024;30(10):e70053.
58. Dai Y, Wang H, Lian A, Li J, Zhao G, Hu S, et al. A comprehensive perspective of Huntington's disease and mitochondrial dysfunction. Mitochondrion. 2023;70:8–19.
59. Kumar P, Choudhary A, Kinger S, Jagtap YA, Dubey AR, Gutti RK, et al. Proteostasis defects: medicinal challenges of imperfect aging & neurodegeneration. Transl Med Aging. 2023;7:87–97.
60. Lee J, Giordano S, Zhang J. Autophagy, mitochondria and oxidative stress: cross-talk and redox signalling. Biochem J. 2012;441(2):523–40.
61. Robbins PD, Jurk D, Khosla S, Kirkland JL, LeBrasseur NK, Miller JD, et al. Senolytic drugs: reducing senescent cell viability to extend health span. Annu Rev Pharmacol Toxicol. 2021;61:779–803.
62. Jia Y, Han L, Ramage CL, Wang Z, Weng CC, Yang L, et al. Co-targeting BCL-XL and BCL-2 by PROTAC 753B eliminates leukemia cells and enhances efficacy of chemotherapy by targeting senescent cells. Haematologica. 2023;108(10):2626–38.
63. de Godoy MCX, Macedo JA, Gambero A. Researching new drug combinations with senolytic activity using senescent human lung fibroblasts MRC-5 cell line. Pharmaceuticals (Basel, Switzerland). 2024;17(1)
64. Zhang L, Pitcher LE, Prahalad V, Niedernhofer LJ, Robbins PD. Targeting cellular senescence with senotherapeutics: senolytics and senomorphics. FEBS J. 2023;290(5):1362–83.
65. Byun S, Lee M, Kim M. Gene therapy for Huntington's disease: the final strategy for a cure? J Mov Disord. 2022;15(1):15–20.
66. Ouvrier B, Ismael S, Bix GJ. Senescence and SASP are potential therapeutic targets for ischemic stroke. Pharmaceuticals (Basel, Switzerland). 2024;17(3)
67. Russo L, Babboni S, Andreassi MG, Daher J, Canale P, Del Turco S, et al. Treating metabolic dysregulation and senescence by caloric restriction: killing two birds with one stone? Antioxidants (Basel, Switzerland). 2025;14(1)
68. Wang Y, Li J, Zhang Z, Wang R, Bo H, Zhang Y. Exercise improves the coordination of the mitochondrial unfolded protein response and mitophagy in aging skeletal muscle. Life (Basel, Switzerland). 2023;13(4)

69. Leslie EM, Deeley RG, Cole SP. Multidrug resistance proteins: role of P-glycoprotein, MRP1, MRP2, and BCRP (ABCG2) in tissue defense. Toxicol Appl Pharmacol. 2005;204(3):216–37.
70. Machiela E, Rudich PD, Traa A, Anglas U, Soo SK, Senchuk MM, et al. Targeting mitochondrial network disorganization is protective in C. Elegans models of Huntington's disease. Aging Dis. 2021;12(7):1753–72.
71. Bell MC, Meier SE, Ingram AL, Abisambra JF. PERK-opathies: an endoplasmic reticulum stress mechanism underlying neurodegeneration. Curr Alzheimer Res. 2016;13(2):150–63.
72. Rozpędek-Kamińska W, Siwecka N, Wawrzynkiewicz A, Wojtczak R, Pytel D, Diehl JA, et al. The PERK-dependent molecular mechanisms as a novel therapeutic target for neurodegenerative diseases. Int J Mol Sci. 2020;21(6)
73. Garret M, Du Z, Chazalon M, Cho YH, Baufreton J. Alteration of GABAergic neurotransmission in Huntington's disease. CNS Neurosci Ther. 2018;24(4):292–300.
74. Guo X, Disatnik MH, Monbureau M, Shamloo M, Mochly-Rosen D, Qi X. Inhibition of mitochondrial fragmentation diminishes Huntington's disease-associated neurodegeneration. J Clin Invest. 2013;123(12):5371–88.
75. Häusl AS, Brix LM, Hartmann J, Pöhlmann ML, Lopez JP, Menegaz D, et al. The co-chaperone Fkbp5 shapes the acute stress response in the paraventricular nucleus of the hypothalamus of male mice. Mol Psychiatry. 2021;26(7):3060–76.
76. Bailus BJ, Scheeler SM, Simons J, Sanchez MA, Tshilenge KT, Creus-Muncunill J, et al. Modulating FKBP5/FKBP51 and autophagy lowers HTT (huntingtin) levels. Autophagy. 2021;17(12):4119–40.
77. Sahu MR, Rani L, Subba R, Mondal AC. Cellular senescence in the aging brain: a promising target for neurodegenerative diseases. Mech Ageing Dev. 2022;204:111675.
78. Beck J, Horikawa I, Harris C. Cellular senescence: mechanisms, morphology, and mouse models. Vet Pathol. 2020;57(6):747–57.
79. von Kobbe C. Targeting senescent cells: approaches, opportunities, challenges. Aging. 2019;11(24):12844–61.
80. Goswami A, Dikshit P, Mishra A, Mulherkar S, Nukina N, Jana NR. Oxidative stress promotes mutant huntingtin aggregation and mutant huntingtin-dependent cell death by mimicking proteasomal malfunction. Biochem Biophys Res Commun. 2006;342(1):184–90.
81. Nousis L, Kanavaros P, Barbouti A. Oxidative stress-induced cellular senescence: is labile iron the connecting link? Antioxidants (Basel, Switzerland). 2023;12(6)
82. Oda T, Gotoh N, Kasamatsu T, Handa H, Saitoh T, Sasaki N. DNA damage-induced cellular senescence is regulated by 53BP1 accumulation in the nuclear foci and phase separation. Cell Prolif. 2023;56(6):e13398.
83. Shreeya T, Ansari MS, Kumar P, Saifi M, Shati AA, Alfaifi MY, et al. Senescence: a DNA damage response and its role in aging and neurodegenerative diseases. Front Aging. 2023;4:1292053.
84. Zhang Y, Zou M, Wu H, Zhu J, Jin T. The cGAS-STING pathway drives neuroinflammation and neurodegeneration via cellular and molecular mechanisms in neurodegenerative diseases. Neurobiol Dis. 2024;202:106710.
85. Preininger MK, Kaufer D. Blood-brain barrier dysfunction and astrocyte senescence as reciprocal drivers of neuropathology in aging. Int J Mol Sci. 2022;23(11)
86. Lin J, Epel E. Stress and telomere shortening: insights from cellular mechanisms. Ageing Res Rev. 2022;73:101507.
87. Stefanidis I, Voliotis G, Papanikolaou V, Chronopoulou I, Eleftheriadis T, Kowald A, et al. Telomere length in peripheral blood mononuclear cells of patients on chronic hemodialysis is related with telomerase activity and treatment duration. Artif Organs. 2015;39(9):756–64.
88. Brami-Cherrier K, Anzalone A, Ramos M, Forne I, Macciardi F, Imhof A, et al. Epigenetic reprogramming of cortical neurons through alteration of dopaminergic circuits. Mol Psychiatry. 2014;19(11):1193–200.
89. Lossi L, Castagna C, Merighi A. An overview of the epigenetic modifications in the brain under normal and pathological conditions. Int J Mol Sci. 2024;25(7)

90. Guedes-Dias P, de Proença J, Soares TR, Leitão-Rocha A, Pinho BR, Duchen MR, et al. HDAC6 inhibition induces mitochondrial fusion, autophagic flux and reduces diffuse mutant huntingtin in striatal neurons. Biochim Biophys Acta. 2015;1852(11):2484–93.
91. Ivanova MM, Changsila E, Iaonou C, Goker-Alpan O. Impaired autophagic and mitochondrial functions are partially restored by ERT in Gaucher and Fabry diseases. PLoS One. 2019;14(1):e0210617.

Cellular Senescence in Amyotrophic Lateral Sclerosis (ALS)

9

Rahamat Unissa Syed, S. Afsar, and Fulvia di Pillo

Introduction

The neurodegenerative condition known as ALS progresses until it becomes fatal as it affects brain and spinal cord motor neurons. The medical condition ALS and its common name Lou Gehrig's disease shows symptoms due to proteostasis dysfunction [1, 2]. The motor neurons then are unable to send messages down the axons, and patients gradually lose muscle strength, and also being paralyzed lose their ability to move, speak swallow, or even breathe [3]. The development of ALS depends on four different pathological processes, which include misfolded protein aggregation together with oxidative stress as well as mitochondrial failure alongside neuroinflammation. Numerous mutations in the SOD1 gene strongly affect the pathophysiology of ALS through Sacrosin [4, 5]. The result is toxic protein aggregation and selective motor neuron death. Worldwide, approximately 2–3 per

R. U. Syed (✉)
Department of Pharmaceutics, College of Pharmacy, University of Hai'l, Hai'l, Saudi Arabia
e-mail: ru.syed@uoh.edu.sa

S. Afsar
Department of Virology, Sri Venkateswara University, Tirupathi, Andhra Pradesh, India
e-mail: mammoth.afsar1@gmail.com

F. di Pillo
Centre for Medicinal Chemistry (CQMED), Campinas, SP, Brazil
e-mail: fulviadipillo@unicamp.br

G. Gupta (ed.), *Cellular Senescence and Brain Aging*,
https://doi.org/10.1007/978-981-96-8873-9_9

100,000 people develop Als each year, and about 1 in an average of every 500 humans will die of ALS [6]. It usually affects persons between the ages of 40 and 70, with no specifically distinguished factors except a higher frequency in men compared to women. While relatively rare, the disease causes dramatic suffering for those who contract it, with severe physical disability and an average survival time after diagnosis of 3–5 years [7, 8].

Longevity assurance genes regulate the signal transduction of stress responses, proteostasis, and metabolism leading to cellular homeostasis and organismal lifespan. These genes are categorized into public (shared among species) and private (unique to species) longevity assurance mechanisms. Public genes of interest are those that are part of antioxidant defense, DNA repair, and autophagy, such as SOD1, which is involved in the attenuation of oxidative stress [6]. But, SOD1 mutations disrupt this balance, causing misfolding of the protein, and increased oxidative burden, resulting in neuronal toxicity. In addition to accelerating neurodegeneration, mutant SOD1 in ALS also interacts with age-related processes that accelerate disease progression. In ALS pathology, however, longevity assurance mechanisms are imperfect: misfolded proteins accumulate, mitochondria are dysfunctional, clearance systems are impaired, and the cells thus become damaged [4, 5]. There may be sex differences, with impedance to ALS depending on sex, perhaps in response to hormonal effects on stress response pathways. Despite very similar metabolic timescales among individuals, biological aging is highly variable with increasing biological age and the transition from cellular homeostasis to dysfunction. Celebrating their success, Friedman notes that aging and oxidative stress pose special regulatory challenges in ALS, characteristic of the fact that motor neurons go from a functional to a degenerative state prematurely due to age-related mitochondrial defects, with disease onset and progression accelerated [3].

Critical to the mechanism of ALS pathogenesis is the concept of hemodynamic space, the organism's ability to withstand stress until hemodynamics can be regained. Aging and cellular senescence reduce hemodynamic capacity, and this limits motor neurons' ability to fend off oxidative damage, protein aggregation, and neuroinflammation, all of which are characteristic of the disease [9]. The heterogeneity in disease progression is attributable, in part, to this hemodynamic collapse that contributes to the observable variation in functional loss among patients. There is equally such variability at the molecular and cellular level, where variations of mitochondrial resilience, proteostasis efficiency, and immune responses regulate disease severity. For therapeutic interventions, it is essential to distinguish between harmful, beneficial, and neutral changes in the aging process. Some cellular changes, for example, increased autophagy, may be compensatory, while others like chronic neuroinflammation are neurodegeneration-promoting [10]. In ALS research, it is important to understand what cells can tolerate, and how they can adapt and bypass age-related insults (paths that make cells more resilient, such as proteostasis, the immune response, and mitochondria) may be involved in therapeutics. These aspects will address key issues in aging, senescence, and longevity assurance in ALS and thereby obtain deeper insights into disease mechanisms and interventions, which may help extend neuronal survival and delay disease progression.

The importance of mutant SOD1 in ALS pathogenesis is imperative for designing successful therapies and ultimately enhancing patient survival [9]. One of the best-characterized genetic causes of ALS is due to mutations in the SOD1 gene. Mutations in this gene account for 20% of familial cases, and a smaller fraction of <5% of sporadic cases [11, 12]. This includes mutations that cause SOD1 misfolding and for the protein to aggregate, in turn driving a cascade of cellular dysfunctions [13]. These pathological lead to the gradual degeneration of motor neurons, which ultimately causes muscle weakness and atrophy due to apparent ALS instances [10]. The molecular mechanisms by which mutant SOD1 leads to neurodegeneration will enable researchers to determine potential therapeutic targets, allowing them to create evidence-based interventions for inducing or slowing disease progression [14, 15]. Knowledge of these molecules will also be essential to discovering new biomarkers for early diagnosis, prognosis, or monitoring the course of the disease leading to an improvement in welfare quality for ALS patients [16, 17].

Aging and cellular senescence are also very relevant to the status of ALS because they aggravate neurodegenerative processes [18]. Motor neuron vulnerability results from a combination of cellular damage and impaired regenerative capacity, oxidative stress, and accumulated motor workloads due to aging. Together, all of these factors speed up the development and advancement of ALS [19]. This comprehensive overview of the relationship between aging, senescence, and ALS can help develop specific treatments to reduce their effects to slow down disease progression (Fig. 9.1).

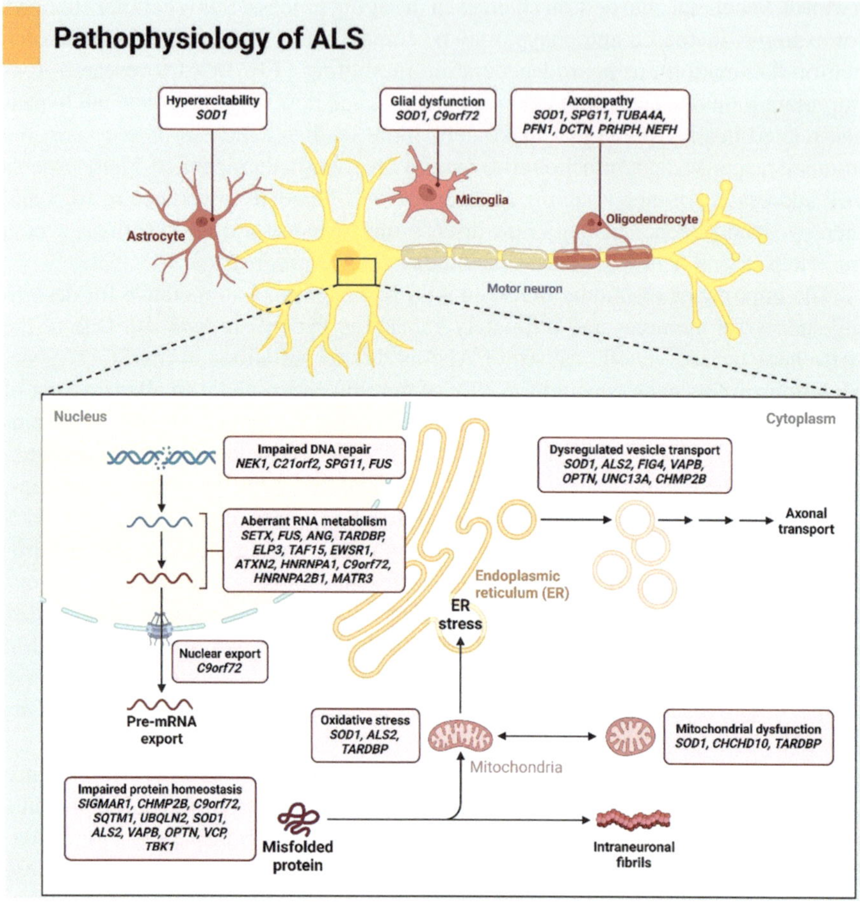

Fig. 9.1 Cellular and molecular pathways in ALS. This image illustrates the pathophysiology of amyotrophic lateral sclerosis (ALS). It highlights various cellular dysfunctions such as hyperexcitability, glial dysfunction, and axonopathy, involving multiple genes like SOD1 and C9orf72. Within the nucleus and cytoplasm, processes including impaired DNA repair, aberrant RNA metabolism, ER stress, oxidative stress, mitochondrial dysfunction, and dysregulated vesicle transport are depicted, leading to motor neuron degeneration

Mutant SOD1 and ALS

Structure and Function of the SOD1

The SOD1 enzyme appears as one of the most frequently seen antioxidant proteins throughout all cells. The main function of SOD1 involves eliminating reactive oxygen species by reducing superoxide radicals that form during cellular respiration [20, 21]. SOD1 is a homodimeric enzyme; each monomer has 153 amino acids and includes an active site where catalytic copper and zinc ions are bound [22]. The catalytic cycle of the enzyme involves a double reduction and oxidation of its copper ion leading to the dismutation of two superoxide radicals into hydrogen peroxide and molecular oxygen [23]. Apart from the antioxidant role of SOD1, it is involved in several cellular processes, including signaling cascades, transcriptional regulation, and apoptosis [24]. The enzyme mostly present in the cytoplasm can also exist in mitochondrial intermembrane space and nuclei, which indicates supplementary cell compartment tasks [25, 26]. Proper folding and metalation are critically important for the structural integrity, and function of SOD1. These processes are frequently disrupted, usually through mutations in the SOD1 gene, causing protein misfolding, aggregation, and toxicity bestowed by a gain-of-function that underlies pathogenesis in amyotrophic lateral sclerosis (ALS) and other neurodegenerative disorders [27, 28].

Mechanisms of SOD1 Mutations in Causing ALS

SOD1 is a widely known genetic risk cause of ALS, which is an intensifying segmental motor neuron degenerative disease. The SOD1 gene is mutated in familial and also sporadic forms of ALS, albeit with differential frequency related to population stratifications [29]. The mechanisms that lead to ALS due to SOD1 mutations are complex and multiple. One of which is the blocking effect on the due function of the SOD1 enzyme, which mediates cell detoxification from ROS reactive oxygen species [30, 31]. Changes in SOD1 gene structure generate protein molecules with improper structures, which become toxic to cells while remaining inside cells [32]. Many scientific studies have proven how damaging SOD1 aggregation is in the development of ALS [22, 33]. Misfolded protein aggregates may interact with other cellular organelles such as mitochondria and affect the normal functioning of the cell. This disruption leads to the activation of cellular stress pathways, including the unfolded protein response (UPR), which can further sensitize cells and exacerbate damage [34]. Mutant SOD1 proteins (SOD1mut) interact specifically with Derlin-1, which is a key component of the ERAD machinery. ER stress in turn activates the ASK1-dependent cell death pathway by perturbing interaction. This process has been demonstrated to be essential for the development of ALS in mutant SOD1-expressing models [35]. An additional major way that SOD1 mutations cause ALS is by disrupting the usual control over regulation of the expression of SOD1. It is the case when a mutated SOD1 can create even more of this protein, hence worsening cellular toxicity induced by those misfolded accumulations [36]. The role of SOD1 mutations in contributing to ALS is complex and highly diverse with disruption of normal cellular mechanisms, protein aggregates accumulation, as well as activation of cellular stress pathways. More studies will be required for additional understanding of pathogenesis and identifying the targeted therapeutic approaches to alleviate this devastating disease [37, 38].

Aging and Senescence in ALS

Aging and Cellular Senescence

The complexity of aging reflects that this is a gradual time-dependent biological process during which the function slowly depresses over time, and individuals become more susceptible to disease and ultimately die [39]. Aging at the cellular level is also accompanied by the appearance of a phenomenon called cellular senescence, which refers to a state of irreversible arrest cell division (proliferation), and thus cells acquire specific phenotype [40, 41]. Cellular senescence can be induced by a variety of stimuli, such as telomere shortening, oxidative stress, DNA damage, activation of oncogene, and many other stresses. Senescent cells exhibit characteristic morphological and functional modifications, like enlarged and flattened morphology, and elevated generation of inflammatory mediators [42, 43]. The SASP functions as a core characteristic of senescent cells through their production of pro-inflammatory cytokines with chemokines and growth factors together with proteases [44]. Although initially advantageous as a tumor-suppressive barrier, chronic exposure to senescent cells and the SASP can promote deleterious effects on surrounding cells and tissues that generate diseases of aging and tissue dysfunction [45]. In addition, senescent cells express several specific molecular markers, including the expression of cell cycle inhibitors (p16INK4a or p21), the upregulation of an enzyme known as SA-β-gal, and focal perinucleolar heterochromatin foci, which can be detected by specific histochemical stains [46, 47]. Many tissues and organs show the increased formation of senescent cells that affect both the natural aging process and diseases of age-related onset, particularly neurodegeneration [48].

Relevance of Aging and Senescence in Neurodegenerative Diseases

Relating cellular senescence and aging mechanisms directly affect the development and worsening of three major neurological diseases: Alzheimer's disease, Parkinson's disease, and multiple sclerosis. Deleterious compounds that senescent cells generate including reactive oxygen species (ROS) inflict harm to cellular structures and accelerate disease development [49, 50]. Senescence further underlies the pathogenesis of neurodegenerative diseases, promoting alterations in normal cellular functions like protein metabolism and autophagic activity [51]. The literature demonstrates that senescent cells block LC3 autophagy processes which leads to cellular breakdown and increases cellular toxicity [52]. Additionally, aging/senescence may further accelerate the neurodegenerative disease progression by disturbing the regular functioning of dynamic neural networks in the brain. This is because aging also affects the ability of brain neural networks, such as the hippocampus and prefrontal cortex, which are essential for memory and cognitive operation [53, 54]. Aging and senescence are large contributors to the onset and spread of NDs [55]. It remains to be seen how aging and senescence are mechanistically linked to these conditions, which requires more research for progress to be elaborated as much-needed therapeutic options for the treatment of such devastating diseases [56, 57].

Evidence Linking Aging and Senescence to ALS Progression

Age and cell senescence have recently emerged as critical factors in the development of ALS. Aging is by far the most significant risk factor for ALS as it gradually causes diminishing cell repair mechanisms while strengthening oxidative stress, giving rise to a more extensive susceptibility of motor neurons [29, 58]. This vulnerability is further exacerbated by the phenomenon of cellular senescence, wherein cells undergo an irreversible state of cell cycle arrest and secrete pro-inflammatory cytokines, among other factors as paracrine signaling agents [31, 59, 60]. SASP factors maintain a toxic microenvironment that exacerbates neuroinflammation, contributing to additional motor neuron degeneration [61]. Studies have revealed that senescent glial cells, astrocytes, and microglia are central players in the pathogenesis of ALS. They release neurotoxic substances and deregulate their function to stop supporting neuron activity [62]. Cellular senescence and associated pathways thus serve as promising targets for developing therapeutic strategies that delay ALS progression and improve patient survival [63]. Taken together, these research works provide persuasive arguments that the unique complexity of aging is weakly connected with cellular senescence, resulting in the progression of ALS [64, 65]. Increased senescent cell burden was detected in the spinal cords of ALS patients and SOD1 mutant mouse models, where increased numbers of senescent cells were observed with increasing disease severity [66]. In the motor cortex and spinal cord of ALS patients and animal models, senescence markers such as p16 INK4a expression levels of senescence-associated β-galactosidase are increased [67]. Using senolytic agents to clear these cells improved motor function, extended lifespan, and reduced neuroinflammation in SOD1 mutant ALS mouse models [68]. Transplantation of senescent cells in the spinal cords of wild-type mice led to motor deficits and neuroinflammation that recapitulated an ALS-like phenotype [69]. In spinal cord from ALS patients and animal models, age-related accumulation of senescent astrocytes and microglia also promotes neuroinflammation as well as neurodegeneration [70]. Transcriptome analysis on differentiated motor neurons generated from ALS-patients showed a deregulated gene-pattern including upregulation of senescence-association genes and pathways, suggesting the involvement of cellular senescence in disease development [71]. Together, these studies delineate the central role of aging and cellular senescence in driving ALS progression and have strong implications for therapeutic interventions aimed at eliminating senescent cells or addressing the broader alterations to the senescence response seen in ALS.

Therapeutic Implications and Strategies

Targeting Mutant SOD1 and Its Associated Pathways

Amyotrophic lateral sclerosis (ALS) is a neurodegenerative condition that takes many years to cause the death of motor neurons [22]. Familial ALS frequently arises due to mutations in the superoxide dismutase 1 (SOD1) gene. Toxic protein aggregation and oxidative stress resulting from mutant SOD1 results in the degeneration of motor neurons [72]. It is possible that targeting mutant SOD1 and its associated pathways could provide a promising therapeutic approach. The approaches include antisense oligonucleotides (ASOs) to reduce SOD1 expression, small molecules to promote its degradation, and gene therapies designed to silence the mutant gene [21, 73]. In addition, we are working on targeting oxidative stress by antioxidants and modulating autophagy to eliminate SOD1 aggregates. Therapeutic interventions targeting these pathways have the potential to decelerate disease progression and promote better clinical outcomes for ALS, indicating further attention is still required [22, 74]. Oxidized SOD1 refers to the superoxide dismutase 1 enzyme, which has been oxidatively modified [75]. Collectively, the modified state is normally referred to as decreased function and an inclination to aggregate. This also plays a part in neurodegenerative diseases like amyotrophic lateral sclerosis (ALS). Oxidized SOD1 has been associated with oxidative stress inducement and cellular lesions [12, 76]. Guo et al. studied the involvement of the oxidized from ALS-linked mutations in SOD1 in cellular senescence and aging. Namely, NSC34 cells expressing hSOD1G93A mutant displayed premature senescence, and the transgenic mice with the same mutation revealed early aging phenotypes.

Additionally, selective elimination of oxidized from SOD1 decelerated aging because oxidization from SOD1 is an aging cause. Mitochondrial dysfunction and elevated MDVs were typical both for the senescent cells and the mice. Thus, they were both characterized by an impaired mitochondrial response to oxidation stress [77] (Fig. 9.2).

The Drosophila SOD1, abbreviated as dSOD1, is the fruit fly's counterpart of the human superoxide dismutase 1 enzyme. It performs a vital function in safeguarding the cells against oxidative stress by transforming the superoxide radicals into oxygen and hydrogen peroxide [78]. The studies on dSOD1 help uncover the ways of oxidative stress and the development of such neurodegenerative disorders as amyotrophic lateral sclerosis [79]. Agudelo et al. present an analysis of age-dependent degeneration in a single motor neuron using a Drosophila SOD1 model for ALS. Mutant dSOD1 flies showed progressive deterioration of walking behavior and selective loss of motor neurons, as well as more prominent polyubiquitinated proteins. This study, therefore, describes SOD loss-of-function-dependent, age-related morphological and molecular alterations that resemble human ALS [80].

On the other hand, Şahin et al. utilized the knock-in approach with the Drosophila SOD1 gene targeted to add the human ALS-causing mutation of the SOD1 gene. The SOD1 mutants showed severe phenotypes, such as neurodegeneration or locomotor abnormalities, although without overexpression of the targeted protein.

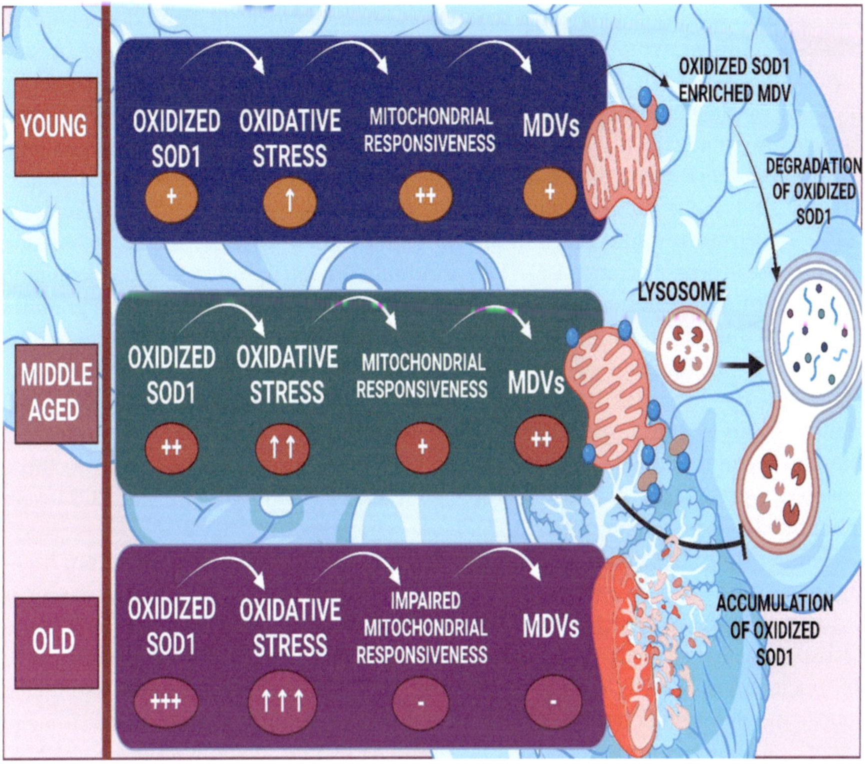

Fig. 9.2 Age-related oxidative stress and mitochondrial decline. This picture shows groups of different ages of ALS, and the effect of oxidative stress, and mitochondrial malfunction in the progression. Greater mitochondrial responsiveness was associated with lower oxidized SOD1 and oxidative stress in young subjects. Oxidative stress also increases but mitochondrial responsiveness decreases in prepare of middle-aged individuals. Oxidized SOD1 accumulates significantly in older individuals; this is associated with a high level of oxidative stress and impaired mitochondrial responsiveness leading to motor neuron degeneration

Moreover, the knock-in shows both gain- and loss-of-function parts, which is why this model is useful for SOD1-based ALS studies [81]. ATXN2 and FUS are two genes involved in amyotrophic lateral sclerosis (ALS) disease. While mutations in ATXN2 cause protein aggregation and neurodegeneration, the case of FUS mutated causes disturbance in RNA processing and transport [82]. They both cause motor neuron degeneration in ALS [83, 84]. Liu et al. analyzed the mutation spectrum within Central South China genes.

SOD1 was the most commonly mutated gene in 753 ALS patients, followed by ATXN2 and FUS for autosomal dominant cases and NEK1 CACNA1H for sporadic cases. The study emphasizes SOD1, not only as a frequent mutation in Chinese ALS patients but also that sex and genetic modifiers concur to modify the phenotype of ALS [85]. Presymptomatic SOD1 is defined as the period before the emergence of symptoms in ALS patients with SOD1 gene mutations. At this point, the primary and cellular levels provide a basis for the commencement of neurodegeneration, while no clinical symptoms are manifest [86]. Bame et al. characterized plasma metabolites in ALS's SOD1 model mouse. As a result of the in-depth metabolic profiling of plasma from presymptomatic SOD1 mice, there was a significant difference in amino acid levels, primarily aspartate and cystine/cysteine, when compared to WT. Metabolic changes specific to gender were also found. Based on the results of this study, plasma metabolites could be used as diagnostic biomarkers for ALS. Moreover, it might assist in the early detection of disease and the evaluation of drug treatment efficacy [87]. The accumulation of lipofuscin is one common hallmark of aging and senescent cells. These are yellow-brown pigment granules that show lipid-containing residues of lysosomal digestion. All this accumulation will lead to dysfunction of the cells, a mark of the neurodegenerative disease as exhibited by oxidative stress and autophagy activity deteriorating [88, 89]. Bandyopadhyay et al. found that the levels of lipofuscin within motor neurons from these presymptomatic SOD1 (G85R and G93A) ALS mice were similar to those observed in wild-type animals. Autophagy markers (SQSTM1/p62 and LC3) were downregulated in mutant mice.

Furthermore, chloroquine treatment resulted in lipofuscin accumulation, highlighting overactive autophagy within ALS motor neurons. This hyperactivity is associated with the MTORC1 pathway, which sheds light on how the autophagy/lysosome pathway could be involved in ALS [90]. Another term for a sequence of six nucleotides in DNA or RNA is hexanucleotide. However, they can also create loops, and due to this nature, these types of repeats have been implicated in many genetic diseases. Some expansions in a hexanucleotide repeat within the C9orf72 gene cause both ALS and frontotemporal dementia [91, 92]. Laaksovirta et al. examined the most common genetic variants and clinical features of ALS patients in Finland, especially those associated with the C9orf72 hexanucleotide repeat expansion. 815/2682 (31%) patients had C9orf72 expansion. OMIN Note: The C9pos patients had an earlier onset and shorter survival compared with the ALS pts of unknown cause; in addition, other alleles were more frequent among population controls. ALS-FTD was significantly more frequent among C9pos patients [93].

Erythropoietin (EPO) is a hormone that the kidneys secrete. The kidney releases this hormone in response to low levels of oxygen due to anemia and other conditions that stimulate red blood cell production in the bone marrow. Erythropoietin, an essential hormone, is required to maintain proper oxygen transfer. It is currently used clinically to treat anemia [94–96]. Noh et al. show the immunomodulatory and anti-inflammatory activity of EPO on SOD1(G93A) mice. EPO treatment strongly decreased the levels of pro-inflammatory cytokines and augmented anti-inflammatory cytokine contents, which led to the delayed onset time of symptoms and conserved MNCs [97].

Furthermore, Koyama et al. investigated how the solubility of mutant SOD1 associated with ALS changes during disease progression. Instead, the mutant SOD1 was converted to an insoluble form that built up as the disease went on. The decreased insoluble SOD1 accumulation after overexpression of Hsp70 suggests that both the proteasome and molecular chaperones control SOD1 solubility and, therefore, have potential as therapeutic targets [98]. The G93A mutation, located on the SOD1 gene as mentioned above, is a particular type of mutation linked to amyotrophic lateral sclerosis (ALS). Substitutions at position 93 by alanine to glycines. The result is the production of a misfolded, toxic version of the SOD1 protein, which drives continued and progressive motor neuron loss [99]. Kuźma et al. investigated tau expression and alternative splicing in the central nervous system of SOD1(G93A) mutant transgenic mice at several ages, as these processes have not been previously studied. Finally, the study proved early alterations in tau expression among young homozygous transgenic mice hippocampus where total tau-mRNA was decreased but expressed higher during the entire life period on the cerebellum [100]. Similarly, Wier et al. also studied muscle contractility and motor unit integrity in the SOD1(G93A) mouse model. Muscle contractility dysfunction, also noted before motor unit loss, was identified even more ahead of the clinical affection of the model. In female mice, this dysfunction was delayed, but the loss of muscle contractility still preceded motor unit death in both genders [101]. A protein whose function for protein degradation and quality control paths is route ubiquilin 2 (UBQLN2) is involved in keeping cell homeostasis [102]. A recent study identified mutations in the UBQLN2 gene associated with amyotrophic lateral sclerosis (ALS) and frontotemporal dementia, which interfere with protein clearance and cause neurodegeneration [103, 104]. Deng et al. discovered that mutations of UBQLN2 result in dominantly inherited X-linked juvenile and adult-onset ALS, as well as ALS/dementia. Since the mutations all impact protein degradation, this suggests that defects in ubiquilin 2 can be attributed to neurodegeneration. Therefore, a single pathogenic mechanism is involved that may potentially be targeted for treatment [105].

FUS encodes a protein that functions in DNA repair, RNA processing, and transport. FUS mutations have been identified in patients with ALS and FTD. they can interfere with cellular processes and thus eventually result in neurodegeneration [106]. Zou et al. discovered de novo FUS gene mutations in juvenile-onset sporadic ALS cases from the Chinese population. After screening 11 juvenile-onset SALS patients, 105 adult-onset ALS patients, and 245 healthy controls, the researchers identified several new FUS mutations. These findings indicated that FUS gene mutations may be linked to juvenile-onset SALS, and young ALS patients with rapid progression should also be verified for the mutation of this gene [107]. Another study by Potenza et al. investigated the impact of long-term consumption of caffeine in SOD1(G93A) ALS rodents. In ALS mice, caffeine curtailed survival significantly but resulted in a nonsignificant earlier onset of the disease (ALSTDRG, 1999). Changes in adenosine A2A receptor, glial glutamate transporter expression, and GFAP levels were found to be affected within the spinal cords of ALS mice but not as a consequence of caffeine administration [108]. GLP-1 receptor agonists are a class of medications that are indicated for type 2 diabetes as well as obesity [109]. Li et al. found that Ex-4, a GLP-1 receptor agonist, behaves as a neuroprotector using cellular and animal models of the disease. Our results show that Ex-4 neuroprotective/trophically in NSC-19 cells ameliorated glucose tolerance and enhanced behavior and neuronal survival in SOD1 mutants. These data indicate that GLP-1 receptor agonists are candidate therapeutic agents for ALS [110] (Table 9.1).

Table 9.1 This table summarizes studies on mutant SOD1's role in ALS, focusing on aging, senescence implications, and pathways involved, highlighting potential therapeutic targets for modulating ALS progression and neurodegeneration

Model	Pathway involved	Mutant SOD1 and aging	Senescence implications	Reference
NSC34 cells, transgenic mice	Oxidized SOD1, mitochondrial malfunction	↑Oxidized SOD1, premature senescence	↑Mitochondrial malfunction	[77]
Drosophila (dsod1)	SOD loss-of-function, polyubiquitination	Age-dependent motor neuron degeneration	↑Polyubiquitinated proteins	[80]
Drosophila (knock-in)	Gain- and loss-of-function components	Severe neurodegeneration, locomotor deficits	Gain- and loss-of-function	[81]
ALS patients (Central South China)	Genetic mutations (SOD1, ATXN2, FUS, NEK1)	SOD1's most frequent mutation	Early disease onset (SOD1, FUS)	[85]
SOD1(G93A) mice	Metabolomic pathways, amino acid changes	Differences in plasma metabolites	Gender-specific metabolite changes	[87]
SOD1(G85R, G93A) mice	Autophagy/ lysosome (MTORC1 pathway)	Lack of lipofuscin accumulation	Hyperactive autophagy	[90]
ALS patients (Finland)	C9orf72 expansion, ALS-FTD	C9orf72 hexanucleotide repeat expansion	Early onset, shorter survival	[93]
SOD1(G93A) mice	Immune-inflammatory response	Reduced pro-inflammatory cytokines	↑anti-inflammatory cytokines	[97]
Familial ALS-linked mutant SOD1	Proteasome, molecular chaperones	Solubility changes with disease progression	Insoluble SOD1 accumulation	[98]
SOD1(G93A) mice	Tau expression, alternative splicing	Early tau expression alterations	Age-related tau changes	[100]
SOD1(G93A) mice	Muscle contractility, motor unit connectivity	Early muscle contractility dysfunction	Preceding motor unit loss	[101]
ALS patients	Protein degradation (UBQLN2 mutations)	Mutations in UBQLN2	Impaired protein degradation	[105]
Juvenile-onset SALS patients	FUS gene mutations	De novo FUS gene mutations	Aggressive progression	[107]
SOD1(G93A) mice	Adenosine receptor (A2A), glutamate transporter	Shortened survival with caffeine	Alterations in adenosine receptors	[108]
SOD1 mutant mice, NSC-19 cells	GLP-1 receptor agonist	Neuroprotective effects of Ex-4	Improved neuronal survival	[110]

Modulating Aging and Senescence Processes in ALS

Modulation of aging and senescence processes is achieved by targeting the pathways leading to cellular damage and decline in ALS [111]. Candidate approaches range from boosting autophagy (the cellular process of cleaning up damaged proteins and organelles) to selective removal of senescent cells using drugs known as senolytics to manipulation of several key signaling pathways (notably mTOR, AMPK and the sirtuins) thought to be able to increase healthspan [112, 113]. Furthermore, antioxidants and anti-inflammatory agents might also play a neuroprotective/neuroprotection role in ALS since oxidative stress and inflammation are characteristic features of the disease [114–116]. The gene therapy strategies, in principle, are directed to correct or silence the mutant SOD1 and, thus, finally eliminate its deleterious effects [117]. Taken together, these strategies could provide a more holistic approach to slowing the progression of ALS by targeting aspects of aging and senescence underlying disease pathogenesis [118, 119]. Trias et al. study investigates when senescence markers first appear in microglia during the development of paralysis. This study also supported the fact that senescent microglia vigorously promotes cell death of motor neurons and neuroinflammation. The findings indicate that cellular senescence is an important aspect of ALS pathology and provides further evidence to associate it with aging and neurodegeneration [120] (Fig. 9.3).

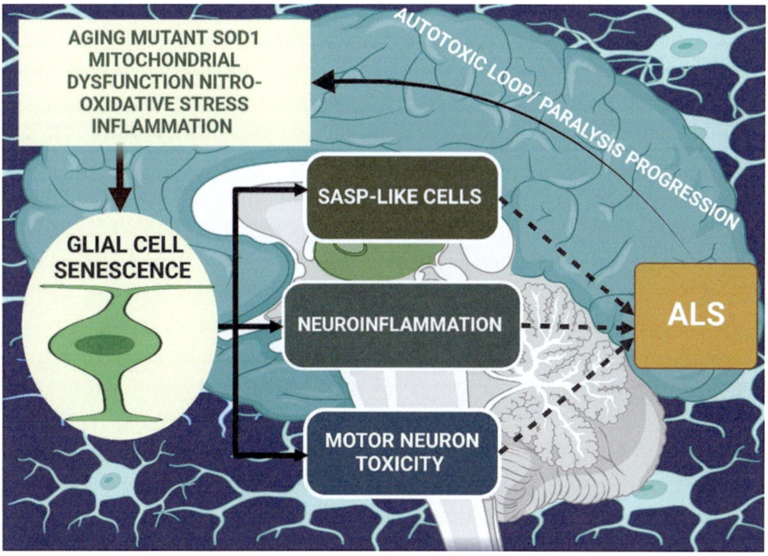

Fig. 9.3 Glial senescence-driven autotoxic loop in ALS. This figure provides an overview of the development of ALS in humans depicting all shown accelerators involved: aging mutant SOD1, mitochondrial dysfunction, nitro-oxidative stress, and inflammatory mechanisms. Together, these changes culminate in glial cell senescence and the subsequent emergence of SASP-like cells, neuroinflammation, and motor neuron toxicities. This leads to an autotoxic loop, driving further paralysis progression and potentially leading to the development of ALS

The impact of ALCAT1 dysregulation on mitochondrial integrity and cellular energy metabolism results in diverse diseases, including heart failure, obesity, diabetes, as well as neurodegenerative disorders [121]. Liu et al., discovering that ALCAT1 is involved in ALS pathogenesis, identified pharmacological inhibition of ALCAT1 alleviates ALS by blocking SOD1 protein aggregation and mitochondrial malfunction. Furthermore, this intervention mitigates motor neuron impairment and prolongs lifespan in SOD1G93A transgenic mice, thus suggesting ALCAT1 as a potential candidate for ALS treatment [122]. G93A SOD1 is a classical mutant form of the superoxide dismutase 1(SOD1) gene and serves as a component leading to ALS. G93A SOD1 studies provide insights into the understanding of ALS pathology and serve as a platform for potential therapies [123]. Trumbullet al. tested the ability of these two analogs, in chronic treatments with parenteral administration, to increase the lifespan of G93A SOD1 ALS mice. However, both of these approaches not only failed to improve overall survival as much as expected from the initial data but also were associated with several potential safety concerns [124]. CD4 T-cell aging is characterized by decreased immune function, such as reduced proliferation, altered cytokine production, and increased susceptibility to apoptosis [125]. Hence, they have a weakened immune response, which makes elders more prone to infections and less responsive to vaccines. Zaccai et al. demonstrated the role of CD4 T cell aging in neuroinflammation in an ALS mice model. We found it results in changes in what happens to the CD4 T-cell compartment over time as animals age, which turns into a chronic inflammatory response that makes everything worse. Taken together, these findings imply that the ambition of CD4 T-cell aging may be a potential healing scheme for ALS [126]. Furthermore, Filho et al. provide a comparative lipidomic analysis of the motor cortex and spinal cord tissues in a well-established rodent model of ALS, which is embedded with significant alterations in lipid metabolism playing crucial roles during the disease progression. They found decreased cardiolipin and increased cholesteryl esters, indicating progressive mitochondrial abnormalities and lipid droplet accumulation in astrocytes [127]. In another study, Chio et al. analyzed the clinical features of Italian ALS patients carrying intermediate ATXN2 polyQ repeats. Patients with ≥31 polyQ repeats had a significantly more rapid progression of the disease, shorter survival, and greater presence of frontotemporal dementia. Their findings reveal a relationship between ATXN2 polyQ repeats and the ALS phenotype, as well as with disease progression, which will have implications for prognosis and clinical trial stratification [128].

The ubiquitin-proteasome system (UPS) is an essential cellular route used to degrade and recycle proteins. It labels damaged or incorrectly folded proteins with ubiquitin molecules, which then target them for degradation by the proteasome [129, 130]. These processes regulate protein quality control, cell cycle, and many other cellular processes that maintain the homeostasis of a living cell. Various diseases can arise due to defects in the UPS, including neurodegenerative disorders and cancer [131]. Keskinet al. tested the impact of cellular pathway perturbations on misfolded SOD1 in fibroblasts taken from ALS patients. The study concluded that proteasome inhibition did indeed elevate levels of misfolded SOD1. This would indicate that the proteasome activity may also be implicated in ALS pathogenicity [132].

Furthermore, Dangoumau et al. examined a patient presenting with a slowly progressive form of ALS. They used a selection of these mutations, which are in non-overlapping common mutation regions of the protein, to show that different SOD1 mutations lead to different ALS phenotypes. The other four mutants caused aggregate formation and decreased cell viability, similar to those induced by D125T [133]. The ARHGEF28 gene (Rho guanine nucleotide exchange factor 28) encodes a protein that is part of the intracellular signaling pathways controlling cytoskeletal dynamics [134]. Rho family GTPases are proteins that play a critical role in cell migration, adhesion, and morphogenesis; they are turned on by this gene [135].

Furthermore, variations in the ARHGEF28 gene have been implicated in many types of cancers and developmental ailments [136]. Song et al. screened ARHGEF28 gene mutations in another Chinese sporadic ALS cohort. While all rare coding variants and even only the subset of three LoF alleles were slightly overrepresented among familial ALS cases, these findings applied more to a genetic contribution of rare variation rather than pointing toward a common goal in gene function [137]. Motor neuron degeneration is a condition wherein motor neurons are lost over a period, which causes muscle weakness and, eventually, atrophy of the muscles [138]. It disrupts how the nervous system and muscles talk to one another, which can impair movement and coordination. It has a feature of diseases such as ALS and spinal muscular atrophy [139, 140]. Angenstein et al. used MRI to follow the degeneration of motor neurons in G93A-SOD1 mice, showing age-dependent increases in MRI signal intensities (likely indicative for hyperintense areas) of motor brain stem nuclei beginning around day 90, where it parallels disease manifestations due to motor neuron dysfunction [141].

Interferons are proteins that are released by host cells in response to a pathogen. They are also important in the immune response as they prevent viral replication and are used to regulate a variety of other cells of the immune system [142]. These compounds are used to treat other diseases; for example, human interferon is used therapeutically in some types of cancer and viral infections [143, 144]. Wang et al. showed activation of interferon (IFN) signaling with SOD1(G93A)) mouse spinal cord astrocytes. Rather, we find that IFN-stimulated genes are upregulated at presymptomatic ages in astrocytes surrounding motor neurons. Reducing or deleting IFNα receptor 1 shut off the signaling of IFNs and extended the lives of ALS mice [145]. Impairments of speech, swallowing, and other functions in the muscles controlled by the brain stem are known as bulbar motor deficits [146]. Such deficits are typical of neurodegenerative diseases, such as ALS, making clear enunciation spontaneous, followed by swallowing and especially the handling of secretions become impossible to carry out properly, considerably impairing the quality of air [147, 148]. Early detection and appropriate supportive care are essential for improving the prognosis of patients discovered or diagnosed with OPG [149]. Smittkamp et al. showed that bulbar involvement followed spinal involvement, and the most important point is how robustly we could characterize those deficits in these mice. Tongue motility and force were reduced during the later symptomatic phase of the disease. Our study sheds light on motor signs onset, particularly regarding the relation between bulbar and spinal involvement in patients with ALS [150]. The blood-spinal cord barrier is a diffusion-biological barrier that keeps the harmful materials from entering into the SC interstitial against that allows necessary molecules like nutrients and small particles to pass across [151–153]. It is essential for proper homeostasis and the functioning of the central nervous system [154]. Davis et al. evaluated the impact of the transplantation of human bone marrow stem cells into symptomatic ALS mice on the blood-spinal cord barrier (BSCB). Profile improvements included capillary morphology, intact basement membrane, coherence of axonal myelin, and capillary density following transplantation. Our findings indicate that stem cell treatment might heal the injured BSCB in ALS [155].

Synergistic Pathway Targeting in ALS Management

Synergistic pathway targeting in ALS treatment. In the therapeutic schemes used today, several cellular and molecular pathways are targeted simultaneously to increase their efficiency [156]. Several investigators have proposed that ALS is associated with progressive motor neuron degeneration, which is mediated through means such as oxidative stress, mitochondrial dysfunction, and protein misfolding [157, 158]. Its significance largely emanates from the fact that targeting several different pathways, including autophagy, neuroinflammation, and synaptic function together, among others, instead of a single target, ensures much more efficient disease progression attenuation [159]. It is a promising possibility that such combined administrations might decelerate the progression of neuronal loss and, therefore, yield improved outcomes in patients by joining, for example, antioxidant therapy with anti-inflammatory drugs or agents promoting protein homeostasis [160, 161]. Overall, this multifacet approach will need a detailed analysis of intricate cross talks between these signaling pathways and the identification of important modulatory nodes to obtain maximal therapeutic potential and minimize adverse effects [162, 163]. Research is needed to substantiate the efficacy of combined treatments in clinical situations [164]. Kupershmidt et al. studied novel multi-target iron-chelators on neuroprotective efficacy in ALS, AD, and healthy aging models. Thus, the findings showed that both compounds display neuroprotective properties via reducing oxidative stress, driving positive behavioral changes, and inducing neural protective pathways and hence may act as potential therapeutics for NDDs [165].

Superoxide dismutase 1 (SOD1) WT is the normal form of superoxide dismutase. This enzyme protects cells from oxidative stress by breaking down various types of superoxide radicals into either oxygen or hydrogen peroxide [166]. It is critical for cellular immunity, obstructing reactive oxygen species and redox universality [167]. Prudencio et al. investigate the interaction of mutant SOD1 with wild-type (WT) SOD1 in this ALS model. This suggests that it can modify the course of disease by altering mutant SOD1 toxicity and aggregation. The results shed further light on the complex relationship between WT and mutant SOD1 in ALS [168].

Furthermore, Martin et al. demonstrate that transgenic mice expressing human SOD1 specifically in their skeletal muscle soon acquire an ALS-like syndrome. The existence of this finding suggests that skeletal muscle can be an independent originator of ALS pathogenesis, contributing to motor neuron degeneration. These results, therefore, support a non-autonomous basis of motor neuron death in this disease and highlight the complexity involved in ALS [169]. The study of age-dependent gene expression is defined as the changes in levels of gene regulation that occur at different stages across an organism's lifespan [170, 171]. These changes can affect multiple biological processes, such as metabolism, cell cycle regulation, and stress responses, leading to aging and age-related diseases [172]. Lobsiger et al. identified global, age-dependent gene expression change in motor neurons of SOD1 mutant mice. This study emphasized dysregulation of the complement system and neuronal regenerative response, offering novel ideas for potential targets in ALS treatment. The findings shed light on the molecular processes at work in ALS and promising points of therapeutic intervention [173]. The UNC13A rs12608932 is a single nucleotide polymorphism that relates to other neurodegenerative diseases such as ALS. Taken together, these first results point toward the AD-specific effects of this variant on gene expression or protein function, which may ultimately lead to enhanced neurotoxicity and vulnerability besides promoting disease progression [174, 175]. Calvo et al. reported on the clinical and metabolic phenotype of patients with ALS who bear the UNC13A rs12608932 variant. It increased the risk of ALS, was tied to specific motor and cognitive deficits, and shortened survival. However, the findings from this study would support including this variant in ALS genetic panels to facilitate prognostic prediction and stratification for clinical trials [176].

Motor impairment is the incomplete or full lack of ability to work with one's body elements, and people will be unable to take activities they desire. Causes can range from neurological disorders, injuries, or diseases that cause problems with the functioning of muscles, which ultimately affects one's ability to perform daily activities [177, 178]. Li et al. produced a double transgenic mouse line that expresses SOD1(G93A) and amyloid precursor protein (APP)-C100, which leads to an earlier onset of motor dysfunction compared with a single SOD1(G93A). The results indicate that amyloid-lighting may promote SOD1 aggregation, suggesting a possible effect of amyloid-beta lighters in the generation of ALS and reinforcing abnormal accumulation of SOD1 protein as one cause for ALS [179]. Importantly, the RNFL is a key layer of the retina and consists of axons that belong to ganglion cells. Functions to carry essential information from the eye to pass through the brain [180, 181]. Measurement of RNFL thickness is commonly used to evaluate the optic nerve status and contributes to the identification of glaucomatous eyes [182]. Zhang et al. examined changes in the RNFL and RGCs in patients with ALS. It showed RNFL thickening in the first year of disease and thinning thereafter, which was associated with ALS progression. The inverse U-shaped curve of RNFL changes may represent an early-stage motor neuron axonopathy. This phenomenon may comprise part of the pathophysiological basis for its use in ALS diagnosis and monitoring in the future or even as a marker for potential treatments targeting MNC function [183]. Riluzole is an FDA-approved drug for the treatment of amyotrophic lateral sclerosis (ALS). By blocking the release of glutamate, a neurotransmitter can become toxic to neurons when present at high levels [184]. Riluzole is capable of slightly prolonging life and slowing the course of action of signs or symptoms in

ALS. Its specific mode of action is not clearly defined, but it is supposed to help motor neurons protect against excitotoxicity [185]. Hogg et al. tested the effect of riluzole treatment in three different ALS mouse models (SOD1G93A, TDP-43A315T, FUS (1-359)) as well. However, riluzole offered no benefit to lifespan or motor function in any model, indicating that such a compound would almost certainly not proceed beyond preclinical assessment if tested today and challenging existing standards for preclinical evaluation of ALS therapeutics [186]. In another study, Aydemiret et al. measured hematological and serum biochemistry parameters during the presymptomatic and symptomatic stages of ALS (SOD1G93A rats). Some parameters, such as weight loss, cholesterol or glucose, and procalcitonin, were recorded that may be potential biomarkers for early diagnosis in ALS patients, for predicting the outcome, and for a better quality of life with increased survival time [187].

IL-10 is an anti-inflammatory cytokine that can be produced by a host of different cell types, particularly T cells, macrophages, and dendritic cells, among others. It is essential to limit immune responses to avoid tissue damage through the suppression of pro-inflammatory cytokine production and its activity on regulatory T cells. Strickland et al. demonstrated the impact of IL-10-mediated immunomodulation in SOD1-G93A mouse models'. The expression of IL-10 not only extends lifespan but also simultaneously modulates inflammatory processes, suggesting a benefit to targeting multiple inflammatory pathways for ALS treatment [188]. It is a protein fragment that's snipped from an amyloid precursor protein (APP). One of the most well-known conditions linked to amyloid is probably Alzheimer's disease, under which it gathers brain plaques [189]. Those plaques interfere with cell function, thus causing cognitive deficits and memory impairment [185]. Alzheimer's pathology manifests as an accumulation of beta-amyloid, which is one of the targets for interventions [190]. Turner et al. used a model to analyze the impact of ALS-linked mutant SOD1 on beta-amyloid aggregation and toxicity. These results demonstrated that overexpression of mutant SOD1 does not enhance amyloid pathology in ALS models, as shown by the failure of mutant SOD1 to stimulate beta-amyloid or gradual reduction in toxicity [191]. Astrocytes from induced pluripotent stem cells are a subtype of the glial cell type, which is prepared by reprogramming somatic cells [180]. It offers an in vitro tool to investigate astrocyte activities, effects, and attack modes of different disease types and ways that can be developed, mostly neurodegenerative diseases as well as ALS [192, 193]. Birger et al. employed iPSC-derived astrocytes from C9orf72-mutant ALS patients to study neurotoxicity. The outcomes indicated that C9-astrocytes were toxic to motor neurons and had higher oxidative stress, implying that therapeutic interventions would be necessary for both motor neurons as well as astrocyte populations [194]. Similarly, Das et al. described how the astrocytes become less supportive of motor neurons in aging and SOD1(G93A) ALS mice. GDNF easily and partially reversed the senescent phenotype of astrocytes. If such aged astrocytes could be replaced with young GDNF-producing cells, this would allow the survival-promoting actions of GDNF on motor neurons to also have a significant impact on diseases such as ALS [195].

Acetylcholine is a neurotransmitter in the central and peripheral nervous systems. Virtually, it is essential for the activation of muscles and vital to memory and learning [196]. Arousal, attention, and motivation are affected by the neurotransmitter acetylcholine at levels in the brain. Its malfunction has been linked to diseases, among others, Alzheimer's disease as well as Myasthenia gravis, which is an illustration of its role in body cognition and movement [182]. Sugita et al. examined the impact of such an increase in acetylcholine on neuromuscular junctions in (NMJs) in aging and ALS. The acceleration of NMJ degeneration and motor deficits in SOD1G93A mice by overexpressing vesicular acetylcholine transporter, specifically for male mice, revealed that normal cholinergic signaling is protective against NMJ degeneration and loss of function [197]. Heat shock proteins (HSPs) are a family of proteins expressed by cells in response to exposure to stressful conditions. These proteins act like molecular chaperones that stabilize, refold, and prevent the aggregation of misfolded proteins [198, 199]. HSPs protect cells against damage from stress, and they take part in many other cellular activities like protein folding, protein repair, and protein destruction [200]. Cellular stress granules are crucial not only for cell survival in times of need but also have potential roles to play in diseases, including cancer and neurodegeneration [201]. Krishnan et al. investigated the consequences of overexpression of Hsp27 in SOD1(G93A) mice.

Nevertheless, despite protection against acute motor neuron injury by Hsp27, it did not extend lifespan or delay the onset of disease in the SOD1 mouse model of ALS. Following all, the study showed that solely Hsp27 was unable to protect against chronic damage of motor nerve cells in ALS [202]. Primary glia are non-neuronal central nervous system cells that support and protect neurons, such as astrocytes, oligodendrocytes, and microglia. On the other hand, astrocytes maintain the blood–brain barrier and supply nutrients to its myelin sheaths by oligodendrocytes on neurons [203]. Microglia serve as defense cells, removing waste and pathogens from the central nervous system (CNS) [204, 205]. Hensley et al. showed that primary glia expressing the G93A-SOD1 mutation displayed a neuro-inflammatory phenotype. G93A-SOD1 glia had increased proinflammatory eicosanoids, inducible nitric oxide synthase, and protein carbonyls. Our results suggest glial involvement in EAE and identify a new tool for examining glial inflammation. Moreover, they support the notion of targeting that system to treat ALS [206]. The inner mitochondrial membrane is an important structure of the mitochondria, which contains proteins that are required for oxidative phosphorylation. It is undrained to the vast majority of particles and particles, which allows it to retain the proton slope imperative for ATP generation [207, 208]. The many folds, called cristae, add surface area and help the mitochondria work more efficiently so they can generate more energy [209–211]. Kirkinezos et al. showed that there is the impaired association of cytochrome c to the inner mitochondrial membrane (IMM) in CNS isolated from G93A-SOD1 mice. Further, this was associated with increased mitochondrial lipid peroxidation and decreased complex IV respiration before clinical symptom presentation. This lends a mechanism by which mutant SOD1 displaces cytochrome c interaction, leading to an apoptotic predisposition in ALS [212] (Table 9.2).

Table 9.2 This table summarizes studies on mutant SOD1 in ALS, highlighting age-related effects, molecular impacts, and therapeutic strategies. It emphasizes the significance of targeting oxidative stress, mitochondrial dysfunction, inflammation, and other pathways to mitigate neurodegeneration and improve patient outcomes

Model	Age-related effects	Molecular effects	Therapeutic strategy	Reference
ALS, AD, aging models	Accelerated aging, neuroprotection	↓ Oxidative stress, ↑ Neuroprotection	Multitarget iron-chelation	[165]
ALS models	Interaction of WT and mutant SOD1	Influences toxicity, aggregation	Modulating SOD1 interactions	[168]
Transgenic mice (skeletal muscle)	Muscle-specific aging effects	Motor neuron degeneration	Targeting muscle-specific SOD1 expression	[169]
SOD1 mutant mice	Age-dependent gene expression changes	Dysregulated complement, ↓ Regeneration	Modulating complement and regenerative pathways	[173]
ALS patients (UNC13A variant)	Earlier onset, motor/cognitive deficits	↑ ALS risk, motor/cognitive deficits	Genetic screening and personalized therapy	[176]
Double transgenic mice (SOD1, APP)	Earlier motor impairment	Earlier motor impairment	Targeting amyloid-beta to reduce SOD1 aggregation	[179]
ALS patients	RNFL and RGCs alterations with age	RNFL thickening, decline with progression	Monitoring RNFL for early diagnosis	[183]
ALS mouse models (SOD1G93A, TDP-43A315T, FUS)	No significant improvement with riluzole	No significant improvement	Reevaluating preclinical standards	[186]
SOD1G93A rats	Hematological/serum biochemistry with aging	Changes in biomarkers with progression	Biomarkers for early diagnosis and prognosis	[187]
SOD1-G93A mice	IL-10 modulates aging-related inflammation	Extended lifespan, ↓ inflammation	Targeting IL-10 to modulate inflammation	[188]
ALS models	No promotion of beta-amyloid toxicity	No promotion of beta-amyloid toxicity	Independent targeting of SOD1 and beta-amyloid	[191]
iPSC-derived astrocytes	Increased oxidative stress with age	↑ Oxidative stress, toxic to motor neurons	Combined motor neuron and astrocyte targeting	[194]

(continued)

Table 9.2 (continued)

Model	Age-related effects	Molecular effects	Therapeutic strategy	Reference
SOD1G93A mice	Accelerated NMJ degeneration with aging	Accelerated NMJ degeneration	Maintaining cholinergic signaling	[197]
SOD1(G93A) mice	Hsp27's limited effect on chronic injury	No lifespan extension	Combined therapeutic approaches	[202]
SOD1(G93A) ALS mice	Aging astrocytes and motor neuron survival	↓ Astrocyte support with aging	Replacing aging astrocytes with young GDNF-producers	[195]
Primary glia (G93A-SOD1 mutation)	Neuroinflammatory phenotype with aging	↑ Proinflammatory markers	Targeting glial inflammation	[206]
SOD1(G93A) mice	Impaired cytochrome c with age	Impaired cytochrome c association	Protecting mitochondrial integrity	[206]

Conclusion and Future Perspective

This review highlights that mutant SOD1 plays a vital part in the ALS pathological process and also focuses on the delicate connections between mutant SOD1, aging, and cellular senescence. Mutant SOD1 exerts neurotoxicity through diverse mechanisms, including oxidative stress, protein misfolding, and disrupted proteostasis. These findings implicate that both targeting mutant SOD1 and modulation of senescence-related pathways are critical for developing effective therapeutic interventions. The interrelationship of the above factors with mutant SOD1, aging, and senescence is important for further progression in ALS studies and therapies. Aging and cell senescence not only worsen the effect of mutant SOD1, but they also create a toxic microenvironment that ultimately supports neurodegeneration.

Further research is needed to investigate the precise role of aging and senescence in ALS pathogenesis. This also involves investigating whether senolytic therapies could be used to eliminate senescent cells, banishing their pro-inflammatory tendencies and revitalizing the health of our neurons. Moreover, given the role of mitochondrial dynamics and autophagy in aging as well as ALS, a study involving these parameters may give more hints about the potential targets for therapy. These findings suggest that targeting the combined effects of mutant SOD1, aging and cellular senescence may provide novel avenues for more effective therapeutic strategies in treating ALS. In the future, the scientific community must utilize a multidimensional approach to comprehend these complicated interactions among genetic mutations, biological senescence and aging processes. Such strategies will be key in enhancing the quality of life and increasing the lifespan of ALS patients. Despite the unpreventable nature of understanding complex ALS, ongoing research is making the future brighter with innovative and individualized therapeutic outlooks.

Data Availability Statement Data sharing does not apply to this article as no new data were created or analyzed in this study.

Conflict of Interest The authors declare that there are no competing interests.

References

1. Brown RH, Al-Chalabi A. Amyotrophic lateral sclerosis. N Engl J Med. 2017;377(2):162–72.
2. Feldman EL, Goutman SA, Petri S, Mazzini L, Savelieff MG, Shaw PJ, et al. Amyotrophic lateral sclerosis. Lancet. 2022;400(10360):1363–80.
3. Goutman SA, Hardiman O, Al-Chalabi A, Chió A, Savelieff MG, Kiernan MC, et al. Recent advances in the diagnosis and prognosis of amyotrophic lateral sclerosis. Lancet Neurol. 2022;21(5):480–93.
4. Grad LI, Rouleau GA, Ravits J, Cashman NR. Clinical spectrum of amyotrophic lateral sclerosis (ALS). Cold Spring Harb Perspect Med. 2017;7(8)
5. Hardiman O, Al-Chalabi A, Chio A, Corr EM, Logroscino G, Robberecht W, et al. Amyotrophic lateral sclerosis. Nat Rev Dis Primers. 2017;3:17071.
6. Hulisz D. Amyotrophic lateral sclerosis: disease state overview. Am J Manag Care. 2018;24(15 Suppl):S320–6.
7. Kiernan MC, Vucic S, Cheah BC, Turner MR, Eisen A, Hardiman O, et al. Amyotrophic lateral sclerosis. Lancet. 2011;377(9769):942–55.
8. Meyer T. Amyotrophic lateral sclerosis (ALS) - diagnosis, course of disease and treatment options. Dtsch Med Wochenschr (1946). 2021;146(24–25):1613–8.
9. Abati E, Bresolin N, Comi G, Corti S. Silence superoxide dismutase 1 (SOD1): a promising therapeutic target for amyotrophic lateral sclerosis (ALS). Expert Opin Ther Targets. 2020;24(4):295–310.
10. Fang T, Je G, Pacut P, Keyhanian K, Gao J, Ghasemi M. Gene therapy in amyotrophic lateral sclerosis. Cells. 2022;11(13)
11. Brown CA, Lally C, Kupelian V, Flanders WD. Estimated prevalence and incidence of amyotrophic lateral sclerosis and SOD1 and C9orf72 genetic variants. Neuroepidemiology. 2021;55(5):342–53.
12. Bunton-Stasyshyn RK, Saccon RA, Fratta P, Fisher EM. SOD1 function and its implications for amyotrophic lateral sclerosis pathology: new and renascent themes. Neuroscientist. 2015;21(5):519–29.
13. Choi ES, Dokholyan NV. SOD1 oligomers in amyotrophic lateral sclerosis. Curr Opin Struct Biol. 2021;66:225–30.
14. Goutman SA, Hardiman O, Al-Chalabi A, Chió A, Savelieff MG, Kiernan MC, et al. Emerging insights into the complex genetics and pathophysiology of amyotrophic lateral sclerosis. Lancet Neurol. 2022;21(5):465–79.
15. Kaur SJ, McKeown SR, Rashid S. Mutant SOD1 mediated pathogenesis of amyotrophic lateral sclerosis. Gene. 2016;577(2):109–18.
16. Philips T, Rothstein JD. Rodent models of amyotrophic lateral sclerosis. Curr Protoc Pharmacol. 2015;69(5.67.21):5.67.1.
17. Witzel S, Mayer K, Oeckl P. Biomarkers for amyotrophic lateral sclerosis. Curr Opin Neurol. 2022;35(5):699–704.

18. Eleutherio ECA, Silva Magalhães RS, de Araújo Brasil A, Monteiro Neto JR, de Holanda Paranhos L. SOD1, more than just an antioxidant. Arch Biochem Biophys. 2021;697:108701.
19. Holecek V, Rokyta R. Possible etiology and treatment of amyotrophic lateral sclerosis. Neuro Endocrinol Lett. 2018;38(8):528–31.
20. Gomez M, Germain D. Cross talk between SOD1 and the mitochondrial UPR in cancer and neurodegeneration. Mol Cell Neurosci. 2019;98:12–8.
21. Hilton JB, White AR, Crouch PJ. Metal-deficient SOD1 in amyotrophic lateral sclerosis. J Mol Med (Berl). 2015;93(5):481–7.
22. Pansarasa O, Bordoni M, Diamanti L, Sproviero D, Gagliardi S, Cereda C. SOD1 in amyotrophic lateral sclerosis: "ambivalent" behavior connected to the disease. Int J Mol Sci. 2018;19(5)
23. Papa L, Manfredi G, Germain D. SOD1, an unexpected novel target for cancer therapy. Genes Cancer. 2014;5(1–2):15–21.
24. Shanq K, Sanghai N, Guo Y, Kong J. Implication of post-translationally modified SOD1 in pathological aging. GeroScience. 2021;43(2):507–15.
25. Sheng Y, Chattopadhyay M, Whitelegge J, Valentine JS. SOD1 aggregation and ALS: role of metallation states and disulfide status. Curr Top Med Chem. 2012;12(22):2560–72.
26. Tafuri F, Ronchi D, Magri F, Comi GP, Corti S. SOD1 misplacing and mitochondrial dysfunction in amyotrophic lateral sclerosis pathogenesis. Front Cell Neurosci. 2015;9:336.
27. Xu J, Su X, Burley SK, Zheng XFS. Nuclear SOD1 in growth control, oxidative stress response, amyotrophic lateral sclerosis, and cancer. Antioxidants (Basel, Switzerland). 2022;11(2)
28. Zelko IN, Mariani TJ, Folz RJ. Superoxide dismutase multigene family: a comparison of the CuZn-SOD (SOD1), Mn-SOD (SOD2), and EC-SOD (SOD3) gene structures, evolution, and expression. Free Radic Biol Med. 2002;33(3):337–49.
29. Beers DR, Appel SH. Immune dysregulation in amyotrophic lateral sclerosis: mechanisms and emerging therapies. Lancet Neurol. 2019;18(2):211–20.
30. Brenner D, Freischmidt A. Update on genetics of amyotrophic lateral sclerosis. Curr Opin Neurol. 2022;35(5):672–7.
31. Chia R, Chiò A, Traynor BJ. Novel genes associated with amyotrophic lateral sclerosis: diagnostic and clinical implications. Lancet Neurol. 2018;17(1):94–102.
32. Corcia P, Lejeune P, Vourc'h P, Beltran S, Piegay AS, Blasco H, et al. Comparison between PFN1 and SOD1 mutations in amyotrophic lateral sclerosis. Eur J Neurol. 2023;30(2):552–4.
33. Gois AM, Mendonça DMF, Freire MAM, Santos JR. In vitro and in vivo models of amyotrophic lateral sclerosis: an updated overview. Brain Res Bull. 2020;159:32–43.
34. Berdyński M, Miszta P, Safranow K, Andersen PM, Morita M, Filipek S, et al. SOD1 mutations associated with amyotrophic lateral sclerosis analysis of variant severity. Sci Rep. 2022;12(1):103.
35. Nishitoh H, Kadowaki H, Nagai A, Maruyama T, Yokota T, Fukutomi H, et al. ALS-linked mutant SOD1 induces ER stress- and ASK1-dependent motor neuron death by targeting Derlin-1. Genes Dev. 2008;22(11):1451–64.
36. Forsberg KM, Graffmo KS, Stenvall E, Tabikh N, Marklund SL, Brännström T, et al. Widespread CNS pathology in amyotrophic lateral sclerosis homozygous for the D90A SOD1 mutation. Acta Neuropathol. 2023;145(1):13–28.
37. Saberi S, Stauffer JE, Schulte DJ, Ravits J. Neuropathology of amyotrophic lateral sclerosis and its variants. Neurol Clin. 2015;33(4):855–76.
38. Suzuki N, Nishiyama A, Warita H, Aoki M. Genetics of amyotrophic lateral sclerosis: seeking therapeutic targets in the era of gene therapy. J Hum Genet. 2023;68(3):131–52.
39. Calcinotto A, Kohli J, Zagato E, Pellegrini L, Demaria M, Alimonti A. Cellular senescence: aging, cancer, and injury. Physiol Rev. 2019;99(2):1047–78.
40. Campisi J. Aging, cellular senescence, and cancer. Annu Rev Physiol. 2013;75:685–705.
41. Csekes E, Račková L. Skin aging, cellular senescence and natural polyphenols. Int J Mol Sci. 2021;22(23)

42. Hernandez-Segura A, Nehme J, Demaria M. Hallmarks of cellular senescence. Trends Cell Biol. 2018;28(6):436–53.
43. López-Otín C, Blasco MA, Partridge L, Serrano M, Kroemer G. Hallmarks of aging: an expanding universe. Cell. 2023;186(2):243–78.
44. Miwa S, Kashyap S, Chini E, von Zglinicki T. Mitochondrial dysfunction in cell senescence and aging. J Clin Invest. 2022;132(13)
45. Mohamad Kamal NS, Safuan S, Shamsuddin S, Foroozandeh P. Aging of the cells: insight into cellular senescence and detection methods. Eur J Cell Biol. 2020;99(6):151108.
46. Ogrodnik M. Cellular aging beyond cellular senescence: markers of senescence prior to cell cycle arrest in vitro and in vivo. Aging Cell. 2021;20(4):e13338.
47. Regulski MJ. Cellular senescence: what, why, and how. Wounds. 2017;29(6):168–74.
48. Zhang L, Pitcher LE, Yousefzadeh MJ, Niedernhofer LJ, Robbins PD, Zhu Y. Cellular senescence: a key therapeutic target in aging and diseases. J Clin Invest. 2022;132(15)
49. Baker DJ, Petersen RC. Cellular senescence in brain aging and neurodegenerative diseases: evidence and perspectives. J Clin Invest. 2018;128(4):1208–16.
50. Bao H, Cao J, Chen M, Chen M, Chen W, Chen X, et al. Biomarkers of aging. Sci China Life Sci. 2023;66(5):893–1066.
51. Gonzales MM, Garbarino VR, Pollet E, Palavicini JP, Kellogg DL Jr, Kraig E, et al. Biological aging processes underlying cognitive decline and neurodegenerative disease. J Clin Invest. 2022;132(10)
52. Guo J, Huang X, Dou L, Yan M, Shen T, Tang W, et al. Aging and aging-related diseases: from molecular mechanisms to interventions and treatments. Signal Transduct Target Ther. 2022;7(1):391.
53. Kritsilis M, Rizou SV, Koutsoudaki PN, Evangelou K, Gorgoulis VG, Papadopoulos D. Ageing, cellular senescence and neurodegenerative disease. Int J Mol Sci. 2018;19(10)
54. Li Z, Zhang Z, Ren Y, Wang Y, Fang J, Yue H, et al. Aging and age-related diseases: from mechanisms to therapeutic strategies. Biogerontology. 2021;22(2):165–87.
55. Liu RM. Aging, cellular senescence, and Alzheimer's disease. Int J Mol Sci. 2022;23(4)
56. Wissler Gerdes EO, Zhu Y, Weigand BM, Tripathi U, Burns TC, Tchkonia T, et al. Cellular senescence in aging and age-related diseases: implications for neurodegenerative diseases. Int Rev Neurobiol. 2020;155:203–34.
57. Zia A, Farkhondeh T, Pourbagher-Shahri AM, Samarghandian S. The role of curcumin in aging and senescence: molecular mechanisms. Biomed Pharmacother = Biomedecine & Pharmacotherapie. 2021;134:111119.
58. Al-Chalabi A, Hardiman O, Kiernan MC, Chiò A, Rix-Brooks B, van den Berg LH. Amyotrophic lateral sclerosis: moving towards a new classification system. Lancet Neurol. 2016;15(11):1182–94.
59. Chiò A, Logroscino G, Traynor BJ, Collins J, Simeone JC, Goldstein LA, et al. Global epidemiology of amyotrophic lateral sclerosis: a systematic review of the published literature. Neuroepidemiology. 2013;41(2):118–30.
60. Gendron TF, Petrucelli L. Immunological drivers of amyotrophic lateral sclerosis. Sci Transl Med. 2023;15(721):eadj9332.
61. Izenberg A. Amyotrophic lateral sclerosis and other motor neuron diseases. Continuum (Minneap Minn). 2023;29(5):1538–63.
62. Mercadante S, Al-Husinat L. Palliative care in amyotrophic lateral sclerosis. J Pain Symptom Manag. 2023;66(4):e485–99.
63. Morelot-Panzini C, Bruneteau G, Gonzalez-Bermejo J. NIV in amyotrophic lateral sclerosis: the 'when' and 'how' of the matter. Respirology. 2019;24(6):521–30.
64. Quinn C, Elman L. Amyotrophic lateral sclerosis and other motor neuron diseases. Continuum (Minneap Minn). 2020;26(5):1323–47.
65. Thomson CG, Hutchinson PR, Stern PJ. Misdiagnosis in amyotrophic lateral sclerosis. J Hand Surg Am. 2023;48(8):822–6.
66. Maximova A, Werry EL, Kassiou M. Senolytics: a novel strategy for neuroprotection in ALS? Int J Mol Sci. 2021;22(21)

67. Torres P, Anerillas C, Ramírez-Núñez O, Fernàndez A, Encinas M, Povedano M, et al. A motor neuron disease mouse model reveals a non canonical profile of senescence biomarkers. Dis Model Mech. 2022;15(8)
68. Thadathil N, Selvarani R, Mohammed S, Nicklas EH, Tran AL, Kamal M, et al. Senolytic treatment reduces cell senescence and necroptosis in Sod1 knockout mice that is associated with reduced inflammation and hepatocellular carcinoma. Aging Cell. 2022;21(8):e13676.
69. Clarke BE, Patani R. The microglial component of amyotrophic lateral sclerosis. Brain J Neurol. 2020;143(12):3526–39.
70. Geerligs L, Campbell KL. Age-related differences in information processing during movie watching. Neurobiol Aging. 2018;72:106–20.
71. D'Erchia AM, Gallo A, Manzari C, Raho S, Horner DS, Chiara M, et al. Massive transcriptome sequencing of human spinal cord tissues provides new insights into motor neuron degeneration in ALS. Sci Rep. 2017;7(1):10046.
72. Arnold FJ, Nguyen AD, Bedlack RS, Bennett CL, La Spada AR. Intercellular transmission of pathogenic proteins in ALS: exploring the pathogenic wave. Neurobiol Dis. 2023;184:106218.
73. Benatar M. Lost in translation: treatment trials in the SOD1 mouse and in human ALS. Neurobiol Dis. 2007;26(1):1–13.
74. Sangwan S, Eisenberg DS. Perspective on SOD1 mediated toxicity in amyotrophic lateral sclerosis. Postepy Biochem. 2016;62(3):362–9.
75. Shadfar S, Parakh S, Jamali MS, Atkin JD. Redox dysregulation as a driver for DNA damage and its relationship to neurodegenerative diseases. Transl Neurodegener. 2023;12(1):18.
76. Tak YJ, Park JH, Rhim H, Kang S. ALS-related mutant SOD1 aggregates interfere with mitophagy by sequestering the autophagy receptor optineurin. Int J Mol Sci. 2020;21(20)
77. Guo Y, Guan T, Yu Q, Sanghai N, Shafiq K, Li M, et al. ALS-linked SOD1 mutations impair mitochondrial-derived vesicle formation and accelerate aging. Redox Biol. 2024;69:102972.
78. Parkes TL, Elia AJ, Dickinson D, Hilliker AJ, Phillips JP, Boulianne GL. Extension of Drosophila lifespan by overexpression of human SOD1 in motorneurons. Nat Genet. 1998;19(2):171–4.
79. De Lazzari F, Agostini F, Doni D, Malacrida S, Zordan MA, Costantini P, et al. DJ-1 and SOD1 act independently in the protection against anoxia in Drosophila melanogaster. Antioxidants. 2022;11(8):1527.
80. Agudelo A, St Amand V, Grissom L, Lafond D, Achilli T, Sahin A, et al. Age-dependent degeneration of an identified adult leg motor neuron in a Drosophila SOD1 model of ALS. Biol Open. 2020;9(10)
81. Şahin A, Held A, Bredvik K, Major P, Achilli TM, Kerson AG, et al. Human SOD1 ALS mutations in a Drosophila knock-in model cause severe phenotypes and reveal dosage-sensitive gain- and loss-of-function components. Genetics. 2017;205(2):707–23.
82. Thapa R, Afzal M, Goyal A, Gupta G, Bhat AA, Almalki WH, et al. Exploring ncRNA-mediated regulation of EGFR signalling in glioblastoma: from mechanisms to therapeutics. Life Sci. 2024;345:122613.
83. Blokhuis AM, Groen EJ, Koppers M, van den Berg LH, Pasterkamp RJ. Protein aggregation in amyotrophic lateral sclerosis. Acta Neuropathol. 2013;125(6):777–94.
84. Wang H, Guan L, Deng M. Recent progress of the genetics of amyotrophic lateral sclerosis and challenges of gene therapy. Front Neurosci. 2023;17:1170996.
85. Liu Z, Yuan Y, Wang M, Ni J, Li W, Huang L, et al. Mutation spectrum of amyotrophic lateral sclerosis in Central South China. Neurobiol Aging. 2021;107:181–8.
86. Martinelli I, Zucchi E, Simonini C, Gianferrari G, Zamboni G, Pinti M, et al. The landscape of cognitive impairment in superoxide dismutase 1-amyotrophic lateral sclerosis. Neural Regen Res. 2023;18(7):1427–33.
87. Bame M, Grier RE, Needleman R, Brusilow WS. Amino acids as biomarkers in the SOD1(G93A) mouse model of ALS. Biochim Biophys Acta. 2014;1842(1):79–87.
88. Brunk UT, Terman A. Lipofuscin: mechanisms of age-related accumulation and influence on cell function. Free Radic Biol Med. 2002;33(5):611–9.

89. Różanowska MB. Lipofuscin, its origin, properties, and contribution to retinal fluorescence as a potential biomarker of oxidative damage to the retina. Antioxidants (Basel). 2023;12(12):2111.
90. Bandyopadhyay U, Nagy M, Fenton WA, Horwich AL. Absence of lipofuscin in motor neurons of SOD1-linked ALS mice. Proc Natl Acad Sci USA. 2014;111(30):11055–60.
91. Haeusler AR, Donnelly CJ, Periz G, Simko EA, Shaw PG, Kim MS, et al. C9orf72 nucleotide repeat structures initiate molecular cascades of disease. Nature. 2014;507(7491):195–200.
92. Liu C, Geng Y, Miao H, Shi X, You Y, Xu N, et al. G-quadruplex structures formed by human telomeric DNA and C9orf72 hexanucleotide repeats. Biophys Rev. 2019;11(3):389–93.
93. Laaksovirta H, Launes J, Jansson L, Traynor BJ, Kaivola K, Tienari PJ. ALS in Finland: major genetic variants and clinical characteristics of patients with and without the C9orf72 hexanucleotide repeat expansion. Neurol Genet. 2022;8(2):e665.
94. Jelkmann W. Regulation of erythropoietin production. J Physiol. 2011;589(Pt 6):1251–8.
95. Tang P, Wang H. Regulation of erythropoiesis: emerging concepts and therapeutic implications. Hematology. 2023;28(1):2250645.
96. Thapa R, Goyal A, Gupta G, Bhat AA, Singh SK, Subramaniyan V, et al. Recent developments in the role of protocatechuic acid in neurodegenerative disorders. EXCLI J. 2023;22:595.
97. Noh MY, Cho KA, Kim H, Kim SM, Kim SH. Erythropoietin modulates the immune-inflammatory response of a SOD1(G93A) transgenic mouse model of amyotrophic lateral sclerosis (ALS). Neurosci Lett. 2014;574:53–8.
98. Koyama S, Arawaka S, Chang-Hong R, Wada M, Kawanami T, Kurita K, et al. Alteration of familial ALS-linked mutant SOD1 solubility with disease progression: its modulation by the proteasome and Hsp70. Biochem Biophys Res Commun. 2006;343(3):719–30.
99. Méndez-López I, Sancho-Bielsa FJ, Engel T, García AG, Padín JF. Progressive mitochondrial SOD1(G93A) accumulation causes severe structural, metabolic and functional aberrations through OPA1 down-regulation in a mouse model of amyotrophic lateral sclerosis. Int J Mol Sci. 2021;22(15)
100. Barańczyk-Kuźma A, Usarek E, Kuźma-Kozakiewcz M, Kaźmierczak B, Gajewska B, Schwalenstocker B, et al. Age-related changes in tau expression in transgenic mouse model of amyotrophic lateral sclerosis. Neurochem Res. 2007;32(3):415–21.
101. Wier CG, Crum AE, Reynolds AB, Iyer CC, Chugh D, Palettas MS, et al. Muscle contractility dysfunction precedes loss of motor unit connectivity in SOD1(G93A) mice. Muscle Nerve. 2019;59(2):254–62.
102. Renaud L, Picher-Martel V, Codron P, Julien JP. Key role of UBQLN2 in pathogenesis of amyotrophic lateral sclerosis and frontotemporal dementia. Acta Neuropathol Commun. 2019;7(1):103.
103. Zhang KY, Yang S, Warraich ST, Blair IP. Ubiquilin 2: a component of the ubiquitin-proteasome system with an emerging role in neurodegeneration. Int J Biochem Cell Biol. 2014;50:123–6.
104. Thapa R, Gupta G, Dave P, Singh SK, Raizaday A, Almalki WH, et al. Current update on the protective effect of epicatechin in neurodegenerative diseases. EXCLI J. 2022;21:897–903.
105. Deng HX, Chen W, Hong ST, Boycott KM, Gorrie GH, Siddique N, et al. Mutations in UBQLN2 cause dominant X-linked juvenile and adult-onset ALS and ALS/dementia. Nature. 2011;477(7363):211–5.
106. López-Erauskin J, Tadokoro T, Baughn MW, Myers B, McAlonis-Downes M, Chillon-Marinas C, et al. ALS/FTD-linked mutation in FUS suppresses intra-axonal protein synthesis and drives disease without nuclear loss-of-function of FUS. Neuron. 2018;100(4):816–30.e7.
107. Zou ZY, Cui LY, Sun Q, Li XG, Liu MS, Xu Y, et al. De novo FUS gene mutations are associated with juvenile-onset sporadic amyotrophic lateral sclerosis in China. Neurobiol Aging. 2013;34(4):1312.e1–8.
108. Potenza RL, Armida M, Ferrante A, Pèzzola A, Matteucci A, Puopolo M, et al. Effects of chronic caffeine intake in a mouse model of amyotrophic lateral sclerosis. J Neurosci Res. 2013;91(4):585–92.

109. Hinnen D. Glucagon-like peptide 1 receptor agonists for type 2 diabetes. Diabetes Spectr. 2017;30(3):202–10.
110. Li Y, Chigurupati S, Holloway HW, Mughal M, Tweedie D, Bruestle DA, et al. Exendin-4 ameliorates motor neuron degeneration in cellular and animal models of amyotrophic lateral sclerosis. PLoS One. 2012;7(2):e32008.
111. Caballero-Hernandez D, Toscano MG, Cejudo-Guillen M, Garcia-Martin ML, Lopez S, Franco JM, et al. The 'omics' of amyotrophic lateral sclerosis. Trends Mol Med. 2016;22(1):53–67.
112. de Carvalho M. Electrodiagnosis of amyotrophic lateral sclerosis: a review of existing guidelines. J Clin Neurophysiol. 2020;37(4):294–8.
113. Donini L, Tanel R, Zuccarino R, Basso M. Protein biomarkers for the diagnosis and prognosis of amyotrophic lateral sclerosis. Neurosci Res. 2023;197:31–41.
114. Kiaei L, Kiaei M. RNA as a source of biomarkers for amyotrophic lateral sclerosis. Metab Brain Dis. 2022;37(6):1697–702.
115. Krishnamurthy K, Pasinelli P. Synaptic dysfunction in amyotrophic lateral sclerosis/frontotemporal dementia: therapeutic strategies and novel biomarkers. J Neurosci Res. 2021;99(6):1499–503.
116. Lisle S, Tennison M. Amyotrophic lateral sclerosis: the role of exercise. Curr Sports Med Rep. 2015;14(1):45–6.
117. Rinaldi F, Motti D, Ferraiuolo L, Kaspar BK. High content analysis in amyotrophic lateral sclerosis. Mol Cell Neurosci. 2017;80:180–91.
118. Sethi A, Everett E, Mehta A, Besbris J, Burke C, Pedowitz E, et al. The role of specialty palliative care for amyotrophic lateral sclerosis. Am J Hosp Palliat Care. 2022;39(7):865–73.
119. Vu LT, Bowser R. Fluid-based biomarkers for amyotrophic lateral sclerosis. Neurotherapeutics. 2017;14(1):119–34.
120. Trias E, Beilby PR, Kovacs M, Ibarburu S, Varela V, Barreto-Núñez R, et al. Emergence of microglia bearing senescence markers during paralysis progression in a rat model of inherited ALS. Front Aging Neurosci. 2019;11:42.
121. Legaki AI, Moustakas II, Sikorska M, Papadopoulos G, Velliou RI, Chatzigeorgiou A. Hepatocyte mitochondrial dynamics and bioenergetics in obesity-related non-alcoholic fatty liver disease. Curr Obes Rep. 2022;11(3):126–43.
122. Liu X, Zhang J, Li J, Song C, Shi Y. Pharmacological inhibition of ALCAT1 mitigates amyotrophic lateral sclerosis by attenuating SOD1 protein aggregation. Mol Metab. 2022;63:101536.
123. Kim BW, Ryu J, Jeong YE, Kim J, Martin LJ. Human motor neurons with SOD1-G93A mutation generated from CRISPR/Cas9 gene-edited iPSCs develop pathological features of amyotrophic lateral sclerosis. Front Cell Neurosci. 2020;14:604171.
124. Trumbull KA, McAllister D, Gandelman MM, Fung WY, Lew T, Brennan L, et al. Diapocynin and apocynin administration fails to significantly extend survival in G93A SOD1 ALS mice. Neurobiol Dis. 2012;45(1):137–44.
125. Lefebvre JS, Haynes L. Aging of the CD4 T cell compartment. Open Longev Sci. 2012;6:83–91.
126. Zaccai S, Nemirovsky A, Lerner L, Alfahel L, Eremenko E, Israelson A, et al. CD4 T-cell aging exacerbates neuroinflammation in a late-onset mouse model of amyotrophic lateral sclerosis. J Neuroinflammation. 2024;21(1):17.
127. Chaves-Filho AB, Pinto IFD, Dantas LS, Xavier AM, Inague A, Faria RL, et al. Alterations in lipid metabolism of spinal cord linked to amyotrophic lateral sclerosis. Sci Rep. 2019;9(1):11642.
128. Chiò A, Moglia C, Canosa A, Manera U, Grassano M, Vasta R, et al. Association of copresence of pathogenic variants related to amyotrophic lateral sclerosis and prognosis. Neurology. 2023;101(1):e83–93.
129. Calvo AC, Manzano R, Mendonça DM, Muñoz MJ, Zaragoza P, Osta R. Amyotrophic lateral sclerosis: a focus on disease progression. Biomed Res Int. 2014;2014:925101.

130. Eisen A, Kiernan M, Mitsumoto H, Swash M. Amyotrophic lateral sclerosis: a long preclinical period? J Neurol Neurosurg Psychiatry. 2014;85(11):1232–8.
131. Dantuma NP, Bott LC. The ubiquitin-proteasome system in neurodegenerative diseases: precipitating factor, yet part of the solution. Front Mol Neurosci. 2014;7:70.
132. Keskin I, Forsgren E, Lange DJ, Weber M, Birve A, Synofzik M, et al. Effects of cellular pathway disturbances on misfolded superoxide dismutase-1 in fibroblasts derived from ALS patients. PLoS One. 2016;11(2):e0150133.
133. Dangoumau A, Verschueren A, Hammouche E, Papon MA, Blasco H, Cherpi-Antar C, et al. Novel SOD1 mutation p.V31A identified with a slowly progressive form of amyotrophic lateral sclerosis. Neurobiol Aging. 2014;35(1):266.e1–4.
134. Mulinari S, Häcker U. Rho-guanine nucleotide exchange factors during development: force is nothing without control. Small GTPases. 2010;1(1):28–43.
135. Gagliardi D, Bresolin N, Comi GP, Corti S. Extracellular vesicles and amyotrophic lateral sclerosis: from misfolded protein vehicles to promising clinical biomarkers. Cell Mol Life Sci. 2021;78(2):561–72.
136. Je G, Keyhanian K, Ghasemi M. Overview of stem cells therapy in amyotrophic lateral sclerosis. Neurol Res. 2021;43(8):616–32.
137. Song Y, Lin F, Ye CH, Huang H, Li X, Yao X, et al. Rare, low-frequency and common coding variants of ARHGEF28 gene and their association with sporadic amyotrophic lateral sclerosis. Neurobiol Aging. 2020;87:138.e1–6.
138. Mancuso R, Navarro X. Amyotrophic lateral sclerosis: current perspectives from basic research to the clinic. Prog Neurobiol. 2015;133:1–26.
139. Mathis S, Couratier P, Julian A, Vallat JM, Corcia P, Le Masson G. Management and therapeutic perspectives in amyotrophic lateral sclerosis. Expert Rev Neurother. 2017;17(3):263–76.
140. Pape JA, Grose JH. The effects of diet and sex in amyotrophic lateral sclerosis. Rev Neurol. 2020;176(5):301–15.
141. Angenstein F, Niessen HG, Goldschmidt J, Vielhaber S, Ludolph AC, Scheich H. Age-dependent changes in MRI of motor brain stem nuclei in a mouse model of ALS. Neuroreport. 2004;15(14):2271–4.
142. Sanchez-Tejerina D, Llaurado A, Sotoca J, Lopez-Diego V, Vidal Taboada JM, Salvado M, et al. Biofluid biomarkers in the prognosis of amyotrophic lateral sclerosis: recent developments and therapeutic applications. Cells. 2023;12(8)
143. Tard C, Defebvre L, Moreau C, Devos D, Danel-Brunaud V. Clinical features of amyotrophic lateral sclerosis and their prognostic value. Rev Neurol. 2017;173(5):263–72.
144. Wang Y, Patani R. Novel therapeutic targets for amyotrophic lateral sclerosis: ribonucleoproteins and cellular autonomy. Expert Opin Ther Targets. 2020;24(10):971–84.
145. Wang R, Yang B, Zhang D. Activation of interferon signaling pathways in spinal cord astrocytes from an ALS mouse model. Glia. 2011;59(6):946–58.
146. Blasco H, Guennoc AM, Veyrat-Durebex C, Gordon PH, Andres CR, Camu W, et al. Amyotrophic lateral sclerosis: a hormonal condition? Amyotroph Lateral Scler. 2012;13(6):585–8.
147. Cieślak M, Roszek K, Wujak M. Purinergic implication in amyotrophic lateral sclerosis-from pathological mechanisms to therapeutic perspectives. Purinergic Signal. 2019;15(1):1–15.
148. Du H, Huo Z, Chen Y, Zhao Z, Meng F, Wang X, et al. Induced pluripotent stem cells and their applications in amyotrophic lateral sclerosis. Cells. 2023;12(6)
149. Gangfuß A, Kohl Z. Amyotrophic lateral sclerosis-motor neuron disease with a wide clinical and genetic spectrum. Nervenarzt. 2023;94(6):494–500.
150. Smittkamp SE, Brown JW, Stanford JA. Time-course and characterization of orolingual motor deficits in B6SJL-Tg(SOD1-G93A)1Gur/J mice. Neuroscience. 2008;151(2):613–21.
151. He J, Mangelsdorf M, Fan D, Bartlett P, Brown MA. Amyotrophic lateral sclerosis genetic studies: from genome-wide association mapping to genome sequencing. Neuroscientist. 2015;21(6):599–615.

152. Dahiya R, Thapa R, Kumawat NK, Singh M, Jakhotiya S, Deopa D, et al. Epigenetic mechanisms in inflammation. In: Targeting epigenetics in inflammatory lung diseases. Springer; 2023. p. 43–55.
153. Thapa R, Afzal O, Afzal M, Gupta G, Bhat AA, Kazmi I, et al. From LncRNA to metastasis: the MALAT1-EMT axis in cancer progression. Pathol Res Pract. 2023;253:154959.
154. Hogden A, Crook A. Patient-centered decision making in amyotrophic lateral sclerosis: where are we? Neurodegener Dis Manag. 2017;7(6):377–86.
155. Garbuzova-Davis S, Haller E, Navarro S, Besong TE, Boccio KJ, Hailu S, et al. Transplantation of human bone marrow stem cells into symptomatic ALS mice enhances structural and functional blood-spinal cord barrier repair. Exp Neurol. 2018;310:33–47.
156. Martin Schaff C, Kurent JE, Kolodziejczak S, Milic M, Foster LA, Mehta AK. Neuroprognostication for patients with amyotrophic lateral sclerosis: an updated, evidence-based review. Semin Neurol. 2023;43(5):776–90.
157. McCombe PA, Wray NR, Henderson RD. Extra-motor abnormalities in amyotrophic lateral sclerosis: another layer of heterogeneity. Expert Rev Neurother. 2017;17(6):561–77.
158. Nguyen DKH, Thombre R, Wang J. Autophagy as a common pathway in amyotrophic lateral sclerosis. Neurosci Lett. 2019;697:34–48.
159. Riva N, Agosta F, Lunetta C, Filippi M, Quattrini A. Recent advances in amyotrophic lateral sclerosis. J Neurol. 2016;263(6):1241–54.
160. Chipika RH, Siah WF, McKenna MC, Li Hi Shing S, Hardiman O, Bede P. The presymptomatic phase of amyotrophic lateral sclerosis: are we merely scratching the surface? J Neurol. 2021;268(12):4607–29.
161. Gao FB, Almeida S, Lopez-Gonzalez R. Dysregulated molecular pathways in amyotrophic lateral sclerosis-frontotemporal dementia spectrum disorder. EMBO J. 2017;36(20):2931–50.
162. Gastl R, Ludolph AC. Amyotrophic lateral sclerosis. Nervenarzt. 2007;78(12):1449–57; quiz 58–9.
163. Harms MB, Baloh RH. Clinical neurogenetics: amyotrophic lateral sclerosis. Neurol Clin. 2013;31(4):929–50.
164. Kumar V, Islam A, Hassan MI, Ahmad F. Therapeutic progress in amyotrophic lateral sclerosis-beginning to learning. Eur J Med Chem. 2016;121:903–17.
165. Kupershmidt L, Youdim MBH. The neuroprotective activities of the novel multi-target iron-chelators in models of Alzheimer's disease, amyotrophic lateral sclerosis and aging. Cells. 2023;12(5)
166. Fukai T, Ushio-Fukai M. Superoxide dismutases: role in redox signaling, vascular function, and diseases. Antioxid Redox Signal. 2011;15(6):1583–606.
167. Roy J, Saucier D, O'Connell C, Morin PJ. Extracellular vesicles and their diagnostic potential in amyotrophic lateral sclerosis. Clin Chim Acta. 2019;497:27–34.
168. Prudencio M, Durazo A, Whitelegge JP, Borchelt DR. An examination of wild-type SOD1 in modulating the toxicity and aggregation of ALS-associated mutant SOD1. Hum Mol Genet. 2010;19(24):4774–89.
169. Martin LJ, Wong M. Skeletal muscle-restricted expression of human SOD1 in transgenic mice causes a fatal ALS-like syndrome. Front Neurol. 2020;11:592851.
170. Viñuela A, Brown AA, Buil A, Tsai PC, Davies MN, Bell JT, et al. Age-dependent changes in mean and variance of gene expression across tissues in a twin cohort. Hum Mol Genet. 2018;27(4):732–41.
171. Bhat AA, Thapa R, Goyal A, Subramaniyan V, Kumar D, Gupta S, et al. Curcumin-based nanoformulations as an emerging therapeutic strategy for inflammatory lung diseases. Future Med Chem. 2023;15:583–6.
172. Turner MR, Agosta F, Bede P, Govind V, Lulé D, Verstraete E. Neuroimaging in amyotrophic lateral sclerosis. Biomark Med. 2012;6(3):319–37.
173. Lobsiger CS, Boillée S, Cleveland DW. Toxicity from different SOD1 mutants dysregulates the complement system and the neuronal regenerative response in ALS motor neurons. Proc Natl Acad Sci USA. 2007;104(18):7319–26.

174. Wang L, Zhang L. MicroRNAs in amyotrophic lateral sclerosis: from pathogenetic involvement to diagnostic biomarker and therapeutic agent development. Neurol Sci. 2020;41(12):3569–77.
175. Tan HHG, Westeneng HJ, van der Burgh HK, van Es MA, Bakker LA, van Veenhuijzen K, et al. The distinct traits of the UNC13A polymorphism in amyotrophic lateral sclerosis. Ann Neurol. 2020;88(4):796–806.
176. Calvo A, Canosa A, Moglia C, Manera U, Grassano M, Vasta R, et al. Clinical and metabolic signature of UNC13A rs12608932 variant in amyotrophic lateral sclerosis. Neurol Genet. 2022;8(6):e200033.
177. Woolley SC, Rush BK. Considerations for clinical neuropsychological evaluation in amyotrophic lateral sclerosis. Arch Clin Neuropsychol. 2017;32(7):906–16.
178. Minassian K, Hofstoetter U, Tansey K, Mayr W. Neuromodulation of lower limb motor control in restorative neurology. Clin Neurol Neurosurg. 2012;114(5):489–97.
179. Li QX, Mok SS, Laughton KM, McLean CA, Volitakis I, Cherny RA, et al. Overexpression of Abeta is associated with acceleration of onset of motor impairment and superoxide dismutase 1 aggregation in an amyotrophic lateral sclerosis mouse model. Aging Cell. 2006;5(2):153–65.
180. Robichaud PP, Arseneault M, O'Connell C, Ouellette RJ, Morin PJ. Circulating cell-free DNA as potential diagnostic tools for amyotrophic lateral sclerosis. Neurosci Lett. 2021;750:135813.
181. Bhat AA, Moglad E, Bansal P, Kaur H, Deorari M, Thapa R, et al. Pollutants to pathogens: the role of heavy metals in modulating TGF-β signaling and lung cancer risk. Pathol Res Pract. 2024;256:155260.
182. Valadi N. Evaluation and management of amyotrophic lateral sclerosis. Prim Care. 2015;42(2):177–87.
183. Zhang Y, Liu X, Fu J, Zhang Y, Yang X, Zhang S, et al. Selective and inverse U-shaped curve alteration of the retinal nerve in amyotrophic lateral sclerosis: a potential mirror of the disease. Front Aging Neurosci. 2021;13:783431.
184. Bellingham MC. A review of the neural mechanisms of action and clinical efficiency of riluzole in treating amyotrophic lateral sclerosis: what have we learned in the last decade? CNS Neurosci Ther. 2011;17(1):4–31.
185. Nirmalananthan N, Greensmith L. Amyotrophic lateral sclerosis: recent advances and future therapies. Curr Opin Neurol. 2005;18(6):712–9.
186. Hogg MC, Halang L, Woods I, Coughlan KS, Prehn JHM. Riluzole does not improve lifespan or motor function in three ALS mouse models. Amyotroph Lateral Scler Frontotemporal Degener. 2018;19(5–6):438–45.
187. Aydemir D, Surucu S, Basak AN, Ulusu NN. Evaluation of the hematological and serum biochemistry parameters in the pre-symptomatic and symptomatic stages of ALS disease to support early diagnosis and prognosis. Cells. 2022;11(22)
188. Strickland MR, Ibanez KR, Yaroshenko M, Diaz CC, Borchelt DR, Chakrabarty P. IL-10 based immunomodulation initiated at birth extends lifespan in a familial mouse model of amyotrophic lateral sclerosis. Sci Rep. 2020;10(1):20862.
189. O'Brien RJ, Wong PC. Amyloid precursor protein processing and Alzheimer's disease. Annu Rev Neurosci. 2011;34:185–204.
190. Pampalakis G, Mitropoulos K, Xiromerisiou G, Dardiotis E, Deretzi G, Anagnostouli M, et al. New molecular diagnostic trends and biomarkers for amyotrophic lateral sclerosis. Hum Mutat. 2019;40(4):361–73.
191. Turner BJ, Li QX, Laughton KM, Masters CL, Lopes EC, Atkin JD, et al. Brain beta-amyloid accumulation in transgenic mice expressing mutant superoxide dismutase 1. Neurochem Res. 2004;29(12):2281–6.
192. Phatnani H, Maniatis T. Astrocytes in neurodegenerative disease. Cold Spring Harb Perspect Biol. 2015;7(6)

193. Thapa R, Afzal O, Bhat AA, Goyal A, Alfawaz Altamimi AS, Almalki WH, et al. New horizons in lung cancer management through ATR/CHK1 pathway modulation. Future Med Chem. 2023;15(19):1807–18.
194. Birger A, Ben-Dor I, Ottolenghi M, Turetsky T, Gil Y, Sweetat S, et al. Human iPSC-derived astrocytes from ALS patients with mutated C9ORF72 show increased oxidative stress and neurotoxicity. EBioMedicine. 2019;50:274–89.
195. Das MM, Svendsen CN. Astrocytes show reduced support of motor neurons with aging that is accelerated in a rodent model of ALS. Neurobiol Aging. 2015;36(2):1130–9.
196. Colović MB, Krstić DZ, Lazarević-Pašti TD, Bondžić AM, Vasić VM. Acetylcholinesterase inhibitors: pharmacology and toxicology. Curr Neuropharmacol. 2013;11(3):315–35.
197. Sugita S, Fleming LL, Wood C, Vaughan SK, Gomes MP, Camargo W, et al. VAChT overexpression increases acetylcholine at the synaptic cleft and accelerates aging of neuromuscular junctions. Skelet Muscle. 2016;6:31.
198. Costa J, de Carvalho M. Emerging molecular biomarker targets for amyotrophic lateral sclerosis. Clin Chim Acta. 2016;455:7–14.
199. Bhat AA, Afzal O, Agrawal N, Thapa R, Almalki WH, Kazmi I, et al. A comprehensive review on the emerging role of long non-coding RNAs in the regulation of NF-κB signaling in inflammatory lung diseases. Int J Biol Macromol. 2023;253(Pt 4):126951.
200. Brent JR, Franz CK, Coleman JM 3rd, Ajroud-Driss S. ALS: management problems. Neurol Clin. 2020;38(3):565–75.
201. Bagyinszky E, Hulme J, An SSA. Studies of genetic and proteomic risk factors of amyotrophic lateral sclerosis inspire biomarker development and gene therapy. Cells. 2023;12(15)
202. Krishnan J, Vannuvel K, Andries M, Waelkens E, Robberecht W, Van Den Bosch L. Overexpression of Hsp27 does not influence disease in the mutant SOD1(G93A) mouse model of amyotrophic lateral sclerosis. J Neurochem. 2008;106(5):2170–83.
203. Allen NJ, Lyons DA. Glia as architects of central nervous system formation and function. Science (New York, NY). 2018;362(6411):181–5.
204. de Almeida JP, Silvestre R, Pinto AC, de Carvalho M. Exercise and amyotrophic lateral sclerosis. Neurol Sci. 2012;33(1):9–15.
205. Eisen A. Amyotrophic lateral sclerosis: a 40-year personal perspective. J Clin Neurosci. 2009;16(4):505–12.
206. Hensley K, Abdel-Moaty H, Hunter J, Mhatre M, Mou S, Nguyen K, et al. Primary glia expressing the G93A-SOD1 mutation present a neuroinflammatory phenotype and provide a cellular system for studies of glial inflammation. J Neuroinflammation. 2006;3:2.
207. Kühlbrandt W. Structure and function of mitochondrial membrane protein complexes. BMC Biol. 2015;13(1):89.
208. Thapa R, Gupta G, Bhat AA, Almalki WH, Alzarea SI, Kazmi I, et al. A review of Glycogen Synthase Kinase-3 (GSK3) inhibitors for cancers therapies. Int J Biol Macromol. 2023;253(Pt 7):127375.
209. Lee JM, Tan V, Lovejoy D, Braidy N, Rowe DB, Brew BJ, et al. Involvement of quinolinic acid in the neuropathogenesis of amyotrophic lateral sclerosis. Neuropharmacology. 2017;112(Pt B):346–64.
210. Marlin E, Viu-Idocin C, Arrasate M, Aragón T. The role and therapeutic potential of the integrated stress response in amyotrophic lateral sclerosis. Int J Mol Sci. 2022;23(14)
211. McGoldrick P, Joyce PI, Fisher EM, Greensmith L. Rodent models of amyotrophic lateral sclerosis. Biochim Biophys Acta. 2013;1832(9):1421–36.
212. Kirkinezos IG, Bacman SR, Hernandez D, Oca-Cossio J, Arias LJ, Perez-Pinzon MA, et al. Cytochrome c association with the inner mitochondrial membrane is impaired in the CNS of G93A-SOD1 mice. J Neurosci Off J Soc Neurosci. 2005;25(1):164–72.

Cellular Senescence in Vascular Dementia (VaD)

10

Muhammad Afzal, Imran Kazmi, and Gaurav Gupta

Introduction

VaD, second only to Alzheimer's disease, is a leading cause of dementia, observed in 15–30% of dementia cases throughout the world [1]. It is due to cerebrovascular diseases that cause a deficiency of cerebral blood flow, which leads to mood swings, executive dysfunction, and progressive cognitive deterioration. VaD is defined by a series of vascular attacks that injure neuronal survival and network integrity. These insults are small vessel disease, infarcts, white matter lesions, and persistent ischemia [2]. Though its clinical significance is irrefutable, there are not many effective treatment methods currently available for VaD, for which new pathogenic

M. Afzal (✉)
Department of Pharmaceutical Sciences, Pharmacy Program, Batterjee Medical College, Jeddah, Saudi Arabia
e-mail: mohmmad.afzal@bmc.edu.sa

I. Kazmi
Department of Biochemistry, Faculty of Science, King Abdulaziz University, Jeddah, Saudi Arabia
e-mail: ikazmi@kau.edu.sa

G. Gupta
Centre for Research Impact & Outcome, Chitkara College of Pharmacy, Chitkara University, Rajpura, Punjab, India
e-mail: Gaurav.gupta@chitkara.edu.in

G. Gupta (ed.), *Cellular Senescence and Brain Aging*,
https://doi.org/10.1007/978-981-96-8873-9_10

pathways must be found. Cellular senescence is a hallmark of aging and age-related disorders [3]. A steady and irreversible standing still of cell division is known as cellular senescence, which is commonly triggered by genotoxic damage, replicative stress, or oncogene activation. Specific molecular markers of senescent cells included upregulation of cyclin dependent kinase inhibitors p16INK4a and p21CIP1, increased enzymatic activity of senescence-associated beta galactosidase (SA-β-gal), activation of DNA damage response, remodeling of chromatin and the pro-inflammatory senescence-associated secretory phenotype (SASP) characterized by growth factors, cytokines, chemokines, and proteases [4]. These secretions diffuse inflammation and tissue damage in the aged brain. Senescent astrocytes and microglia maintain neuroinflammation and oxidative stress in the brain, whereas senescent endothelial cells are responsible for BBB leakage and microvascular rarefaction [5]. These results prove a molecular link between cellular senescence and VaD. Finally, senescent neurons may accumulate due to metabolic stress or long-term ischemia that aggravates cognitive deficits [6]. The convergence of the brain's vascular damage mechanisms and senescence-associated pathways suggests active participation of senescent cells in the pathophysiology of vascular dysfunction and not passive spectators [7, 8]. Interestingly, the SASP variables, such as glial activation and neuronal dysfunction in VaD models, correlate with cerebrovascular lesions, including IL-6, TNF-α, and MMPs. These findings have motivated experimental pursuit of therapeutic senescence, either by modulating the secretion phenotype of senescent cells with senomorphics or by purging them using senolytics [9]. A convincing paradigm for understanding the course of the disease and finding new biomarkers and therapeutic strategies is the introduction of cellular senescence into the model of VaD research. This chapter aims to give a detailed overview of the available knowledge regarding the correlation between cellular senescence and vascular cognitive impairment to gain understanding in the experimental studies on senescence-targeted therapies to prevent VaD [10, 11] (Fig. 10.1).

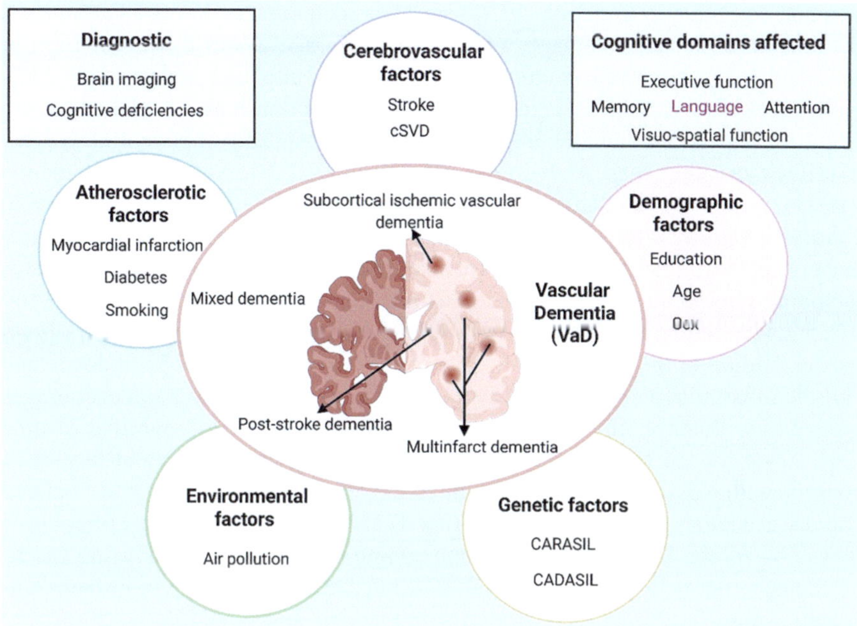

Fig. 10.1 Multifactorial risk landscape of vascular dementia. Figure showing the range of risk variables that influence the genesis of VaD. VaD arises from several risk factors that impact vascular homeostasis, including atherosclerotic, genetic, environmental, and cerebrovascular disorders. The diagnosis of VaD is based on assessing cognitive impairments in conjunction with neuroimaging to identify underlying vascular changes. Vascular dementia (VaD), cerebral autosomal recessive arteriopathy (CARASIL), cerebral autosomal dominant arteriopathy (CADASIL) with subcortical infarcts and leukoencephalopathy, and cerebral small vessel disease (cSVD)

Pathophysiological Basis of Vascular Dementia

VaD caused by complex pathophysiological pathways, such as microvascular injury and chronic cerebral hypoperfusion, will damage neuronal function and connections, resulting in a gradual decline in cognitive function [12]. Immune cells and neurotoxic chemicals can transverse the brain parenchyma via endothelial dysfunction and blood–brain barrier (BBB) disruption, accelerating neuronal injury and cognitive deficits [13, 14]. Activated microglia and increments in reactive oxygen species also contribute to oxidative stress and chronic neuroinflammation, aggravating synaptic dysfunction and neuronal death [15]. In VaD, the integration of these pathways worsens neurodegeneration and compromises brain homeostasis. Our combined results accentuate the complex nature of VaD pathophysiology and identify promising therapeutic options for disease modification and early intervention [16]. Vascular cognitive impairment (VCI) is the vascular cause-induced cognitive decline without dementia. Such an initial stage of vascular cognitive dysfunction (VCD) has been seen to precede VaD. Through the stimulation of vascular damage, VCI significantly aggravates the progression of VaD and has related risk factors that include diabetes and hypertension [17]. Agrawal et al. discussed vascular pathology and its contribution to dementia and cognitive impairment, and the importance of mixed AD and other neurodegenerative diseases other than Alzheimer's disease (AD) [18]. Age-related dementia variations are explored, particularly focusing on findings from community-based cohorts, including others such as the Rush Memory and Aging Project. New data and concepts from neuropathology are also discussed [19]. Cerebrovascular cells, including smooth muscle, pericytes, and endothelial cells, support neuronal activity; regulate cerebral blood flow; and maintain the integrity of the blood–brain barrier. Failure of these cells results in reduced perfusion, barrier breakdown, and chronic inflammation in VaD that disturbs the neurovascular unit and encourages brain tissue damage, thus leading to neurodegeneration and cognitive decline [20]. According to Iadecola et al., the vascular cognitive impairment that often coexists with Alzheimer's disease is one of the primary reasons for age-associated cognitive decline and vascular dementia [21]. They highlight the critical role played by cerebral blood arteries in maintaining brain health by analyzing vascular damage that deteriorates neuronal-glial interactions, particularly in the hemisphere white matter and searching for preventive and remedial measures that focus on midlife vascular risks [22]. The apolipoprotein E (APOE) ε4 allele mediates neuronal repair and lipid transport. It is closely related to Alzheimer's disease, but it is not clear what role it has in VaD. Investigations show that in mixed or VaD pathology, APOE ε4 may amplify vascular damage, impair the integrity of the brain barrier, and increase the risk of cognitive decline among the patients [23]. Govindpani et al. reported that vascular dysfunction plays a key role in AD and perturbs homeostasis, β-amyloid clearance, and brain metabolism. Pathological changes in permeability of the blood–brain barrier, angiogenesis, and vessel dynamics, which are associated with oxidative stress, and amyloid toxicity, create a vicious cycle of progressive neurovascular disease, with potential possibilities for therapy [24]. Neuropeptides are small signaling molecules that look like proteins and have roles in vascular control, neuroinflammation, and neuromodulation [25]. Neural adaptive processes, including regulation of neuropeptide levels from reduced neuropeptide Y (NPY) and vasoactive intestinal peptide (VIP), are important factors that lead to inflammation, blood–brain barrier compromise, and disturbed cerebral blood flow in VaD, which exacerbates the neurovascular events set in motion as VaD hallmark pathology develops [26]. Yan et al. explored emerging perspectives on neuropeptides' use as biomarkers for VaD. Neuropeptides largely control vascular functions, which are lasting chemical messages in the brain. Traditional and new neuropeptides are discussed in the essay, their diagnostic potential and molecular processes for early VaD diagnosis and therapy [27].

Cellular Senescence: Mechanisms and Markers

Oxidative stress, DNA damage, telomere attrition, and mitochondrial dysfunction, common in aging and neurodegenerative diseases, are the primary agents of cellular senescence in brain and vascular cells. Senescent cells adopt a senescence-associated secretory phenotype (SASP) that is characterized by the secretion of matrix metalloproteinases + chemokines + pro-inflammatory cytokines that (e.g., IL-6 + IL-1β) that exacerbates neuroinflammation and compromises vascular integrity and therefore brain dysfunction [28]. Increases in senescent glial, endothelial, and neuronal populations are being observed in the brain and cerebrospinal fluid (CSF), and share key biomarkers such as p16INK4a, p21CIP1, SA-β-gal activity, and SASP factors [29]. Those biomarkers can be employed as targets for diagnosing and treating disorders associated with blood vessels, including vascular dementia [30, 31]. The blood movement to the brain tissue required to provide oxygen and nutrients is called cerebral blood flow (CBF). In VaD, neuronal damage, persistent hypoperfusion, and cognitive impairment result from reduced CBF due to stroke or small vessel disease. Neurovascular connection worsened due to uncontrolled CBF causes the bulk of the pathophysiology and development of VaD [32]. Fang et al. reviewed the importance of vascular dysfunction in aging and neurodegenerative diseases such as AD and VaD, chronic vascular inflammation and vascular dysfunction aggravated by the alteration of the local cells and cerebral blood vessel morphology, which makes cognitive impairment worse [33]. These changes are also consequential for cerebral blood flow, fluid dynamics, and vascular integrity. Understanding these mechanisms might make it easier to identify clinically significant vascular biomarkers in the prevention and treatment of diseases [34].

The rearrangement of chromatin to affect gene accessibility and expression while leaving other DNA sequences intact is termed chromatin rearrangement. Bringing too much chaos with aberrant chromatin remodeling of genes involved in vascular integrity and cognitive function, vasodementia (VaD) results in vascular endothelial failure, neuroinflammation, and neuronal senescence [35]. Such epigenetic modifications may be therapeutic targets, which may intensify the development of VaD [36]. The mechanisms regulating various aspects of senescence phenotype and its functional effects were studied by Herranz et al. Their findings show how important it is to understand these mechanisms to enhance the detection and characterization of senescent cells in vivo to ultimately develop treatment plans to modify the senescence pathway [37]. Restitution of the brain's ability to heal and rearrange neuronal networks after the damage is called reinstating regenerative plasticity. Cognitive impairment in VaD is due to impaired neurogenesis and synaptic plasticity. With strategies that boost regenerative plasticity, such as enhancing neurotrophic factors, reducing inflammation or tackling senescence, neuronal loss may be resisted, and cognitive outcomes in VaD patients enhanced [38]. Based on

Kandasamy et al., the second most common cause of memory loss, VaD is caused by decreased cerebral blood flow (CBF), which is caused by conditions such as atherosclerosis and hypertension. Disrupting TGF-β signaling may lower endothelial cell death, neuroinflammation, and synaptic loss, minimizing vascular disease and cognitive decline [39].

Evidence Linking Cellular Senescence to VaD

As supported by increasing studies, cellular senescence plays an important role in the pathophysiology of VaD. Senescence of endothelial cells of cerebrovasculature causes a hypoxic and inflammatory milieu, which is detrimental to the viability of neurons as it disrupts the integrity of the brain barrier physiologically and disturbs the vascular homeostasis. Similarly, senescent glial and neuronal cells contribute to cognitive decline by induction of chronic neuroinflammation and synaptic dysfunction through SASP factors. Clinical analysis of postmortem tissue from VaD patients identifies senescence markers; these suggest pathological relevance, while preclinical studies using mouse disease models reveal the expression of p16ink4a and SA-β-gal in diseased brain regions [7, 40]. Increased heart rate, blood pressure, and vascular resistance that result from the promotion by hyperthyroidism (which is characterized by excessive production of thyroid hormones) can cause VaD and VCI. Cerebrovascular damage may be accelerated by such changes, leading to cognitive impairment. Proper management of hyperthyroidism could flatten vascular risk factors, which may not allow or can only support a milder development of VaD [41]. Sheng et al. determined nonlinear associations between TSH, FT4, and cognitive scores when examining NHANES 2011–2012 data and using Mendelian randomization.

In contrast, genetically predicted hyperthyroidism reduced the risk of dementia, while hypothyroidism or elevated TSH increased the risk of dementia, VaD, and Alzheimer's disease. In contrast, low TSH and FT4 levels correlated with better test performance [42]. Factors such as age, high blood pressure, diabetes, smoking, and cardiovascular disorders influence the risk of all-cause dementia, including VaD. These have increased the risk of cognitive decline by triggering cerebrovascular injury. Reducing brain vascular injury is of paramount importance in preventing or avoiding the onset of dementia, particularly VaD; vascular risk factor control is necessary [43]. Despite Wennberg et al.'s investigation finding no overall association between the risk of VAlD, AD, or any other cause of dementia and breast cancer survivorship, women diagnosed over the age of 65 had a remarkably high risk for all-cause dementia (SHR 1.30), AD (1.35), and VaD (1.64), which implies segregation by age as an incidental risk factor [44].

Interactions Between Senescence, Neuroinflammation, and Vascular Dysfunction

Vascular dysfunction, neuroinflammation, and senescence represent closely related phenomena significant in the pathophysiology of VaD. A body of research from recent reviews suggests that senescent endothelial, glial, and neuronal cells interact with immune cells in both directions, a body of research gaining new weight as we speak. They alter innate and adaptive immunity by releasing cytokines and chemokines that call in and engage T cells, macrophages and microglia, maintaining a pro-inflammatory environment [45]. The senescence associated secretory phenotype (SASP), which includes matrix metalloproteinases, IL-6, IL-1B and TNF-α, is the culprit for this inflammatory cascade. It reduces the function of the blood–brain barrier, causes oxidative stress, and damages vascular integrity. Progressive neurovascular degeneration is promoted by resultant endothelial dysfunction and sustained inflammation, changing cerebral perfusion and neuronal homeostasis. SASP constant signaling also establishes harmful feedback loops where oxidative destruction caused by inflammation triggers further cellular senescence, resulting in a reinforcing neurovascular destruction and cognitive impairment [46, 47]. Such interrelated processes not only explain the pathophysiological determinants of VaD, but also target important treatment areas. Hopefully, one can interdict this cycle, reduce neuroinflammation, and return vascular function by emphasizing senescent cells and their secretory features. As recent research comprehends, such an enhanced understanding of such cellular relationships will underscore the importance of senescence-immunovascular cross talk as an area where mechanistic knowledge and intervention strategies can be applied to vascular cognitive impairment [48].

Therapeutic Implications and Senescence-Targeting Strategies

In VaD, the approach to therapy that tackles cellular senescence is now getting increasingly common. There are two subsets of drugs that have already proven their effectiveness in cellular senescence elimination, or at least in decreasing the pro-inflammatory SASP profile of senescent cells, senolytics (dasatinib, quercetin), and senomorphics (rapamycin, metformin) [49]. These complementary tactics, which involve antioxidants, anti-inflammatory drugs, and vascular protective tactics such as exercise, resveratrol, and omega-3 fatty acids, reverse/relieve the oxidative stress and endothelial dysfunction associated with VaD. Senotherapeutics may also be targeted selectively to brain tissue by vehicles of delivery based on nanocarriers like liposomes and polymeric nanoparticles under development for better blood–brain barrier penetration. Together, these tactics present a multimodal strategy for decreasing the effects of senescence on neurovascular disease in VaD [50]. Key players in vascular disease on the molecular level are gelatinase (MMP-9), cyclophilin A (Cypa), and extracellular matrix metalloproteinase inducer (EMMPRIN). Gelatinase causes blood–brain barrier breakdown and neurovascular injury in VaD, but CypA and EMMPRIN promote inflammation and matrix remodeling. These interactions impair endothelial function and promote the pathogenesis of VaD through increased neuroinflammation [51].

Along with the CCH mice model, Chai et al. studied this Cypa-EMMPRIN-gelatinase pathway in 48 VaD patients and 36 cognitively unaffected individuals. According to the results, EMMPRIN was higher and CypA lower in the CCH brain and VaD serum. Gelatinase differences—MMP-2 in CCH and MMP-9 in VaD—indicate complex regulatory mechanisms and a promising role as a biomarker and a target of therapy in VaD [52]. SIRT6 is a NAD+-dependent deacetylase that is required for DNA repair and the regulation of inflammation and the cellular stress response. SIRT6 deficiency is associated with neurovascular injury, neuroinflammation, and cognitive impairment in VaD. Maintaining SIRT6 activity may help diminish these processes, suggesting its usefulness as a VaD treatment target [53]. Du et al. demonstrated that SIRT6 prevents cognitive impairment induced by chronic cerebral hypoperfusion (CCH) in a STAT5, PGAM5, and Drp1-dependent fashion. Reduced neuronal SIRT6 expression was found in the CCH utilizing a BCAS mouse model. Neuron-specific Sirt6 ablation worsened neuronal recovery, while MDL-800 therapy decreased the loss of neurons and helped them recover. In addition, a correlation was established between SIRT6 expression in monocytes and cognitive results in human patients, with potential translational importance [54]. Multipotent stem cells known as bone marrow-derived stem cells (BMSCS) possess the properties to develop into many types of cells, such as endothelial and neuronal. By promoting vascular repair, neuroinflammation, and enhancing cognition, BMSCS have shown promise in the neuroprotection and neurogenesis of VaD, giving a promising treatment approach to VaD [55]. Wang et al. demonstrated that in a VaD rat model, introduction of co-transfected BMSCs with NGF and TERT had a significant effect on cognitive function. Their results showed that BMSCs co-transfected with NGF-TERT had better treatment outcomes with improved learning and memory than the BMSCS with no NGF and TERT. This improvement may have resulted from elevated NGF, TrkA, and SYN expression in the hippocampus CA1 region, suggesting potential treatment implications for VaD [56]. Bioactive lipids, such as plasma sphingolipids, play a role in producing cell membranes and signaling pathways that are associated with inflammation and death. Endothelial dysfunction, neuroinflammation, and cognitive loss have all been associated with altered plasma sphingolipid levels (ceramides and sphingosine-1-phosphate) in VaD. These lipids might serve as clinical disease targets and biomarkers in the pathophysiology of VaD [57]. Based on Chua et al., there is a plasma sphingolipid difference between VaD and AD with d16:1 species downregulated in VaD and d18:1 species upregulated in AD. It was demonstrated that some sphingolipid combinations exhibit high sensitivity and specificity for classifying AD and VaD, which bears evidence of their candidacy as targets for treatment and biomarkers [58].

Clinical Trials and Translational Perspectives

The therapeutic potential of senescence-targeting drugs in age-related neurodegeneration, like VaD, has been revealed through new advances in translational research. In preclinical models, metformin and the senolytic combo of dasatinib and quercetin have demonstrated hope of reducing inflammation and cognitive changes associated with senescence. These medicines are now under clinical studies, such as NCT04063124 and NCT04685590, to understand their influence on cognitive outcomes and systemic senescence load in the aged populations [59]. Although there are few direct trials in VaD, newly published data suggest that senescent cell targeting is the desired agent to lessen neurovascular dysfunction. Recent reviews emphasize the translational opportunities of numerous treatments and encourage a focused research project in vascular cognitive impairment [60]. Ischemic brain damage causes a reduction in blood flow; thus, oxygen and nutrients to the brain tissues are denied. It plays a big role in VaD, leading to microinfarcts, white matter lesions, and neuronal death. These vascular insults enable the onset and the course of cognitive loss in VaD by disrupting neuronal networks crucial to cognition [61]. Duy Linh et al. reported that most of the 17 randomized controlled trials of vascular dementia and VCI were either underway or recent, with very little representation across all trial phases. Less research is done in VaD than Alzheimer's, even though pharmaceutical companies still invest less heavily [62].

Based on Duy Linh et al., most of the 17 randomized controlled studies on vascular dementia and VCI were still ongoing and recent. At the same time, there was very little representation in all trial phases. Treatment studies on VaD are fewer than those on Alzheimer's, and pharmaceutical companies still invest little in this area. Three critical European studies that deal with multidomain interventions, including FINGER, predive, and MAPT, are included in the current research on dementia prevention. Following a trend toward preventative approaches toward at-risk high-risk senior populations, the HATICE program was also developed to manage modifiable risk factors through online counselling [63]. The efforts in a global collaborative effort, the International Database on Ageing and Dementia (IDAD), collect longitudinal data on dementia, including VaD, and cognitive aging. In helping to conduct better diagnoses, preventive measures, and more focused therapies in various populations; in making cross-national comparisons, identifying risk factors, and assisting in developing biomarkers, it plays a vital role in VaD research [64]. Even if the precise relationship between neuropathology and cognition is unknown, Solomon et al. recommend further emphasis on preventing cognitive impairment. Mechanisms of multidomain therapies and enrichment relying on biomarkers and genetics are the mechanisms considered in the experiments. International initiatives such as EDPI and IDAD are created to support wide-scale interventions and produce beneficial preventative measures [64]. Randomized controlled trials (RCTS) evaluate to a large extent approaches to the diagnosis and therapy for VaD. Medications like cholinesterase inhibitors, antihypertensives, and antiplatelet medicines are effective and provide excellent evidence. RCTS have indicated that reducing vascular risk factors, specifically hypertension, could slow down cognitive decline and possibly prevent VaD [65]. Based on a study by Fillit et al., cardiovascular risk factors including diabetes, obesity, dyslipidemia, and hypertension are associated with a greater chance of dementia and cognitive impairment. According to clinical studies, controlling these variables, including with statins and antihypertensives, might limit cognitive impairment, which encourages positive cardiovascular risk prevention in dementia preventive measures [66].

Challenges and Knowledge Gaps

The most significant information vacuum in cellular senescence research for VaD is the source of difficulties. First, diagnostic restraints limit the identification of senescent cells in VaD, as the lack of reliable biomarkers prevents timely diagnosis and treatment monitoring. The complex human pathophysiology behind VaD is often not duplicated by preclinical models, thereby creating a gap between findings in the laboratory and the translation of these to clinical use of senescence-targeted therapeutics [67]. That is the second big translational barrier. Safety and long-term effects of such medicines remain a topic in ethical debate, especially in aging cohorts where the role of senescence may be bicorne, both enhancing and posthumous. Improved preclinical models, increased ability to discover biomarkers, and ethical frameworks that will guide clinical trials are all needed to address these gaps [68]. Cerebrospinal fluid (CSF) pro-inflammatory cytokine TNF-alpha (tumor necrosis factor-alpha) is elevated in VaD. Its raised levels, responsible for neuronal injury and cognitive loss, represent neuroinflammation, endothelial dysfunction, and blood–brain barrier disruption. Enhanced CSF-TNF-α reinforces the possibility of one of the strongest chronic inflammation sources as a diagnosis and treatment target, and also bears chronic inflammation involvement in VaD pathophysiology [69]. Román et al. indicate that the global diagnostic category of cognitive impairment due to vascular origin, which encompasses VaD and VCI, is vascular cognitive disorder (VCD). Old age, hypertension, diabetes mellitus, and heart disease are risk factors for VaD. Biomarker studies reported on VaD patients normal CSF-tau values, increased CSF-NFL, and higher levels of VEGF, TGF-beta, and

TSF-TNF-alpha. VaD was responsive to acetylcholinesterase inhibitors and had low levels of CSF acetylcholinesterase, suggesting cholinergic deficiency. To prevent VaD, it is important to control vascular risks and especially hypertension [70]. Cerebrovascular dysfunction is the cause of vascular mild cognitive impairment (VaMCI), which precedes VaD. It presents mild impairments in executive function and memory, evidenced via attention, that do not impact functioning. It is essential to identify and manage early to arrest the tide of the illness. VaMCI is prodrome for VaD [71]. Liu et al.'s narrative review of multimodal neuroimaging in vascular mild cognitive impairment (VaMCI) highlighted the measurement of tissue architecture, function, and vascular dysfunction using magnetic resonance imaging (MRI). The study focuses on the biomarkers for VaMCI early detection, progression monitoring, and optimum individualized treatment strategy [72]. The immunological factor, pro-inflammatory cytokine, interleukin-6 (IL-6) is associated with the etiology of VaD. Cognitive decline due to heightened IL-6 levels also contributes to endothelial dysfunction, neuroinflammation, and blood–brain barrier breakage. It correlates systemic inflammation to cerebral small vessel disease, a key pathology driving VaD, promoting vasculature damage and accelerating white matter lesions [73]. Custohero et al. conducted a meta-analysis to determine the role of inflammatory biomarkers in differentiating AD and VaD. The possibility that IL-6 has the potential qualities of a diagnostic/prognostic biomarker is illustrated by the fact that blood levels of IL-6 in VaD patients vastly exceeded blood levels of IL-6 in AD patients, and that higher levels of IL-6 were related to an increased risk of future VaD [74]. The liver-derived peptide hormone hepcidin suppresses intestinal iron absorption and macrophage iron release, regulating systemic iron homeostasis. Dysregulated hepcidin expression results in oxidative stress, neuroinflammation, and iron accumulation in the brain in VaD, aggravating vascular damage and cognitive loss. It may be therapeutic to target the hepcidin pathways in slowing down the course of VaD [75, 76]. In Alzheimer's disease (AD), blood hepcidin and Nfl did not significantly augment the AUC with p-Tau181 and amyloid composite, according to Delaby et al. However, with NFL, it significantly improved the value of AUC for VaD about NfL alone, implying that they may serve as biomarkers for diagnostic use regarding VaD [77] (Table 10.1).

Table 10.1 The primary cellular causes of VaD by means of senescence-mediated pathways are presented in this table. It describes related molecular pathways including p53/p21, NF-κB, and SASP; it points to endothelial cells, astrocytes, microglia, neurons, pericytes, and vascular smooth muscle cells. The table addresses therapeutic ideas aimed at senescence with the goal to possibly repair or modulate VaD pathology, as well as their effects on neurovascular integrity and cognitive decline which are also described on the basis of preclinical studies

Cellular senescence	Molecular pathways	Impact of vascular dementia (VaD)	Preclinical/clinical	Therapeutic insight	Reference
Senescence in vascular cells	p53/p21, p16INK4a/Rb, mTOR, NF-kB	Accelerated vascular aging, endothelial dysfunction, reduced blood flow	Preclinical studies show endothelial cell senescence in vascular dysfunction	Targeting senescence pathways (e.g., p16INK4a, p53) could improve endothelial function and reduce cardiovascular events	[78]
Senescence in glial cells	STAT3, NF-kB, IL-6, p16INK4a, TGF-β	Contributes to neuroinflammation and neuronal damage	In vivo models show increased senescence markers in glial cells post-injury	Modulating inflammation and glial senescence through NF-kB inhibitors may reduce neurodegeneration	[79, 80]
Senescence in neurons	ATM/ATR, p53, DNA damage response, MAPK	Loss of neuroplasticity, cognitive decline, neurodegeneration	Evidence from models of neurodegeneration showing neuroinflammation and synaptic dysfunction	Senescence clearance strategies (e.g., using senolytics) to reduce neurodegeneration and improve cognition	[81]
Senescence in endothelial cells	VEGF, mTOR, p21, TGF-β	Impaired blood–brain barrier integrity, hypoxia	Preclinical studies show endothelial cell senescence in vascular dementia models	Targeting endothelial cell senescence could improve BBB permeability and reduce vascular cognitive impairment	[82]
Senescence in microglia	NF-kB, p16INK4a, IL-6, TNF-α	Microglial activation, neuroinflammation leading to neuronal death	Preclinical studies in Alzheimer's and VaD show increased microglial senescence and inflammation	Modulating microglial senescence could prevent excessive neuroinflammation and protect against neuronal loss	[83]
Senescence in astrocytes	TGF-β, IL-1β, p16INK4a, NF-kB	Astrocyte dysfunction, impaired neurovascular coupling	Clinical studies indicate astrocyte senescence plays a role in VaD progression	Targeting astrocyte senescence to restore neurovascular coupling and reduce neurodegeneration	[84]

Future Directions

The future tendencies of treating VaD with a point of view of cellular senescence suggest a shift toward precise methods to enhance clinical outcomes and preventative measures. Based on patient-specific senescence signatures, genetic risk profiles, and epigenetic markers, personalized medicine may be able to stratify patients as to their vulnerability to senescence-driven neurovascular disease and personalize their optimal therapy [85]. Some recent studies on neurovascular ageing and senescence have revealed that incorporating senescence profiling into VaD management via expansion of biomarker screening (p16INK4a, SASP components, or single-cell transcriptomics) would augment the potential for early diagnosis, the ability to monitor the course of the diseases, and influence the choice of therapy [86]. In addition, in at-risk aging populations, frailty may be postponed or avoided by proactive efforts at cognitive resilience, including lowering vascular risk, physical exercise, dietary modification, and early pharmacologic intervention with senotherapeutics. These varied strategies focus on the necessity for long-term research and clinical work with confirmation of senescence as a treatment target and expansion of our understanding of cell-specific and time-specific functions in cerebrovascular dysfunction. Combined, these future initiatives represent an attendance shift toward personalized, biologically driven forms of treatment aimed at preventing VaD through the cessation of neurovascular deterioration that occurs with aging [87].

Conclusion

Vascularized dysfunction, chronic neuroinflammation, oxidative stress, and blood–brain barrier collapse are all characteristics of VaD, and cellular senescence is a primary and complex feature in the pathophysiology of VaD. All these factors combined lead to cognitive decline. More comprehensive, while mechanistic knowledge of senescence in different brain cell types (endothelial cells, astrocytes, microglia, and neurons) is urgently needed due to the increasing prevalence of VaD in aging populations. For future studies to map out the senescence markers of the disease that are unique to cells and areas to understand the complexities underlying VaD's pathobiology and determine optimal locations to intervene, certain things must be done. Further, there is excellent promise for developing senescence-associated biomarkers that will help in early means detection, patient stratification, and monitoring therapy outcome from biomarkers of circulating SASP components, senescence gene expression, or new imaging with the potential to become objective biomarkers. While such barriers as low blood–brain barrier permeability, potential off-target effects, and requirements to devise brain-specific drug-delivery methods persist, senotherapeutic strategies, including, for example, senolytics and senomorphics, could create novel avenues for intervention on the translational ground. Such obstacles could be overcome, and target specificity could be enhanced by improving the delivery procedure by nanoparticles and using exosome-mediated transport. Also, the range of single-cell transcriptome analysis, patient-derived induced pluripotent stem cell models, and well-defined longitudinal cohorts will be of critical importance for identifying senescence-associated targets and adjusting treatment algorithms. As the research evolves, converting senescence biology into beneficial therapies will require a cross-disciplinary approach from fundamental science, bioengineering, and clinical neurology. Lastly, the approach allows the utilization of precision medicine strategies which will enable delayed onset, slow progression, and protected cognitive function, and thus can be transformative for the implementation of changes in the course of VaD, and offer hope for alleviating the worldwide burden of vascular cognitive impairment.

References

1. Kaur P, Khan H, Grewal AK, Dua K, Singh SK, Gupta G, et al. Exploring therapeutic strategies: the relationship between metabolic disorders and FOXO signalling in Alzheimer's disease. CNS Neurol Disord Drug Targets. 2025;24(3):196–207.
2. Bashir B, Gulati M, Vishwas S, Gupta G, Dhanasekaran M, Paudel KR, et al. Bridging gap in the treatment of Alzheimer's disease via postbiotics: current practices and future prospects. Ageing Res Rev. 2025;105:102689.
3. Bir SC, Khan MW, Javalkar V, Toledo EG, Kelley RE. Emerging concepts in vascular dementia: a review. J Stroke Cerebrovasc Dis. 2021;30(8):105864.
4. Onorati A, Havas AP, Lin B, Rajagopal J, Sen P, Adams PD, et al. Upregulation of PD-L1 in senescence and aging. Mol Cell Biol. 2022;42(10):e0017122.
5. Vishwas S, Bashir B, Birla D, Khandale N, Chaitanya M, Chellappan DK, et al. Neuroprotective role of phytoconstituents-based nanoemulsion for the treatment of Alzheimer's disease. Curr Top Med Chem. 2024;24(19):1683–98.
6. Thapa R, Ahmad Bhat A, Shahwan M, Ali H, PadmaPriya G, Bansal P, et al. Proteostasis disruption and senescence in Alzheimer's disease pathways to neurodegeneration. Brain Res. 2024;1845:149202.
7. Knopp RC, Erickson MA, Rhea EM, Reed MJ, Banks WA. Cellular senescence and the blood-brain barrier: implications for aging and age-related diseases. Exp Biol Med (Maywood). 2023;248(5):399–411.
8. Che J, Sun Y, Deng Y, Zhang J. Blood-brain barrier disruption: a culprit of cognitive decline? Fluids Barriers CNS. 2024;21(1):63.
9. Kisku A, Nishad A, Agrawal S, Paliwal R, Datusalia AK, Gupta G, et al. Recent developments in intranasal drug delivery of nanomedicines for the treatment of neuropsychiatric disorders. Front Med (Lausanne). 2024;11:1463976.
10. Ouvrier B, Ismael S, Bix GJ. Senescence and SASP are potential therapeutic targets for ischemic stroke. Pharmaceuticals (Basel, Switzerland). 2024;17(3)
11. Carreno G, Guiho R, Martinez-Barbera JP. Cell senescence in neuropathology: a focus on neurodegeneration and tumours. Neuropathol Appl Neurobiol. 2021;47(3):359–78.
12. Chakraborty S, Vishwas S, Harish V, Gupta G, Paudel KR, Dhanasekaran M, et al. Exploring nanoparticular platform in delivery of repurposed drug for Alzheimer's disease: current approaches and future perspectives. Expert Opin Drug Deliv. 2024;21(12):1771–92.
13. Venkat P, Chopp M, Chen J. Models and mechanisms of vascular dementia. Exp Neurol. 2015;272:97–108.
14. Altahrawi AY, James AW, Shah ZA. The role of oxidative stress and inflammation in the pathogenesis and treatment of vascular dementia. Cells. 2025;14(8)
15. Bhat AA, Moglad E, Afzal M, Thapa R, Almalki WH, Kazmi I, et al. Therapeutic approaches targeting aging and cellular senescence in Huntington's disease. CNS Neurosci Ther. 2024;30(10):e70053.

16. Teleanu DM, Niculescu AG, Lungu II, Radu CI, Vladâcenco O, Roza E, et al. An overview of oxidative stress, neuroinflammation, and neurodegenerative diseases. Int J Mol Sci. 2022;23(11)
17. Rundek T, Tolea M, Ariko T, Fagerli EA, Camargo CJ. Vascular cognitive impairment (VCI). Neurotherapeutics. 2022;19(1):68–88.
18. Ashique S, Sirohi E, Kumar S, Rihan M, Mishra N, Bhatt S, et al. Aducanumab in Alzheimer's disease: a critical update. Curr Med Chem. 2024;31(31):5004–26.
19. Agrawal S, Schneider JA. Vascular pathology and pathogenesis of cognitive impairment and dementia in older adults. Cereb Circ Cogn Behav. 2022;3:100148.
20. Procter TV, Williams A, Montagne A. Interplay between brain pericytes and endothelial cells in dementia. Am J Pathol. 2021;191(11):1917–31.
21. Acharya M, Singh N, Gupta G, Tambuwala MM, Aljabali AAA, Chellappan DK, et al. Vitamin D, Calbindin, and calcium signaling: unraveling the Alzheimer's connection. Cell Signal. 2024;116:111043.
22. Iadecola C. The pathobiology of vascular dementia. Neuron. 2013;80(4):844–66.
23. Husain MA, Laurent B, Plourde M. APOE and Alzheimer's disease: from lipid transport to physiopathology and therapeutics. Front Neurosci. 2021;15:630502.
24. Govindpani K, McNamara LG, Smith NR, Vinnakota C, Waldvogel HJ, Faull RL, et al. Vascular dysfunction in Alzheimer's disease: a prelude to the pathological process or a consequence of it? J Clin Med. 2019;8(5)
25. Bhat AA, Gupta G, Afzal O, Kazmi I, Al-Abbasi FA, Alfawaz Altamimi AS, et al. Neuropharmacological effect of risperidone: from chemistry to medicine. Chem Biol Interact. 2023;369:110296.
26. Carniglia L, Ramírez D, Durand D, Saba J, Turati J, Caruso C, et al. Neuropeptides and microglial activation in inflammation, pain, and neurodegenerative diseases. Mediat Inflamm. 2017;2017:5048616.
27. Yan X, Ma Y, Yan L, Li S, Xu Y. Neuropeptides as potential biomarkers in vascular dementia. J Integr Neurosci. 2024;23(3):66.
28. Alharbi KS, Javed Shaikh MA, Imam SS, Alshehri S, Ghoneim MM, Almalki WH, et al. Role of flavonoids in management of various biological targets in Alzheimer's disease: evidence from preclinical to clinical studies. Curr Med Chem. 2023;30(18):2061–74.
29. Pandey SN, Singh G, Semwal BC, Gupta G, Alharbi KS, Almalki WH, et al. Therapeutic approaches of nutraceuticals in the prevention of Alzheimer's disease. J Food Biochem. 2022;46(12):e14426.
30. Barnes RP, Fouquerel E, Opresko PL. The impact of oxidative DNA damage and stress on telomere homeostasis. Mech Ageing Dev. 2019;177:37–45.
31. Qin Y, Liu H, Wu H. Cellular senescence in health, disease, and lens aging. Pharmaceuticals (Basel, Switzerland). 2025;18(2)
32. Rajeev V, Fann DY, Dinh QN, Kim HA, De Silva TM, Lai MKP, et al. Pathophysiology of blood brain barrier dysfunction during chronic cerebral hypoperfusion in vascular cognitive impairment. Theranostics. 2022;12(4):1639–58.
33. Bhardwaj S, Kesari KK, Rachamalla M, Mani S, Ashraf GM, Jha SK, et al. CRISPR/Cas9 gene editing: new hope for Alzheimer's disease therapeutics. J Adv Res. 2022;40:207–21.
34. Fang YC, Hsieh YC, Hu CJ, Tu YK. Endothelial dysfunction in neurodegenerative diseases. Int J Mol Sci. 2023;24(3)
35. Pathak S, Gupta G, Gilhotra RM. The role of diazepam in epigenetics: from the molecular level to clinical implications. Adv Mind Body Med. 2021;35(1):25–33.
36. Sharma H, Koirala S, Chew YL, Konopka A. DNA damage and chromatin rearrangement work together to promote neurodegeneration. Mol Neurobiol. 2025;62(1):1282–90.
37. Herranz N, Gil J. Mechanisms and functions of cellular senescence. J Clin Invest. 2018;128(4):1238–46.
38. Gao L, Zhang Y, Sterling K, Song W. Brain-derived neurotrophic factor in Alzheimer's disease and its pharmaceutical potential. Transl Neurodegener. 2022;11(1):4.

39. Kandasamy M, Anusuyadevi M, Aigner KM, Unger MS, Kniewallner KM, de Sousa DMB, et al. TGF-β signaling: a therapeutic target to reinstate regenerative plasticity in vascular dementia? Aging Dis. 2020;11(4):828–50.
40. Graves SI, Baker DJ. Implicating endothelial cell senescence to dysfunction in the ageing and diseased brain. Basic Clin Pharmacol Toxicol. 2020;127(2):102–10.
41. Ertek S, Cicero AF. Hyperthyroidism and cardiovascular complications: a narrative review on the basis of pathophysiology. Arch Med Sci. 2013;9(5):944–52.
42. Sheng X, Gao J, Chen K, Zhu X, Wang Y. Hyperthyroidism, hypothyroidism, thyroid stimulating hormone, and dementia risk: results from the NHANES 2011-2012 and Mendelian randomization analysis. Front Aging Neurosci. 2024;16:1456525.
43. Gorelick PB, Scuteri A, Black SE, Decarli C, Greenberg SM, Iadecola C, et al. Vascular contributions to cognitive impairment and dementia: a statement for healthcare professionals from the american heart association/american stroke association. Stroke. 2011;42(9):2672–713.
44. Wennberg A, Ding M, Feychting M, Modig K. Risk of all-cause dementia, Alzheimer disease, and vascular dementia in breast cancer survivors: a longitudinal register-based study. Neurol Clin Pract. 2023;13(4):e200173.
45. Tian Z, Ji X, Liu J. Neuroinflammation in vascular cognitive impairment and dementia: current evidence, advances, and prospects. Int J Mol Sci. 2022;23(11)
46. Yue Z, Nie L, Zhao P, Ji N, Liao G, Wang Q. Senescence-associated secretory phenotype and its impact on oral immune homeostasis. Front Immunol. 2022;13:1019313.
47. Coppé JP, Desprez PY, Krtolica A, Campisi J. The senescence-associated secretory phenotype: the dark side of tumor suppression. Annu Rev Pathol. 2010;5:99–118.
48. Valencia I, Vallejo S, Dongil P, Romero A, San Hipólito-Luengo Á, Shamoon L, et al. DPP4 promotes human endothelial cell senescence and dysfunction via the PAR2-COX-2-TP axis and NLRP3 inflammasome activation. Hypertension. 2022;79(7):1361–73.
49. Behfar Q, Ramirez Zuniga A, Martino-Adami PV. Aging, senescence, and dementia. J Prev Alzheimers Dis. 2022;9(3):523–31.
50. Kakoti BB, Hernandez-Ontiveros DG, Kataki MS, Shah K, Pathak Y, Panguluri SK. Resveratrol and omega-3 fatty acid: its implications in cardiovascular diseases. Front Cardiovasc Med. 2015;2:38.
51. Halliday MR, Pomara N, Sagare AP, Mack WJ, Frangione B, Zlokovic BV. Relationship between cyclophilin a levels and matrix metalloproteinase 9 activity in cerebrospinal fluid of cognitively normal apolipoprotein e4 carriers and blood-brain barrier breakdown. JAMA Neurol. 2013;70(9):1198–200.
52. Chai YL, Rajeev V, Poh L, Selvaraji S, Hilal S, Chen CP, et al. Chronic cerebral hypoperfusion alters the CypA-EMMPRIN-gelatinase pathway: implications for vascular dementia. J Cereb Blood Flow Metab. 2023;43(5):722–35.
53. Guo Z, Li P, Ge J, Li H. SIRT6 in aging, metabolism, inflammation and cardiovascular diseases. Aging Dis. 2022;13(6):1787–822.
54. Du Y, He J, Xu Y, Wu X, Cheng H, Yu J, et al. SIRT6 prevent chronic cerebral hypoperfusion induced cognitive impairment by remodeling mitochondrial dynamics in a STAT5-PGAM5-Drp1 dependent manner. J Transl Med. 2024;22(1):788.
55. Zeng CW. Multipotent mesenchymal stem cell-based therapies for spinal cord injury: current progress and future prospects. Biology. 2023;12(5)
56. Wang F, Chang G, Geng X. NGF and TERT co-transfected BMSCs improve the restoration of cognitive impairment in vascular dementia rats. PLoS One. 2014;9(6):e98774.
57. Arsenault EJ, McGill CM, Barth BM. Sphingolipids as regulators of neuro-inflammation and NADPH oxidase 2. NeuroMolecular Med. 2021;23(1):25–46.
58. Chua XY, Torta F, Chong JR, Venketasubramanian N, Hilal S, Wenk MR, et al. Lipidomics profiling reveals distinct patterns of plasma sphingolipid alterations in Alzheimer's disease and vascular dementia. Alzheimers Res Ther. 2023;15(1):214.
59. Zhu M, Meng P, Ling X, Zhou L. Advancements in therapeutic drugs targeting of senescence. Ther Adv Chronic Dis. 2020;11:2040622320964125.

60. Csik B, Nyúl-Tóth Á, Gulej R, Patai R, Kiss T, Delfavero J, et al. Senescent endothelial cells in cerebral microcirculation are key drivers of age-related blood-brain barrier disruption, microvascular rarefaction, and neurovascular coupling impairment in mice. Aging Cell. 2025:e70048.
61. Kalaria RN, Akinyemi R, Ihara M. Stroke injury, cognitive impairment and vascular dementia. Biochim Biophys Acta. 2016;1862(5):915–25.
62. Linh TTD, Hsieh YC, Huang LK, Hu CJ. Clinical trials of new drugs for vascular cognitive impairment and vascular dementia. Int J Mol Sci. 2022;23(19)
63. Imtiaz B, Tolppanen A-M, Kivipelto M, Soininen H. Future directions in Alzheimer's disease from risk factors to prevention. Biochem Pharmacol. 2014;88(4):661–70.
64. Solomon A, Mangialasche F, Richard E, Andrieu S, Bennett DA, Breteler M, et al. Advances in the prevention of Alzheimer's disease and dementia. J Intern Med. 2014;275(3):229–50.
65. Battle CE, Abdul-Rahim AH, Shenkin SD, Hewitt J, Quinn TJ. Cholinesterase inhibitors for vascular dementia and other vascular cognitive impairments: a network meta-analysis. Cochrane Database Syst Rev. 2021;2(2):Cd013306.
66. Fillit H, Nash DT, Rundek T, Zuckerman A. Cardiovascular risk factors and dementia. Am J Geriatr Pharmacother. 2008;6(2):100–18.
67. Baker DJ, Petersen RC. Cellular senescence in brain aging and neurodegenerative diseases: evidence and perspectives. J Clin Invest. 2018;128(4):1208–16.
68. Cummings SR, Lui LY, Aversa Z, Mau T, Fielding RA, Atkinson EJ, et al. Correction to: biomarkers of cellular senescence and major health outcomes in older adults. GeroScience. 2025; https://doi.org/10.1007/s11357-025-01619-4.
69. Hu Y, Zhang F, Ikonomovic M, Yang T. The role of NRF2 in cerebrovascular protection: implications for vascular cognitive impairment and dementia (VCID). Int J Mol Sci. 2024;25(7)
70. Román GC. Vascular dementia. Advances in nosology, diagnosis, treatment and prevention. Panminerva Med. 2004;46(4):207–15.
71. Iadecola C, Duering M, Hachinski V, Joutel A, Pendlebury ST, Schneider JA, et al. Vascular cognitive impairment and dementia: JACC scientific expert panel. J Am Coll Cardiol. 2019;73(25):3326–44.
72. Liu Q, Zhang X. Multimodality neuroimaging in vascular mild cognitive impairment: a narrative review of current evidence. Front Aging Neurosci. 2023;15:1073039.
73. Xia Y, Xu Z, Zhang Y, Jiang D, Zhu Y, Liang X, et al. Circulating cytokines and vascular dementia: a bi-directional Mendelian randomization study. Exp Gerontol. 2024;189:112394.
74. Custodero C, Ciavarella A, Panza F, Gnocchi D, Lenato GM, Lee J, et al. Role of inflammatory markers in the diagnosis of vascular contributions to cognitive impairment and dementia: a systematic review and meta-analysis. GeroScience. 2022;44(3):1373–92.
75. Pal A, Cerchiaro G, Rani I, Ventriglia M, Rongioletti M, Longobardi A, et al. Iron in Alzheimer's disease: from physiology to disease disabilities. Biomol Ther. 2022;12(9)
76. Raha AA, Ghaffari SD, Henderson J, Chakraborty S, Allinson K, Friedland RP, et al. Hepcidin increases cytokines in Alzheimer's disease and down's syndrome dementia: implication of impaired iron homeostasis in neuroinflammation. Front Aging Neurosci. 2021;13:653591.
77. Delaby C, Alcolea D, Busto G, Gabelle A, Ayrignac X, Bennys K, et al. Plasma Hepcidin as a potential informative biomarker of Alzheimer disease and vascular dementia. Alzheimers Res Ther. 2025;17(1):42.
78. Sun X, Feinberg MW. Vascular endothelial senescence: pathobiological insights, emerging long noncoding RNA targets, challenges and therapeutic opportunities. Front Physiol. 2021;12:693067.
79. Maupin EA, Adams KL. Cellular senescence in glial cells: implications for multiple sclerosis. J Neurochem. 2025;169(1):e16301.
80. Alshaebi F, Sciortino A, Kayed R. The role of glial cell senescence in Alzheimer's disease. J Neurochem. 2025;169(3):e70051.
81. Shreeya T, Ansari MS, Kumar P, Saifi M, Shati AA, Alfaifi MY, et al. Senescence: a DNA damage response and its role in aging and neurodegenerative diseases. Front Aging. 2023;4:1292053.

82. Augustin HG, Koh GY. A systems view of the vascular endothelium in health and disease. Cell. 2024;187(18):4833–58.
83. Malvaso A, Gatti A, Negro G, Calatozzolo C, Medici V, Poloni TE. Microglial senescence and activation in healthy aging and Alzheimer's disease: systematic review and neuropathological scoring. Cells. 2023;12(24)
84. Saggu R, Schumacher T, Gerich F, Rakers C, Tai K, Delekate A, et al. Astroglial NF-kB contributes to white matter damage and cognitive impairment in a mouse model of vascular dementia. Acta Neuropathol Commun. 2016;4(1):76.
85. Anstey KJ, Peters R, Zheng L, Barnes DE, Brayne C, Brodaty H, et al. Future directions for dementia risk reduction and prevention research: an international research network on dementia prevention consensus. J Alzheimers Dis. 2020;78(1):3–12.
86. Muthamil S, Kim HY, Jang HJ, Lyu JH, Shin UC, Go Y, et al. Biomarkers of cellular senescence and aging: current state-of-the-art, challenges and future perspectives. Adv Biol. 2024;8(9):e2400079.
87. Angulo J, El Assar M, Álvarez-Bustos A, Rodríguez-Mañas L. Physical activity and exercise: strategies to manage frailty. Redox Biol. 2020;35:101513.

Cellular Senescence in Multiple Sclerosis (MS)

11

Rajiv Dahiya, Sunita Dahiya, Raihan Siddique, Riya Thapa, Asif Ahmad Bhat, Md Sadique Hussain, Thakur Gurjeet Singh, Sachin Kumar Singh, Saurabh Gupta, Pran Kishore Deb, and Moayad Al Shahwan

Introduction

The central nervous system shows three main characteristics during multiple sclerosis (a chronic immune-mediated neurological illness), which include inflammation alongside demyelination and neurodegenerative processes. The various clinical forms of MS include two types of progressive disease, PPMS and SPMS, and the two kinds of relapsing-remitting disease: RRMS and progressive-relapsing MS [1]. The disease begins when immunoreactive T and B lymphocytes enter the blood–brain barrier and then attack oligodendrocytes along with myelin sheaths while causing damage to the axonal connections and generating neurological disabilities. Neuroinflammation and immune dysregulation are the main factors affecting illness

R. Dahiya (✉)
School of Pharmacy, Faculty of Medical Sciences, The University of the West Indies, St. Augustine, Trinidad and Tobago, West Indies

S. Dahiya
Department of Pharmaceutical Sciences, School of Pharmacy, University of Puerto Rico, Medical Sciences Campus, San Juan, PR, USA

R. Siddique · R. Thapa
School of Pharmacy, Suresh Gyan Vihar University, Jaipur, India

A. A. Bhat
BM College of Pharmacy, Farrukhnagar, Gurugram, Haryana, India

M. S. Hussain
Uttaranchal Institute of Pharmaceutical Sciences, Uttaranchal University, Dehradun, India

T. G. Singh
Chitkara College of Pharmacy, Chitkara University, Rajpura, Punjab, India
e-mail: gurjeet.singh@chitkara.edu.in

G. Gupta (ed.), *Cellular Senescence and Brain Aging*,
https://doi.org/10.1007/978-981-96-8873-9_11

development through elevated pro-inflammatory cytokines, while activated microglia and impaired remyelination play crucial roles. Death of cells through premature aging combined with cellular breakdown occurs because of continuous inflammatory responses [2, 3]. The development of new treatments for MS requires knowledge about the multiple pathways, including cellular senescence, which causes permanent neural damage and disability. The worldwide presence of multiple sclerosis involves 2.9 million patients, while regional and demographic distributions lead to significant variations in both incidence and prevalence numbers. The total cases of multiple sclerosis across Asia and sub-Saharan Africa fall lower than North American and European numbers, which exceed 100 cases per 100,000 people [4]. The female population predominates over males (2–3:1), and multiple sclerosis impacts individuals generally between 20 and 40 years of age. Advancements in both knowledge and diagnostic tools about multiple sclerosis have resulted in elevated worldwide cases during the past few decades. The rising disease numbers within regions where MS was formerly regarded as low-risk suggest that lifestyle factors and environmental aspects fundamentally modify how the diseases spread. Young individuals experience substantial neurological impairment because of MS, which leads to major healthcare expenses and significant economic costs [5, 6].

The search for new disease-causing pathways related to cellular senescence illness development and treatment resistance is vital due to increasing disease prevalence and the lack of curative treatments. Cellular senescence represents a durable cell cycle stop mechanism that develops from oxidative stress, DNA damage, and telomere shortening and oncogene activation. During cellular senescence, cells express $p16^{INK4a}$ and $p21^{CIP1}$ while producing a pro-inflammatory phenotype known as SASP and maintaining metabolic functions [7]. The beneficial impact of senescence during tumor suppression and wound healing eventually transforms into age-related pathologies together with neurodegenerative diseases through prolonged accumulation. SASP includes multiple secreted components, such as cytokines,

S. K. Singh
School of Pharmaceutical Sciences, Lovely Professional University, Phagwara, Punjab, India
e-mail: sachin.16030@lpu.co.in

S. Gupta
Department of Pharmacology, Chameli Devi Institute of Pharmacy,
Indore, Madhya Pradesh, India

P. K. Deb
Department of Pharmaceutical Sciences and Technology, Birla Institute of Technology,
Mesra, Ranchi, India

M. Al Shahwan
Centre of Medical and Bio-allied Health Sciences Research, Ajman University,
Ajman, United Arab Emirates

Department of Clinical Sciences, College of Pharmacy and Health Sciences,
Ajman University, Ajman, United Arab Emirates
e-mail: m.shahwan@ajman.ac.ae

chemokines, growth factors, and proteases, which distort tissue conditions while sustaining long-term inflammation and blocking healing processes. The aging state of glial cells elevates their potential to damage neuroinflammatory responses and neuron destruction within the CNS. Drug development for targeting senescent cells demands knowledge about their two-sided aging process and information about their molecular structures [8, 9]. A major, although understudied, framework for inflammation, demyelination, and neurodegeneration progression in MS could be cellular senescence. Multiple sclerosis has an etiological basis linked to cellular senescence structures, according to recently accumulating evidence. Multiple sclerosis displays its two primary manifestations through chronic inflammation and oxidative stress, strongly promoting cell aging in central nervous system-dwelling astrocytes, microglia, oligodendrocyte precursor cells, and endothelial cells. The presence of aged cells in MS lesions deteriorates the blood–brain barrier and delays oligodendrocyte cell development while sustaining continual neuroinflammation through SASP [4, 10]. Research supports the pathogenic significance of $p16^{INK4a}$ and $p21^{CIP1}$ by investigating these proteins' senescence-related pathways in MS brain tissue. Since senescence differs from apoptosis by keeping damaged cells, it may lead to extended impaired central nervous system functioning. An examination of senescence provides fresh insight into MS disease progression because it happens during advanced illness stages that show compartmentalized inflammation. The shift in understanding about senescence creates new opportunities for therapeutic development, enabling better treatment of unmet MS clinical needs and advanced immunomodulatory approaches [11].

Cellular Senescence: Mechanisms and Markers

Hallmarks of Cellular Senescence

Stable arrival at cell cycle stop is the fundamental characteristic of cellular senescence that develops from several triggering agents, including oxidative damage, telomere degradation, oncogenic signals, or chronic inflammation. Survival pathways become modified in senescent cells, causing them to cease cell growth yet continue their metabolic functions while resisting programmed cell death. They show defective mitochondria function, build up persistent DNA damage foci, and generate higher reactive oxygen species (ROS) [12]. The main distinguishing sign of senescent cells manifests through their pro-inflammatory release of cytokines and chemokines with proteases and growth factors that create tissue alterations known as the SASP. The combination of these features relates to glial dysfunction and sustained neuroinflammation, as well as deficient remyelination that occurs in multiple sclerosis patients. The investigation of senescent cells in MS pathogenesis needs complete comprehension of their distinctive qualities to develop strategies for selective cell elimination or modification [13].

Molecular Pathways ($p16^{INK4a}$, p21, p53, SASP)

The cellular pathways that lead to senescence become regulated mainly by tumor suppressors, including $p16^{INK4a}$, $p21^{CIP1/WAF1}$, and p53. Cell cycle arrest occurs because p53-p21 inhibits cyclin-dependent kinases (CDKs) in response to DNA damage and oxidative stress. Rb-based control of the E2F target gene transcription becomes possible when $p16^{INK4a}$-Rb complexes create senescent conditions. Multiple pathways increase their expression levels in glial cells within multiple sclerosis lesions. SASP appears to be controlled through transcriptional and epigenetic regulation, which mainly depends on NF-κB and C/EBPβ factors. The persistent release of SASP factors causes remyelination limitations by increasing neuroinflammation and leading to OPC developmental challenges and immune cell inflammatory responses. The control of MS illness progression and reduced CNS damage becomes possible when researchers concentrate on senescence-associated molecular pathways [14, 15].

Senescence Biomarkers: Tools for Detection in MS

Creating targeted treatments for MS requires precise detection methods to establish the potential harm of senescent cells. Research usually detects CNS tissue senescent phenotypes through biomarkers such as $p16^{INK4a}$, p21, and senescence-associated β-galactosidase (SA-β-gal). SASP components expressed at higher levels through proteins such as IL-6, IL-8, and MMPs serve as diagnostic indicators for senescence activity, and the presence of γH2AX foci represents DNA damage continuance [4]. Several research techniques, including flow cytometry analysis in blood cells, single-cell sequencing analysis, and immunohistochemistry-based tissue labeling for postmortem brains, have supported the study of MS age-related changes. Recent advances in medical imaging offer non-invasive ways to monitor senescent markers through PET tracers aimed at detecting such markers during in vivo imaging studies. The senescence indicators are monitoring tools for assessing therapeutic response to senescence-specific treatments while advancing pathogenesis research on multiple sclerosis [16].

Pathophysiological Role of Senescent Cells in MS

Senescence in CNS Cell Types

Cells within the CNS, including astrocytes, microglia, oligodendrocytes, and endothelial cells, primarily cause multiple sclerosis through senescence processes. The cells of astrocytes show worsened neuroinflammation following senescence because this state breaks down both neuroprotective and cellular repair mechanisms. Chronic inflammation in MS lesions evolves from degenerated microglia cells that generate reactive oxygen species (ROS) through the local release of ROS and pro-inflammatory cytokines. The ability of oligodendrocytes to repair damaged axons diminishes with their senescence development. Endothelial cells that have become senescent function to degrade BBB components, thereby enabling greater entry of immune cells into the brain. Multiple sclerosis pathogenesis intensifies because aging cells of these groups disturb cellular equilibrium while boosting inflammatory responses and limiting myelin recovery and nerve function repair [17, 18].

Senescence-Induced Inflammation and Neurodegeneration

The cells that have become senescent play an essential part in developing neurodegeneration and inflammatory processes in multiple sclerosis. Through SASP mechanisms, senescent cells produce pro-inflammatory cytokines, chemokines, and matrix metalloproteinases that create persistent neuroinflammation. Tissue destruction becomes self-propelling as glial and immune cells get activated by inflammatory conditions present in the surrounding area. The ROS produced by senescent cells intensifies oxidative stress and promotes harmful effects on neurons. The health of the central nervous system suffers from inflammatory conditions because senescent cells continue to survive inside this region, leading to worsening neurodegeneration, the loss of neurons and damage to axons, representing multiple sclerosis pathologies [9, 19].

Contribution to Blood–Brain Barrier Disruption and Demyelination

The breakdown of the blood–brain barrier (BBB) and the development of demyelination requires senescent cells in multiple sclerosis. Endothelial cell dysfunction and BBB disruption occur from elevated levels of inflammatory mediators and matrix metalloproteinase synthesized by aged endothelial cells, thus allowing immune cells to gain entry into the central nervous system. The disruption helps demyelination occur through its inflammatory effect. Myelin repair becomes hindered by ageing cells from both oligodendrocytes and astrocytes, while these cells simultaneously increase the loss of oligodendrocyte progenitor cells and restrict myelin repair mechanisms. The chronic nature of multiple sclerosis develops because poor myelin tissue repair and disruption of the blood–brain barrier end in sustained central nervous system damage [20, 21] (Fig. 11.1).

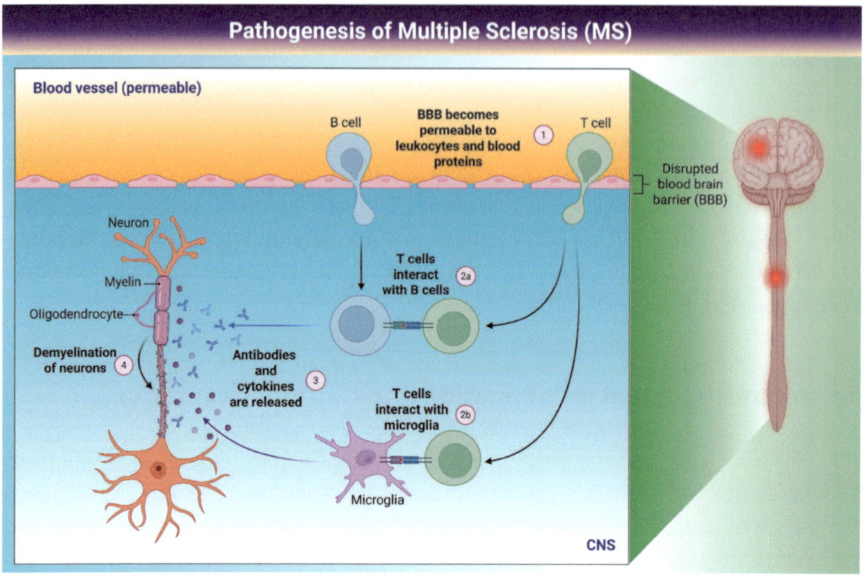

Fig. 11.1 MS pathogenesis: BBB disruption to demyelination. A figure depicts the multiple sclerosis (MS) pathologic processes by showing how blood–brain barrier (BBB) degradation permits leukocyte entry into brain tissue. After T cells initiate contact with B cells and microglia they start immunological responses. Oligodendrocytes experience damage when cells interact with one another because it leads to pro-inflammatory cytokines and antibody production. The complete deterioration of myelin sheaths results in neuronal demyelination. A depiction shows how the CNS immune responses result in MS-related brain damage across multiple steps

Preclinical Evidence of Cellular Senescence in MS Models

Experimental Autoimmune Encephalomyelitis (EAE) and Senescence

Experimental autoimmune encephalomyelitis (EAE) is one of the most common animal models used to explore MS. EAE imitates important aspects of multiple sclerosis, such as demyelination and neuroinflammation. Recent studies examine cellular senescence within EAE because it triggers the escalation of senescent cells throughout the central nervous system as the condition advances. SASP drives inflammatory cytokine production, which worsens tissue degeneration and neuroinflammation while playing a role in creating an inflammatory condition. The pathology of EAE shows definitive participation from astrocytes, oligodendrocytes, and senescent microglia, leading to immune system activation and failure to re-myeline neural tissue. Therapeutic applications of senescent cell eradication seem possible because scientists have identified senescence as an essential contributor to MS disease development [22, 23].

Cellular and Molecular Insights from Rodent Studies

Rodent research has revealed essential knowledge about cellular senescence mechanisms that occur at the molecular and cellular level in MS models. The research studies three essential cell types responsible for neuroinflammation and neurodegeneration among astrocytes, oligodendrocytes, and microglia. Senescence-mediated inflammatory mediator release from microglia cells leads to worse damage throughout the central nervous system [24]. Scientists have established that $p16^{INK4a}$ and p21, along with other pathways, become active in cells when studying cellular senescence at the molecular level. Research has indicated that SASP leads to persistent inflammatory conditions within the body. Studies have established a basis for therapeutic methods to control cellular aging, indicating that CNS molecular aging mechanisms significantly contribute to multiple sclerosis pathogenesis [25, 26].

Anti-Senescence Interventions in Preclinical Models

The research on evaluating anti-senescence therapies for MS models mainly studied senolytic medications that kill specific senescent cells and strategies blocking senescence pathways. The EAE disease models have shown positive results through these treatments since they reduce disease inflammation while minimizing tissue destruction and promoting remyelination. Study results show that the senolytic medications dasatinib and quercetin properly remove senescent cells from the central nervous system, thus reducing inflammation and improving motor function. Genetic techniques focusing on $p16^{INK4a}$ and other cellular senescence proteins have demonstrated effectiveness in managing or stopping the advancement of illnesses. Studies validate the therapeutic value of senescence-focused approaches in MS treatment that show promise for better recovery as well as delaying illness progression [27].

Clinical Evidence and Translational Potential

Senescence Markers in Human MS Tissues and CSF

Multiple senescence markers were discovered in human tissues, especially brain tissue, the spinal cord, and cerebrospinal fluid (CSF), in patients with MS. Identifying senescent cells in MS pathology relies heavily on using important indicators like SA-β-gal and $p16^{INK4a}$ along with others. $p16^{INK4a}$ represents a widely investigated marker for cellular cycle arrest and aging because it functions as a cyclin-dependent kinase inhibitor. SA-β-gal activity is an indicator connecting with the secretory processes of senescence-associated phenotype (SASP). Scientists investigate the diagnostic and prognostic worth of these indicators because they link to multiple sclerosis-related neuroinflammation and neurodegeneration. The clinical evaluation of disease activity through these markers becomes possible because they specifically reflect relevant biological disease processes [28].

Correlation with Disease Severity and Progression

The severity of multiple sclerosis and its disease progression shows direct links to markers of senescence. An increase in both relapse frequency and neurodegenerative effects and disability progression coincides with higher numbers of senescent cells together with their associated markers. Studies show that an increase in senescent cell numbers leads to tissue damage, particularly within areas of MS patient brains which show demyelination and axonal deterioration. A head of medical staffing experts provides an expanded employment opportunity for advanced health service professionals. Scientists consider senescent cell load a suitable biomarker for MS diagnosis since it allows monitoring of disease advancement throughout the CNS [29, 30].

Current Clinical Trials Targeting Senescence Pathways

Multiple clinical studies investigate the therapeutic potential of treating multiple sclerosis through senescence pathway intervention. Such investigations centre on therapeutical compounds that reduce or modify natural cell aging impacts in patients with multiple sclerosis [31, 32]. Ongoing therapeutic research investigates two methods: senolytics eliminate specific senescent cells, and senostatics stop senescence-related adverse effects [33]. Current research evaluates the safety profile, effectiveness, and optimal dosages of these treatments. At the same time, specific investigations focus on three endpoints—neurodegeneration enhancement, inflammation reduction, and remyelination stimulation [34, 35]. The investigation demonstrates strong potential for translation to medical practice because it presents a novel approach to control and possibly halt the progression of MS through studies of senescence-based therapeutic methods [36, 37].

Therapeutic Approaches Targeting Senescence in MS

Senolytics and Senostatics: Mechanisms and Candidates

MS patients experience increased accumulation of senescent cells, leading to tissue structure damage and neuroinflammation. The medicines known as senolytics serve to eliminate these cellular elements [38, 39]. Senolytics eliminate senescent cells through their action on cellular survival pathways, such as Bcl-2 family proteins, while senostatics disrupt the secretory activities of SAPS to stop the damage from senescent cells [40, 41]. The preclinical research on multiple sclerosis suggests that dasatinib, quercetin, and fisetin have the potential as senolotic options for promoting tissue healing while reducing inflammation [42, 43]. Due to their ability to restore cellular homeostasis, these agents serve as promising strategies even though they require further studies. Multiple sclerosis poses specific specificity challenges and possible unintended effects when targeting the disease [19, 44, 45].

Anti-Inflammatory and Immunomodulatory Strategies

Through the SASP MS, senescent cells activate the production of pro-inflammatory cytokines, chemokines, and proteases that create chronic inflammation and immune system breakdown [19, 46]. The inflammatory substances guide neuroinflammation into worsening states while providing acceleration to multiple sclerosis development. The main objective of anti-inflammatory approaches focusing on senescence-associated pathways involves controlling or blocking the SASP to minimize neuroinflammation while supporting tissue healing [47–49]. The assessment continues for using corticosteroids along with interferons specific cytokine inhibitors and other medications to decrease the inflammatory burden that senescent cells generate in patients with MS. These medicinal interventions find their essential role in treating progressive MS because inflammation drives both brain cell deterioration and worsening of the disease condition [19, 50, 51].

Drug Repurposing and Combination Therapies

The established safety records of current medications make them promising candidates for multipurpose use as senescence-targeted medicines to treat multiple sclerosis. Rapamycin and metformin, among other FDA-approved medications, demonstrate the potential to control senescence processes while influencing linked pathways within the context of MS [52, 53]. The combined effects of senolytics with anti-inflammatory and immunomodulatory medications should produce better therapeutic outcomes [54, 55]. The therapy of using senolytic agents like dasatinib with downstream immune modulator medications such as interferon-beta addresses both aging cell origins while targeting their resulting inflammatory effects. Clinical research and ongoing trials evaluate combined medications that provide MS patients with prospects for personalized and effective treatment options [56, 57].

Challenges and Considerations in Senescence-Targeted Therapy

Specificity and Safety of Senescence Modulators

The current major challenge for senescence-targeted multiple sclerosis treatment involves ensuring productive outcomes from senescence-modulating agents. The preventive measures should allow these treatments to target senescent cells selectively without harming healthy, non-senescent cells. Both senescent and regular cells use similar molecular pathways, creating difficulties in maintaining selectivity between these cells [58–60]. When administered throughout the body, these modulators could trigger unintended side effects that might harm patients by causing toxicity combined with immunological problems and other unacceptable outcomes [61, 62]. The extended-term use of senescence modulators gives reason for concern due to their potential unexpected side effects that manifest after regular usage. New treatment development requires thorough testing and adverse effect monitoring because safety and efficacy must be appropriately balanced [63–65].

Timing and Dosing Challenges in MS Context

Multiple obstacles prevent healthcare professionals from identifying suitable administration timing and dosage levels of senescence-targeted treatments for patients with multiple sclerosis. Treatments for various sclerosis deliver different effects based on when they are administered to patients because this condition advances gradually [27, 66]. Too early administration of the therapy can lead to undesirable outcomes, while too late administration might result in irreparable disease progression. The problem of medication side effects interferes with the approach of finding optimal dosage levels which meet therapeutic requirements [67, 68]. The need for detailed dosage planning emerges because improper dosing at either low or high levels generates unfavorable outcomes. The movement of the disease coupled with patient-specific factors increases plan difficulty, requiring customized treatment approaches [69, 70].

Biomarker Development and Patient Stratification

Medical sciences need biomarkers to assess cellular senescence because better patient classification results from these assessments. Identifying patients likely to gain benefits from senescence-focused medication enables doctors to maximize

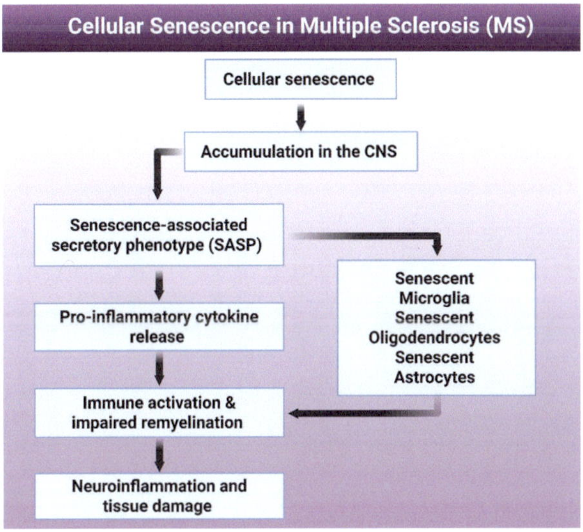

Fig. 11.2 Cellular senescence–driven neuroinflammation in MS. The figure shows the cellular senescence process that occurs when cells experience oxidative stress alongside oncogene activation and encounter DNA damage. When such triggers activate tumor suppressor pathways consisting of the p53/p21 and p16^{INK4a}/Rb axis these elements trigger permanent cell cycle arrest. Living cells in the aging state produce SASP pro-inflammatory substances that help tissue remodeling while also triggering temporary inflammation. SASP release with extended periods of senescence drives both tumor-suppressing properties while contributing to developing age-related diseases and chronic illnesses

therapeutic effects through the trusted utilization of biomarkers [71, 72]. The MS disease structure and the challenging distinction between senescent cells and different types of cells make MS biomarker development a difficult task [73, 74]. Furthermore, a major obstacle still stands in the clinical validation of these biomarkers. Biomarkers developed for medical use will enable practitioners to check therapy responses and choose patients, enhancing therapeutic precision. The achievement of useful biomarkers for MS demands complete scientific study, which follows rigorous clinical trials to establish their reliability for therapeutic applications [75, 76] (Fig. 11.2; Table 11.1).

Table 11.1 The table summarizes key aspects of cellular senescence in multiple sclerosis (MS), detailing its mechanisms, biomarkers, associated therapeutic compounds, and implications across CNS cell types. It highlights the interplay between senescence and MS pathology, emphasizing potential targets for intervention and the translational relevance of senescence-based therapies

Cell type	Mechanism of senescence	Biomarkers	Therapeutic compounds	Implications in MS	Reference
Astrocytes	DNA damage, oxidative stress, SASP activation via NF-κB/C/EBPβ	p16INK4a, p21, SA-β-gal, IL-6, MMPs	Metformin, rapamycin, dasatinib + quercetin	Loss of neuroprotection, impaired repair, SASP-driven neuroinflammation	[77]
Microglia	Chronic activation, ROS production, p16/p21 upregulation	p16INK4a, SA-β-gal, TNF-α, IL-1β	Fisetin, quercetin, senolytic peptides	Exacerbation of neuroinflammation, immune dysregulation, neuronal injury	[9, 78]
Oligodendrocytes	Telomere attrition, DNA damage, Rb pathway activation	γH2AX, p21, SA-β-gal	Dasatinib, FOXO modulators	Impaired remyelination, OPC differentiation failure, myelin loss	[79]
Endothelial cells	Stress-induced senescence, Bcl-2 survival pathway activity	MMP-3, ICAM-1, VCAM-1, $p16^{INK4a}$	Navitoclax, rapamycin	BBB disruption, immune cell infiltration, enhanced lesion formation	[80]
OPC (precursor cells)	Inflammatory cytokine exposure, oxidative stress, impaired cell cycle progression	p16, SA-β-gal, Ki-67 (reduction)	Senomorphics (e.g., resveratrol), JAK inhibitors	Failure to differentiate into oligodendrocytes, remyelination block	[45, 81]
Neurons	Mitochondrial dysfunction, persistent DNA damage, altered autophagy	γH2AX, p53, SA-β-gal	NAD+ boosters (e.g., nicotinamide), senolytics	Axonal degeneration, synaptic dysfunction, cognitive decline	[82, 83]
All CNS cells (SASP)	Activation of NF-κB, chronic stress signaling, paracrine senescence spread	IL-6, IL-8, CCL2, MMPs, SASP transcriptome	Anti-SASP agents, corticosteroids	Chronic inflammation, immune cell activation, feedback loop sustaining MS progression	[84, 85]
General mechanism	$p16^{INK4a}$-Rb pathway, p53-p21 axis, SASP gene regulation by NF-κB and C/EBPβ	SA-β-gal, p16INK4a, p21, γH2AX, SASP markers	Dasatinib, quercetin, fisetin, senomorphics	Neuroinflammation, impaired repair, progressive neurodegeneration, therapeutic target for MS modulation	[86–88]

Future Perspectives

Integrating Senescence into MS Pathogenesis Paradigm

The scientific community better recognizes cellular senescence's important role in MS disease pathology. Senescent cells in the CNS develop into the SASP, which produces inflammatory chemicals that cause neuroinflammation, consequent tissue destruction, and demyelination [89, 90]. MS research, including senescence studies, will help scientists find important treatment targets and biomarkers to understand disease processes better. Observing how senescent cells relate to immune cells and help maintain chronic inflammation allows researchers to discover disease acceleration methods and create new therapeutic strategies. Such an approach will revolutionize MS understanding by bringing cell aging principles into central positions for explaining MS development while surpassing traditional immune-based theories [91, 92].

Personalized Senescence-Modulating Therapies

Personalized medical treatments based on individual patient senescence profiles become possible by developing specific multiple sclerosis remedies that affect senescence [93, 94]. Disease course and effectiveness of therapy can be altered by natural differences in senescence pathways that involve both senescence marker activity and genetic risk factors for cellular aging [95–97]. Researchers need to identify biomarkers that indicate senescence levels in MS patients because this knowledge enables them to select the treatments that target cellular aging for specific individuals. These treatments, made of senolytic and senostatic agents, show promise in decreasing symptoms and stopping disease progression while producing better long-term effects. The development of personalized therapies represents a new direction in multiple sclerosis treatment because it provides better and more successful therapeutic interventions, elevated therapeutic benefit, and reduced adverse reactions [10, 98, 99].

Potential of Regenerative and Remyelination Therapies Linked to Senescence Control

MS treatment may benefit from the fusion of senescence-modulating medications with remyelination and regenerative procedures. Multiple sclerosis shows declining tissue regeneration, including remyelination, because senescent cells restrict the activity of oligodendrocyte precursor cells (OPCs) [100, 101]. Scientists believe that senolytic medicines that destroy senescent cells and senostatics to prevent growth stoppage could improve tissue repair, increase OPC differentiation, and simplify myelin repair processes [29, 102, 103]. Through restoration of CNS regeneration ability, this approach could potentially lead to better clinical benefits for patients with MS. Scientists need to solve some remaining challenges before translating preclinical findings into human treatments because of safety aspects, dosage determination, and long-term treatment effectiveness validation requirements. Increasing clinical outcomes and reducing neurodegenerative damage become possible when senescence control methods integrate with regenerative medicine strategies for multiple sclerosis therapy [4, 104, 105].

Conclusion

Multiple sclerosis disease progression is strongly affected by cellular senescence because senescent cells located in central nervous system tissue drive the evolution of the condition. The disease process intensifies due to inflammatory substances released by these cells that cause persistent inflammation, damage to neurons, and loss of myelin sheaths. The pathogenesis of MS shows deterioration due to senescence because this aging process accelerates tissue breakdown while blocking healing mechanisms. The therapeutic potential of targeting senescence in multiple sclerosis appears promising because senolytics and other senescence-modulating drugs have emerged as viable new therapies. Numerous challenges stand in the way of this research, involving specific treatment protocols, effective side effect mitigation, and dosage determination. Understanding senescence in MS can lead to better treatment options despite existing research challenges. In the future, research needs to develop dependable biomarkers that monitor MS patient senescence because they will allow physicians to implement precise individual therapy plans. More research is required to advance senescence-targeted medicines and determine their long-term safety and effectiveness. Cellular senescence represents a promising fresh therapeutic target in multiple sclerosis, which can help patients while slowing down disease progression. Future MS treatment will likely benefit significantly from research about senescence, biomarker development, and individualized therapy creation, which could help physicians control this harmful disease.

References

1. Oh J, Vidal-Jordana A, Montalban X. Multiple sclerosis: clinical aspects. Curr Opin Neurol. 2018;31(6):752–9.
2. Wu GF, Alvarez E. The immunopathophysiology of multiple sclerosis. Neurol Clin. 2011;29(2):257–78.
3. Derada Troletti C, de Goede P, Kamermans A, de Vries HE. Molecular alterations of the blood–brain barrier under inflammatory conditions: the role of endothelial to mesenchymal transition. Biochim Biophys Acta. 2016;1862(3):452–60.
4. Maupin EA, Adams KL. Cellular senescence in glial cells: implications for multiple sclerosis. J Neurochem. 2025;169(1):e16301.
5. Portaccio E, Magyari M, Havrdova EK, Ruet A, Brochet B, Scalfari A, et al. Multiple sclerosis: emerging epidemiological trends and redefining the clinical course. Lancet Reg Health Eur. 2024;44:100977.

6. Walton C, King R, Rechtman L, Kaye W, Leray E, Marrie RA, et al. Rising prevalence of multiple sclerosis worldwide: insights from the Atlas of MS, third edition. Mult Scler. 2020;26(14):1816–21.
7. Liang R, Qi X, Cai Q, Niu L, Huang X, Zhang D, et al. The role of NLRP3 inflammasome in aging and age-related diseases. Immun Ageing. 2024;21(1):14.
8. Childs BG, Gluscevic M, Baker DJ, Laberge RM, Marquess D, Dananberg J, et al. Senescent cells: an emerging target for diseases of ageing. Nat Rev Drug Discov. 2017;16(10):718–35.
9. Huang W, Hickson LJ, Eirin A, Kirkland JL, Lerman LO. Cellular senescence: the good, the bad and the unknown. Nat Rev Nephrol. 2022;18(10):611–27.
10. Papadopoulos D, Magliozzi R, Mitsikostas DD, Gorgoulis VG, Nicholas RS. Aging, cellular senescence, and progressive multiple sclerosis. Front Cell Neurosci. 2020;14:178.
11. Wissler Gerdes EO, Zhu Y, Weigand BM, Tripathi U, Burns TC, Tchkonia T, et al. Cellular senescence in aging and age-related diseases: implications for neurodegenerative diseases. Int Rev Neurobiol. 2020;155:203–34.
12. Gorgoulis V, Adams PD, Alimonti A, Bennett DC, Bischof O, Bishop C, et al. Cellular senescence: defining a path forward. Cell. 2019;179(4):813–27.
13. Li X, Li C, Zhang W, Wang Y, Qian P, Huang H. Inflammation and aging: signaling pathways and intervention therapies. Signal Transduct Target Ther. 2023;8(1):239.
14. Kumari R, Jat P. Mechanisms of cellular senescence: cell cycle arrest and senescence associated secretory phenotype. Front Cell Dev Biol. 2021;9:645593.
15. Colucci M, Zumerle S, Bressan S, Gianfanti F, Troiani M, Valdata A, et al. Retinoic acid receptor activation reprograms senescence response and enhances anti-tumor activity of natural killer cells. Cancer Cell. 2024;42(4):646–61.e9.
16. Serafini B, Rosicarelli B, Veroni C, Aloisi F. Tissue-resident memory T cells in the multiple sclerosis brain and their relationship to Epstein-Barr virus infected B cells. J Neuroimmunol. 2023;376:578036.
17. Ziar R, Tesar PJ, Clayton BLL. Astrocyte and oligodendrocyte pathology in Alzheimer's disease. Neurotherapeutics. 2025;22(3):e00540.
18. Afridi R, Rahman MH, Suk K. Implications of glial metabolic dysregulation in the pathophysiology of neurodegenerative diseases. Neurobiol Dis. 2022;174:105874.
19. Oost W, Talma N, Meilof JF, Laman JD. Targeting senescence to delay progression of multiple sclerosis. J Mol Med (Berl). 2018;96(11):1153–66.
20. Takata F, Nakagawa S, Matsumoto J, Dohgu S. Blood-brain barrier dysfunction amplifies the development of neuroinflammation: understanding of cellular events in brain microvascular endothelial cells for prevention and treatment of BBB dysfunction. Front Cell Neurosci. 2021;15:661838.
21. Brandl S, Reindl M. Blood-brain barrier breakdown in neuroinflammation: current in vitro models. Int J Mol Sci. 2023;24(16)
22. Constantinescu CS, Farooqi N, O'Brien K, Gran B. Experimental autoimmune encephalomyelitis (EAE) as a model for multiple sclerosis (MS). Br J Pharmacol. 2011;164(4):1079–106.
23. Melamed E, Palmer JL, Fonken C. Advantages and limitations of experimental autoimmune encephalomyelitis in breaking down the role of the gut microbiome in multiple sclerosis. Front Mol Neurosci. 2022;15:1019877.
24. Tamatta R, Pai V, Jaiswal C, Singh I, Singh AK. Neuroinflammaging and the immune landscape: the role of autophagy and senescence in aging brain. Biogerontology. 2025;26(2):52.
25. Wagner KD, Wagner N. The senescence markers p16INK4A, p14ARF/p19ARF, and p21 in organ development and homeostasis. Cells. 2022;11(12)
26. Grootaert MOJ. Cell senescence in cardiometabolic diseases. npj Aging. 2024;10(1):46.
27. Sutter PA, McKenna MG, Imitola J, Pijewski RS, Crocker SJ. Therapeutic opportunities for targeting cellular senescence in progressive multiple sclerosis. Curr Opin Pharmacol. 2022;63:102184.
28. Gao C, Jiang J, Tan Y, Chen S. Microglia in neurodegenerative diseases: mechanism and potential therapeutic targets. Signal Transduct Target Ther. 2023;8(1):359.

29. Gross PS, Durán-Laforet V, Ho LT, Melchor GS, Zia S, Manavi Z, et al. Senescent-like microglia limit remyelination through the senescence associated secretory phenotype. Nat Commun. 2025;16(1):2283.
30. Correale J, Ysrraelit MC. Multiple sclerosis and aging: the dynamics of demyelination and remyelination. ASN Neuro. 2022;14:17590914221118502.
31. Stankiewicz JM, Neema M, Ceccarelli A. Iron and multiple sclerosis. Neurobiol Aging. 2014;35(Suppl 2):S51–8.
32. Stern M, Sorkin L, Milton K, Sperber K. Aging with multiple sclerosis. Phys Med Rehabil Clin N Am. 2010;21(2):403–17.
33. Virley DJ. Developing therapeutics for the treatment of multiple sclerosis. NeuroRx. 2005;2(4):638–49.
34. Enoka RM, Duchateau J. Translating fatigue to human performance. Med Sci Sports Exerc. 2016;48(11):2228–38.
35. Frischer JM, Weigand SD, Guo Y, Kale N, Parisi JE, Pirko I, et al. Clinical and pathological insights into the dynamic nature of the white matter multiple sclerosis plaque. Ann Neurol. 2015;78(5):710–21.
36. Zhang J, Zhang Y, Wang J, Xia Y, Zhang J, Chen L. Recent advances in Alzheimer's disease: mechanisms, clinical trials and new drug development strategies. Signal Transduct Target Ther. 2024;9(1):211.
37. Allanach JR, Farrell JW 3rd, Mésidor M, Karimi-Abdolrezaee S. Current status of neuroprotective and neuroregenerative strategies in multiple sclerosis: a systematic review. Mult Scler. 2022;28(1):29–48.
38. Larochelle C, Uphaus T, Prat A, Zipp F. Secondary progression in multiple sclerosis: neuronal exhaustion or distinct pathology? Trends Neurosci. 2016;39(5):325–39.
39. Schmitt A, Falkai P. DGPPN guideline on anxiety disorders and cognitive dysfunction in the elderly or patients with multiple sclerosis. Eur Arch Psychiatry Clin Neurosci. 2015;265(5):361–2.
40. Choi IY, Lee C, Longo VD. Nutrition and fasting mimicking diets in the prevention and treatment of autoimmune diseases and immunosenescence. Mol Cell Endocrinol. 2017;455:4–12.
41. Sanai SA, Saini V, Benedict RH, Zivadinov R, Teter BE, Ramanathan M, et al. Aging and multiple sclerosis. Mult Scler. 2016;22(6):717–25.
42. Khalil M, Teunissen CE, Otto M, Piehl F, Sormani MP, Gattringer T, et al. Neurofilaments as biomarkers in neurological disorders. Nat Rev Neurol. 2018;14(10):577–89.
43. Louapre C, Papeix C, Lubetzki C, Maillart E. Multiple sclerosis and aging. Geriatr Psychol Neuropsychiatr Vieil. 2017;15(4):402–8.
44. Lavorgna L, Brigo F, Moccia M, Leocani L, Lanzillo R, Clerico M, et al. e-Health and multiple sclerosis: an update. Mult Scler. 2018;24(13):1657–64.
45. Kritsilis M, Rizou SV, Koutsoudaki PN, Evangelou K, Gorgoulis VG, Papadopoulos D. Ageing, cellular senescence and neurodegenerative disease. Int J Mol Sci. 2018;19(10)
46. Banwell B. Are children with multiple sclerosis really "old" adults. Mult Scler. 2019;25(7):888–90.
47. Zhu X, Chen Z, Shen W, Huang G, Sedivy JM, Wang H, et al. Inflammation, epigenetics, and metabolism converge to cell senescence and ageing: the regulation and intervention. Signal Transduct Target Ther. 2021;6(1):245.
48. Neumann B, Baror R, Zhao C, Segel M, Dietmann S, Rawji KS, et al. Metformin restores CNS remyelination capacity by rejuvenating aged stem cells. Cell Stem Cell. 2019;25(4):473–85.e8.
49. Hammond TR, Dufort C, Dissing-Olesen L, Giera S, Young A, Wysoker A, et al. Single-cell RNA sequencing of microglia throughout the mouse lifespan and in the injured brain reveals complex cell-state changes. Immunity. 2019;50(1):253–71.e6.
50. Yeung MSY, Djelloul M, Steiner E, Bernard S, Salehpour M, Possnert G, et al. Dynamics of oligodendrocyte generation in multiple sclerosis. Nature. 2019;566(7745):538–42.
51. Nguyen AL, Eastaugh A, van der Walt A, Jokubaitis VG. Pregnancy and multiple sclerosis: clinical effects across the lifespan. Autoimmun Rev. 2019;18(10):102360.

52. Kular L, Jagodic M. Epigenetic insights into multiple sclerosis disease progression. J Intern Med. 2020;288(1):82–102.
53. Koutsoudaki PN, Papadopoulos D, Passias PG, Koutsoudaki P, Gorgoulis VG. Cellular senescence and failure of myelin repair in multiple sclerosis. Mech Ageing Dev. 2020;192:111366.
54. Choi S, Guo L, Cordeiro MF. Retinal and brain microglia in multiple sclerosis and neurodegeneration. Cells. 2021;10(6)
55. Moccia M, Lanzillo R, Brescia Morra V, Bonavita S, Tedeschi G, Leocani L, et al. Assessing disability and relapses in multiple sclerosis on tele-neurology. Neurol Sci. 2020;41(6):1369–71.
56. Aliper A, Jellen L, Cortese F, Artemov A, Karpinsky-Semper D, Moskalev A, et al. Towards natural mimetics of metformin and rapamycin. Aging. 2017;9(11):2245–68.
57. Romashkan S, Chang H, Hadley EC. National Institute on Aging workshop: repurposing drugs or dietary supplements for their senolytic or senomorphic effects: considerations for clinical trials. J Gerontol A Biol Sci Med Sci. 2021;76(6):1144–52.
58. Bousset L, Gil J. Targeting senescence as an anticancer therapy. Mol Oncol. 2022;16(21):3855–80.
59. Naseri A, Nasiri E, Sahraian MA, Daneshvar S, Talebi M. Clinical features of late-onset multiple sclerosis: a systematic review and meta-analysis. Mult Scler Relat Disord. 2021;50:102816.
60. Horton DB, Reder AT. Medications for multiple sclerosis and risk of malignancy: what next? Neurotherapeutics. 2021;18(3):1650–3.
61. Fox RJ, Kryscio RJ. A new way to identify promising therapies for progressive multiple sclerosis. Neurology. 2021;96(18):833–4.
62. Dema M, Eixarch H, Villar LM, Montalban X, Espejo C. Immunosenescence in multiple sclerosis: the identification of new therapeutic targets. Autoimmun Rev. 2021;20(9):102893.
63. Li Z, Zhang Z, Ren Y, Wang Y, Fang J, Yue H, et al. Aging and age-related diseases: from mechanisms to therapeutic strategies. Biogerontology. 2021;22(2):165–87.
64. Adamczyk-Sowa M, Nowak-Kiczmer M, Jaroszewicz J, Berger T. Immunosenescence and multiple sclerosis. Neurol Neurochir Pol. 2022;56(3):220–7.
65. Ostolaza A, Corroza J, Ayuso T. Multiple sclerosis and aging: comorbidity and treatment challenges. Mult Scler Relat Disord. 2021;50:102815.
66. Engelhard J, Oleske DM, Schmitting S, Wells KE, Talapala S, Barbato LM. Multiple sclerosis by phenotype in Germany. Mult Scler Relat Disord. 2022;57:103326.
67. International Multiple Sclerosis Genetics Consortium, MultipleMS Consortium. Locus for severity implicates CNS resilience in progression of multiple sclerosis. Nature. 2023;619(7969):323–31.
68. Yong VW. Microglia in multiple sclerosis: protectors turn destroyers. Neuron. 2022;110(21):3534–48.
69. Filippi M, Amato MP, Centonze D, Gallo P, Gasperini C, Inglese M, et al. Early use of high-efficacy disease-modifying therapies makes the difference in people with multiple sclerosis: an expert opinion. J Neurol. 2022;269(10):5382–94.
70. Wiendl H, Gold R, Berger T, Derfuss T, Linker R, Mäurer M, et al. Multiple sclerosis therapy consensus group (MSTCG): position statement on disease-modifying therapies for multiple sclerosis (white paper). Ther Adv Neurol Disord. 2021;14:17562864211039648.
71. Graves JS, Krysko KM, Hua LH, Absinta M, Franklin RJM, Segal BM. Ageing and multiple sclerosis. Lancet Neurol. 2023;22(1):66–77.
72. Graves JS. Cell-specific aging in multiple sclerosis. Neurology. 2023;101(7):285–6.
73. Kuhlmann T, Moccia M, Coetzee T, Cohen JA, Correale J, Graves J, et al. Multiple sclerosis progression: time for a new mechanism-driven framework. Lancet Neurol. 2023;22(1):78–88.
74. Holloway K, Neherin K, Dam KU, Zhang H. Cellular senescence and neurodegeneration. Hum Genet. 2023;142(8):1247–62.
75. Cummings SR, Lui LY, Zaira A, Mau T, Fielding RA, Atkinson EJ, et al. Biomarkers of cellular senescence and major health outcomes in older adults. GeroScience. 2024; https://doi.org/10.1007/s11357-024-01474-9.

76. Bulbiankova D, Díaz-Puertas R, Álvarez-Martínez FJ, Herranz-López M, Barrajón-Catalán E, Micol V. Hallmarks and biomarkers of skin senescence: an updated review of skin senotherapeutics. Antioxidants (Basel). 2023;12(2)
77. Thakur S, Dhapola R, Sarma P, Medhi B, Reddy DH. Neuroinflammation in Alzheimer's disease: current progress in molecular signaling and therapeutics. Inflammation. 2023;46(1):1–17.
78. Chaib S, Tchkonia T, Kirkland JL. Cellular senescence and senolytics: the path to the clinic. Nat Med. 2022;28(8):1556–68.
79. Peng S, Sen B, Mazumdar T, Byers LA, Diao L, Wang J, et al. Dasatinib induces DNA damage and activates DNA repair pathways leading to senescence in non-small cell lung cancer cell lines with kinase-inactivating BRAF mutations. Oncotarget. 2016;7(1):565–79.
80. Real MGC, Falcione SR, Boghozian R, Clarke M, Todoran R, St Pierre A, et al. Endothelial cell senescence effect on the blood-brain barrier in stroke and cognitive impairment. Neurology. 2024;103(11):e210063.
81. Alshaebi F, Sciortino A, Kayed R. The role of glial cell senescence in Alzheimer's disease. J Neurochem. 2025;169(3):e70051.
82. Formentini L, Moroni F, Chiarugi A. Detection and pharmacological modulation of nicotinamide mononucleotide (NMN) in vitro and in vivo. Biochem Pharmacol. 2009;77(10):1612–20.
83. Zhu J, Wu C, Yang L. Cellular senescence in Alzheimer's disease: from physiology to pathology. Transl Neurodegener. 2024;13(1):55.
84. Prata L, Ovsyannikova IG, Tchkonia T, Kirkland JL. Senescent cell clearance by the immune system: emerging therapeutic opportunities. Semin Immunol. 2018;40:101275.
85. Sun Y, Wang X, Liu T, Zhu X, Pan X. The multifaceted role of the SASP in atherosclerosis: from mechanisms to therapeutic opportunities. Cell Biosci. 2022;12(1):74.
86. Schmitt CA, Wang B, Demaria M. Senescence and cancer - role and therapeutic opportunities. Nat Rev Clin Oncol. 2022;19(10):619–36.
87. Ngoi NYL, Liew AQX, Chong SJF, Davids MS, Clement M-V, Pervaiz S. The redox-senescence axis and its therapeutic targeting. Redox Biol. 2021;45:102032.
88. Kudlova N, De Sanctis JB, Hajduch M. Cellular senescence: molecular targets, biomarkers, and senolytic drugs. Int J Mol Sci. 2022;23(8)
89. Ransohoff RM. Multiple sclerosis: role of meningeal lymphoid aggregates in progression independent of relapse activity. Trends Immunol. 2023;44(4):266–75.
90. Ostolaza Ibáñez A, Corroza Laviñeta J, Ayuso Blanco T. Immunosenescence: the role of age in multiple sclerosis. Neurologia (Engl Ed). 2023;38(4):284–90.
91. Lei S-Y, Qu Y, Yang Y-Q, Liu J-C, Zhang Y-F, Zhou S-Y, et al. Cellular senescence: a novel therapeutic target for central nervous system diseases. Biomed Pharmacother. 2024;179:117311.
92. Deng S, Xie H, Xie B. Cell-based regenerative and rejuvenation strategies for treating neurodegenerative diseases. Stem Cell Res Ther. 2025;16(1):167.
93. Censi ST, Mariani-Costantini R, Granzotto A, Tomassini V, Sensi SL. Endogenous retroviruses in multiple sclerosis: a network-based etiopathogenic model. Ageing Res Rev. 2024;99:102392.
94. Zhang Y, Atkinson J, Burd CE, Graves J, Segal BM. Biological aging in multiple sclerosis. Mult Scler. 2023;29(14):1701–8.
95. Griñán-Ferré C, Bellver-Sanchis A, Guerrero A, Pallàs M. Advancing personalized medicine in neurodegenerative diseases: the role of epigenetics and pharmacoepigenomics in pharmacotherapy. Pharmacol Res. 2024;205:107247.
96. Duarte LF, Villalobos V, Farías MA, Rangel-Ramírez MA, González-Madrid E, Navarro AJ, et al. Asymptomatic herpes simplex virus brain infection elicits cellular senescence phenotypes in the central nervous system of mice suffering multiple sclerosis-like disease. Commun Biol. 2024;7(1):811.
97. DiMauro KA, Swetlik C, Cohen JA. Management of multiple sclerosis in older adults: review of current evidence and future perspectives. J Neurol. 2024;271(7):3794–805.

98. Fettig NM, Pu A, Osborne LC, Gommerman JL. The influence of aging and the microbiome in multiple sclerosis and other neurologic diseases. Immunol Rev. 2024;325(1):166–89.
99. Fernández Ó, Sörensen PS, Comi G, Vermersch P, Hartung HP, Leocani L, et al. Managing multiple sclerosis in individuals aged 55 and above: a comprehensive review. Front Immunol. 2024;15:1379538.
100. Goyne CE, Fair AE, Sumowski PE, Graves JS. The impact of aging on multiple sclerosis. Curr Neurol Neurosci Rep. 2024;24(4):83–93.
101. Filippi M, Preziosa P, Barkhof F, Ciccarelli O, Cossarizza A, De Stefano N, et al. The ageing central nervous system in multiple sclerosis: the imaging perspective. Brain. 2024;147(11):3665–80.
102. Ionescu RB, Nicaise AM, Reisz JA, Williams EC, Prasad P, Willis CM, et al. Increased cholesterol synthesis drives neurotoxicity in patient stem cell-derived model of multiple sclerosis. Cell Stem Cell. 2024;31(11):1574–90.e11.
103. Gross RH, Corboy J. De-escalation and discontinuation of disease-modifying therapies in multiple sclerosis. Curr Neurol Neurosci Rep. 2024;24(9):341–53.
104. Wu L, Lu J, Lan T, Zhang D, Xu H, Kang Z, et al. Stem cell therapies: a new era in the treatment of multiple sclerosis. Front Neurol. 2024;15:1389697.
105. Zhu W, Xia Z. Treatment discontinuation in older people with multiple sclerosis. Curr Opin Neurol. 2024;37(3):220–7.

Future Prospects and Challenges

12

Gaurav Gupta

Introduction

Cellular senescence has emerged as a central mechanism driving age-related dysfunction across multiple organ systems, and its role in the aging brain is attracting growing attention for its potential impact on neurodegenerative disease. Diverse cell types in the central nervous system, including astrocytes, microglia, oligodendrocyte progenitors, and even neurons, can enter a state of irreversible growth arrest and adopt a pro-inflammatory secretory phenotype. This senescence-associated secretory phenotype (SASP) comprises cytokines, chemokines, matrix-remodeling enzymes, and growth factors that alter the tissue microenvironment. In the brain, accumulation of senescent cells and their SASP can compromise synaptic function, disrupt the blood–brain barrier, propagate neuroinflammation, and impair endogenous repair mechanisms, thereby accelerating the onset and progression of disorders such as Alzheimer's disease, Parkinson's disease, and amyotrophic lateral sclerosis [1–3].

The discovery that targeted removal of senescent cells, "senolysis," can restore tissue function and extend health span in peripheral organs has galvanized efforts to translate similar strategies to the brain. Preclinical studies have shown that genetic or pharmacological elimination of senescent astrocytes and microglia reduces pathological protein aggregates, dampens inflammatory networks, and improves cognitive performance in animal models. At the same time, approaches aimed at modulating the SASP (so-called senomorphics) hold promise for

G. Gupta (✉)
Centre for Research Impact & Outcome, Chitkara College of Pharmacy,
Chitkara University, Rajpura, Punjab, India
e-mail: Gaurav.gupta@chitkara.edu.in

G. Gupta (ed.), *Cellular Senescence and Brain Aging*,
https://doi.org/10.1007/978-981-96-8873-9_12

attenuating deleterious inflammatory signaling while preserving any beneficial functions of senescent cells, such as roles in wound healing and tumor suppression [4–7].

Despite these encouraging findings, translating senescence-targeted therapies into clinical practice faces multiple challenges. The heterogeneity of senescent cell populations in the brain raises questions about optimal targeting strategies: which cell types should be eliminated, and at what stage of disease progression? The blood–brain barrier presents an additional hurdle for delivering senolytic compounds or gene-based interventions. Furthermore, current diagnostic tools for detecting senescence in vivo remain rudimentary. While fluid biomarkers in cerebrospinal fluid or blood offer noninvasive windows into brain health, their sensitivity and specificity for cellular senescence require rigorous validation. Novel imaging tracers that bind senescence-associated markers may one day enable real-time monitoring of therapeutic effects, but these tools are still in early development [8–11].

Regulatory landscapes add another layer of complexity. Early-phase clinical trials must balance safety and efficacy, defining appropriate end points that capture biochemical changes and meaningful functional outcomes. Given the ethical implications of intervening in aging processes, especially in vulnerable elderly populations, robust patient selection criteria and long-term safety monitoring will be essential. Equally important are access and equity considerations: as life expectancy increases, ensuring that breakthroughs in senescence-targeted therapies benefit diverse patient populations will require coordinated efforts among researchers, industry partners, and healthcare systems [12–15].

Looking ahead, interdisciplinary collaboration will be key to overcoming these obstacles. Advances in nanotechnology may yield delivery platforms capable of breaching the blood–brain barrier with high specificity. At the same time, cutting-edge organoid and 3D culture models of human brain aging could provide more faithful preclinical testing grounds. Artificial intelligence and machine learning promise to accelerate biomarker discovery and patient stratification by integrating multi-omics datasets. At the same time, public–private consortia and open-science initiatives already demonstrate the power of shared data and resources to drive innovation [16, 17].

A proactive, forward-thinking approach is essential as the field stands at this pivotal juncture. Anticipating technical roadblocks, regulatory challenges, ethical considerations, and embedding solutions within research design will expedite the path from bench to bedside. Ultimately, harnessing the full therapeutic potential of senescence clearance or modulation could transform our ability to prevent or slow neurodegenerative diseases, preserving cognitive health and improving quality of life for an increasingly aging global population [18].

Advances in Senescence-Targeting Therapeutics

Over the past decade, directly targeting senescent cells to treat age-related pathologies has shifted from theoretical to translational reality. In peripheral tissues, the first generation of senolytics, small molecules that selectively induce apoptosis in senescent cells, demonstrated striking rejuvenative effects in rodents. Harnessing this same approach in the central nervous system presents unique opportunities and distinct hurdles. Below, we explore four primary therapeutic modalities under active development: senolytic agents, senomorphic strategies to modulate the SASP, gene- and cell-based reprogramming, and advanced delivery platforms leveraging nanotechnology [19, 20].

Senolytic Agents: Next-Generation Small Molecules and Biologics

Early senolytics such as the BCL-2 family inhibitors (e.g., navitoclax) and the flavonoid quercetin in combination with the kinase inhibitor dasatinib paved the way by demonstrating clearance of senescent cells in adipose, lung, and vascular tissues. Translating these compounds to the brain demands molecules that penetrate the blood–brain barrier (BBB) and avoid neurotoxicity. Recent efforts have yielded brain-penetrant derivatives of navitoclax with reduced platelet toxicity and novel inhibitors targeting antiapoptotic proteins upregulated in glial senescence (e.g., selective MCL-1 antagonists) [21, 22].

Beyond small molecules, antibody-based senolytics are emerging. Conjugating pro-apoptotic payloads to antibodies that recognize senescence-associated surface markers such as urokinase-type plasminogen activator receptor (uPAR) or specific lectins enables precision targeting of senescent astrocytes and microglia. Early preclinical studies report efficient clearance of uPAR-positive cells in Alzheimer's models, with concurrent reduction in amyloid burden and improved synaptic plasticity [23, 24].

Senomorphic Strategies: Modulating the SASP

Complete ablation of all senescent cells risks eliminating those that play beneficial roles in tissue repair, wound healing, or tumor suppression. Senomorphics offer an alternative: they attenuate the deleterious components of the SASP without inducing cell death. Small-molecule inhibitors of key SASP regulators such as the JAK/STAT pathway, p38 MAPK, and NF-κB have shown promise in reducing neuroinflammatory cytokines (IL-6, IL-1β, TNFα) secreted by senescent glia [25, 26].

Other approaches exploit epigenetic modulation: inhibitors of bromodomain proteins (BRD4) or histone deacetylases (HDACs) can reprogram the senescent secretome toward a less inflammatory phenotype. In murine models of Parkinson's disease, HDAC inhibition, specifically in microglia, reduced α-synuclein propagation and motor deficits, suggesting that fine-tuning the SASP may slow neurodegenerative cascades without wholesale cell removal [27, 28].

Gene-Based and Cellular Reprogramming Approaches

Gene therapy affords a versatile platform for both senolysis and senomorphics. Viral vectors delivering inducible "suicide genes" under the control of senescence-associated promoters (e.g., $p16^{INK4a}$ or $p21^{CIP1}$) have achieved selective ablation of senescent cells in peripheral tissues; adapting these systems for CNS use has required engineering of viral tropism toward astrocytes and microglia, as well as incorporation of neuron-sparing safeguards. Early work using adeno-associated virus serotype 9 (AAV9) to deliver a caspase-activated construct under the p16 promoter has cleared senescent cells in the hippocampus, restoring neurogenesis in aged rodents [29, 30].

An alternative is in vivo partial reprogramming: transient expression of Yamanaka factors (Oct4, Sox2, Klf4, c-Myc) can reverse senescence hallmarks and rejuvenate cellular function without inducing pluripotency. In the brain, controlled pulses of these factors have improved the electrophysiological properties of aged neurons and reduced glial activation, although long-term safety remains under investigation [31, 32].

Nanotechnology and Delivery Systems

A central barrier to CNS senescence-targeted therapy is achieving sufficient drug concentration behind the BBB. Nanoparticle-based delivery systems have thus taken center stage. Lipid-based nanoparticles (LNPs), polymeric micelles, and exosome-mimetic vesicles functionalized with BBB-transcytosis ligands (e.g., transferrin receptor antibodies or apolipoprotein E mimetic peptides) can ferry both small-molecule senolytics and nucleic acid payloads across endothelial tight junctions [33, 34].

For instance, encapsulating a JAK inhibitor within LNPs decorated with angiopep-2 achieved tenfold higher brain accumulation in murine Alzheimer's models, leading to marked suppression of SASP factors and improved cognitive performance. Similarly, polymeric nanoparticles loaded with siRNAs against Bcl-2 family members have selectively knocked down antiapoptotic genes in senescent microglia, triggering localized cell death. Emerging "smart" nanoparticles incorporate stimuli-responsive elements that release their cargo in response to the acidic or

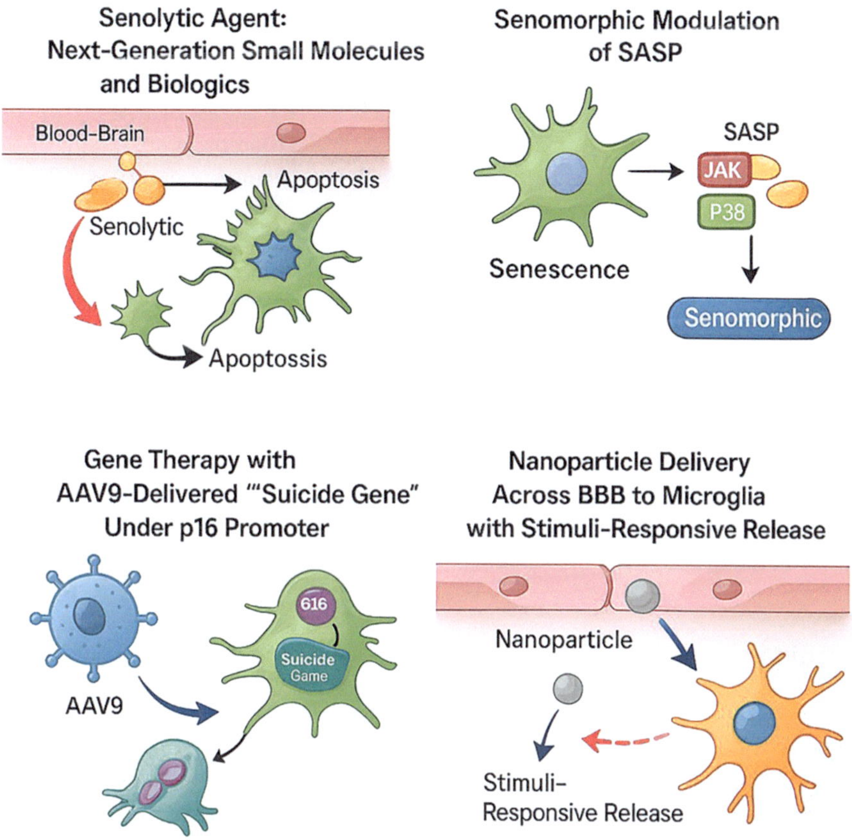

Fig. 12.1 Therapeutic strategies targeting cellular senescence in the aging brain. Biomarkers and diagnostics for brain senescence

protease-rich microenvironment characteristic of senescent cell niches. Such precision could minimize off-target effects in healthy tissue and permit lower dosing regimens [35, 36] (Fig. 12.1).

Detecting and quantifying senescent cells in the living human brain is a critical hurdle for both research and clinical translation. True senescence is defined by persistent DNA damage responses, elevated levels of cyclin-dependent kinase inhibitors ($p16^{INK4a}$, $p21^{CIP1}$), and secretion of a complex pro-inflammatory milieu, the senescence-associated secretory phenotype (SASP). However, sampling brain parenchyma directly is invasive and impractical for routine monitoring. Consequently, three complementary diagnostic avenues have emerged: fluid biomarkers, neuroimaging techniques, and peripheral composite indices. Each offers unique advantages and faces distinct challenges [37–39].

Fluid Biomarkers (CSF and Blood-Based Panels)

Cerebrospinal fluid (CSF) and plasma provide relatively accessible windows into central nervous system health. In CSF, elevated concentrations of SASP components such as interleukin-6, interleukin-1β, matrix metalloproteinases, and growth differentiation factor-15 correlate with chronological age and cognitive decline in neurodegenerative cohorts. Advances in multiplex immunoassays now permit simultaneous quantification of dozens of cytokines and chemokines, enabling the construction of tailored "senescence panels." Parallel efforts focus on cell-free DNA methylation signatures: circulating fragments bearing patterns characteristic of senescent chromatin offer a noninvasive readout of cellular aging [40, 41].

In blood, where SASP factors are more diluted, ultrasensitive detection methods (e.g., single-molecule arrays) are essential to distinguish senescence-derived signals from systemic inflammation. Emerging markers include soluble urokinase-type plasminogen activator receptor (suPAR) and circulating levels of $p16^{INK4a}$ transcripts in peripheral blood mononuclear cells. Although blood-based assays lack the anatomical specificity of CSF measures, they hold promise for large-scale screening and longitudinal monitoring [42, 43].

Imaging Modalities (PET, MRI with Senescence Tracers)

Neuroimaging techniques capable of visualizing senescent cells in situ would revolutionize diagnosis and therapeutic monitoring. Positron emission tomography (PET) tracers targeting β-galactosidase activity, an enzymatic hallmark of senescence, have demonstrated retention in aged rodent brains, indicating feasibility. Other PET probes under development bind to SASP-associated enzymes, such as matrix metalloproteinases, which indirectly measure senescent cell burden [44, 45].

Magnetic resonance imaging (MRI) approaches leverage ultrasmall superparamagnetic iron oxide nanoparticles functionalized to home to cells overexpressing $p16^{INK4a}$. These contrast agents highlight clusters of senescent glia at high spatial resolution. When combined with functional MRI (fMRI), researchers can correlate regions of elevated senescence with alterations in neural connectivity and hemodynamic response. Although still in the early stages, such tracers could enable real-time assessment of therapeutic efficacy in clinical trials [46, 47].

Peripheral Readouts and Composite Scores

Given the invasiveness of CSF collection and the expense of advanced imaging, composite indices derived from peripheral biomarkers are attractive for epidemiological studies and initial patient stratification. These scores integrate circulating SASP factors, inflammatory cytokines, metabolic markers (e.g., lipid profiles, glucose levels), and epigenetic "aging clocks" computed from DNA methylation arrays. Machine-learning models trained on multimodal datasets can infer the likelihood of elevated brain senescence from these systemic signatures. Preliminary data indicate that higher composite senescence scores predict poorer performance on neuropsychological tests and correlate with volumetric atrophy on structural MRI. While such indices cannot replace direct imaging or CSF assays, they offer a scalable, cost-effective tool for screening and longitudinal follow-up [48–51].

Challenges in Clinical Validation

Despite rapid technological advances, validating senescence biomarkers for routine use entails several obstacles. First, senescent cell heterogeneity in the brain means no single marker will capture all relevant cell types. Second, many SASP components are also elevated in systemic inflammatory or metabolic disorders, compromising specificity. Third, longitudinal studies must establish how early biomarker changes forecast clinical outcomes and define the minimal detectable change that reflects meaningful therapeutic impact. To surmount these challenges, concerted efforts are needed to standardize sample collection and assay protocols across centers. Multicenter cohort studies with well-phenotyped participants will be essential to validate biomarker panels against gold-standard end points, such as postmortem histopathology or established neuroimaging measures. A robust armamentarium of fluid, imaging, and composite biomarkers will be indispensable for guiding senescence-targeted interventions and tracking their efficacy in slowing or halting neurodegenerative progression [52, 53].

Translational and Clinical Considerations

Bringing senescence-targeted therapies for neurodegeneration from bench to bedside requires careful orchestration of preclinical and clinical strategies. Three critical domains demand particular attention: the fidelity of experimental models, the design of early-phase trials, and the navigation of regulatory pathways. Each influences the likelihood of success, patient safety, and ultimate therapeutic impact [54, 55].

Preclinical Models: Bridging Rodents to Humans

Traditional rodent models of aging and neurodegeneration have yielded invaluable mechanistic insights, yet they only imperfectly recapitulate human brain aging. Murine lifespans are compressed, and their glial biology, especially microglial activation states, diverges from that of primates. To enhance translational relevance, researchers are developing "humanized" mouse lines expressing human cytokines and cell-cycle regulators, as well as chimeric models in which induced pluripotent stem cell (iPSC)-derived human microglia or astrocytes engraft into the mouse brain. Complementing these approaches, three-dimensional brain organoids and slice cultures derived from aged human donors provide platforms for testing senolytic and senomorphic compounds in human genetic backgrounds. Rigorous comparative studies benchmarking drug responses across rodent, organoid, and nonhuman primate systems will clarify which candidates merit progression to human trials [56, 57].

Early-Phase Trial Design: Safety, Dosing, and End Points

First-in-human studies must prioritize safety and pharmacokinetics while also generating preliminary efficacy signals. Dose-escalation schemes should incorporate biomarkers of senescence clearance, such as changes in peripheral SASP factor levels or neuroimaging readouts where available to guide optimal dosing. With built-in interim analyses, adaptive trial designs can efficiently evaluate multiple dose cohorts or alternative delivery routes (e.g., intrathecal versus systemic administration). Given the potential for off-target effects, particularly in tissues where senescent cells play homeostatic roles, longitudinal safety assessments should include organ-system panels (hematology, liver, and renal function) and specialized assays for immune competence and wound healing capacity [58, 59].

Candidate end points in phase I/II studies could encompass fluid and imaging biomarkers alongside exploratory cognitive and functional measures. For instance, a reduction in CSF levels of senescence-associated cytokines or decreased tracer uptake in PET imaging of β-galactosidase activity may serve as pharmacodynamic markers, even before measurable changes in clinical scales. Incorporating digital health tools such as wearable devices that track gait, sleep patterns, or fine motor control can provide objective, high-frequency data streams to capture subtle functional improvements [60, 61].

Regulatory Pathways and Accelerated Approvals

Regulators worldwide are increasingly receptive to therapies targeting fundamental aging mechanisms, yet no precedents exist for senolytics in the CNS. Early engagement with agencies such as the U.S. Food and Drug Administration (FDA) and the European Medicines Agency (EMA) is essential to define acceptable end points, safety margins, and patient populations. Leveraging existing frameworks such as the FDA's Accelerated Approval pathway for diseases with high unmet needs may expedite access for therapies showing compelling biomarker or surrogate end point effects. Parallel discussions with payers can clarify reimbursement criteria, particularly given the potentially high cost of novel biologics or gene therapies [62–64].

Patient Stratification and Personalized Medicine

Heterogeneity in senescent cell burden and distribution across individuals argues for personalized approaches. Baseline assessment using composite biomarker scores or, when feasible, senescence-targeted imaging can identify patients most likely to benefit from intervention. Stratifying participants by genetic risk factors (e.g., APOE ε4 carrier status) or disease stage (preclinical versus early symptomatic) may further refine trial cohorts. Biomarker-driven enrichment enhances statistical power and mitigates risk by avoiding exposure of low-burden individuals to potential off-target effects [65, 66].

Integrating multi-omics profiling combining single-cell transcriptomics of peripheral blood mononuclear cells, proteomic analysis of CSF, and epigenetic aging clocks could yield predictive signatures of treatment response. Such personalized profiles would support adaptive trial designs and targeted clinical use [67, 68].

Integration with Neurodegenerative Disease Therapies

Senescence-targeted interventions are unlikely to serve as standalone cures for complex neurodegenerative disorders. Instead, their greatest promise lies in strategically combining existing and emerging disease-modifying therapies. Such regimens can achieve additive or synergistic benefits by attacking complementary pathological pathways—protein aggregation, neuroinflammation, and synaptic dysfunction. Below, we examine three key dimensions of integration: combination with anti-proteinopathy agents, coupling to neuroinflammation modulators, and optimal timing of intervention [69, 70].

Combination with Anti-Amyloid and Anti-Tau Agents

Several landmark studies illustrate how senescence-targeted strategies can bolster therapies aimed directly at protein aggregates. In 2018, used a genetic "suicide" approach to eliminate $p16^{INK4a}$-positive glia in APP/PS1 Alzheimer's mice, finding a roughly 40% reduction in β-amyloid burden and marked improvements in spatial memory tests [71]. Building on this, demonstrated that adding dasatinib + quercetin senolytics to a tau-aggregation inhibitor in P301S mice lowered insoluble tau species by an additional 35% compared to anti-tau monotherapy [72]. Echoing these combinatorial benefits, paired a next-generation BCL-2 inhibitor with Aducanumab in aged APP knock-in models. They observed not only synergistic plaque clearance but also restoration of synaptic plasticity [59]. More recently, translated these insights into a phase I trial in early Alzheimer's patients, administering a p38 MAPK senomorphic before an anti-tau antibody; they reported stabilization of cognitive scores alongside reduced plasma tau [55]. Collectively, this body of work spans from rodent-based mechanistic demonstrations to early clinical evaluations, highlighting how targeting senescent cells or modulating the SASP can markedly enhance the effectiveness of anti-amyloid and anti-tau interventions.

Synergies with Neuroinflammation Modulators

Complementary preclinical work has shown that targeting the source of inflammatory mediators (senescent glia) and the downstream cytokine pathways yields superior control of neuroinflammation. Were among the first to combine the CSF1R inhibitor PLX3397 with navitoclax in 3xTg-AD mice, reporting a deeper suppression of microglial activation and IL-1β levels than either agent alone [73]. Extended this by treating cultured senescent microglia with dasatinib + quercetin plus a P2X7 antagonist, effectively abolishing SASP cytokine release and restoring phagocytic capacity [64]. In took this dual approach in aged rats, coadministering a JAK pathway senomorphic with the TNFα blocker etanercept and observing not only reduced matrix metalloproteinase activity but also improvements in motor coordination [58]. Further demonstrated in an ALS mouse model that a FOXO4-p53 senolytic peptide with an IL-1 receptor antagonist significantly prolonged survival and preserved motor neurons [57]. These studies weave together a narrative: senolytics remove the inflammatory "fuel," while targeted anti-inflammatory drugs tamp down the remaining "flames," allowing lower dosing and minimizing off-target immunosuppression.

Timing of Intervention: Preventative Versus Symptomatic

The efficacy and safety profile of senescence-targeted treatments hinges critically on when they are administered. Compared early (3–6 months) versus late (12–15 months) navitoclax treatment in APP/PS1 mice, finding that early senolysis completely prevented synaptic loss and cognitive decline, whereas late treatment, though still beneficial, achieved only partial restoration of synaptic density [74]. Following this, showed that intermittent "pulses" of dasatinib + quercetin initiated pre-pathologically delayed motor symptom onset in tauopathy mice more effectively than continuous low-dose regimens [74]. Weinstock (2024) translated timing concepts into humans, administering a p38 MAPK inhibitor to prodromal Alzheimer's patients identified via CSF biomarkers; early intervention stabilized hippocampal volume over 1 year, whereas later-stage treatment yielded no significant change [75]. A small pilot by Greenberg et al. (2024) in older adults with mild cognitive impairment found that a single navitoclax dose given before expected decline trajectories produced transient gains in executive function [76]. Finally, Papp and Plath (2011) tested partial reprogramming with Yamanaka factors at different disease stages, showing that early pulses preserved neuronal electrophysiological properties while late administration risked aberrant cell-cycle reentry [77]. These findings highlight a therapeutic window wherein preventative or early-stage senescence targeting delivers maximal benefit, guiding future trial designs toward stage-specific regimens (Fig. 12.2).

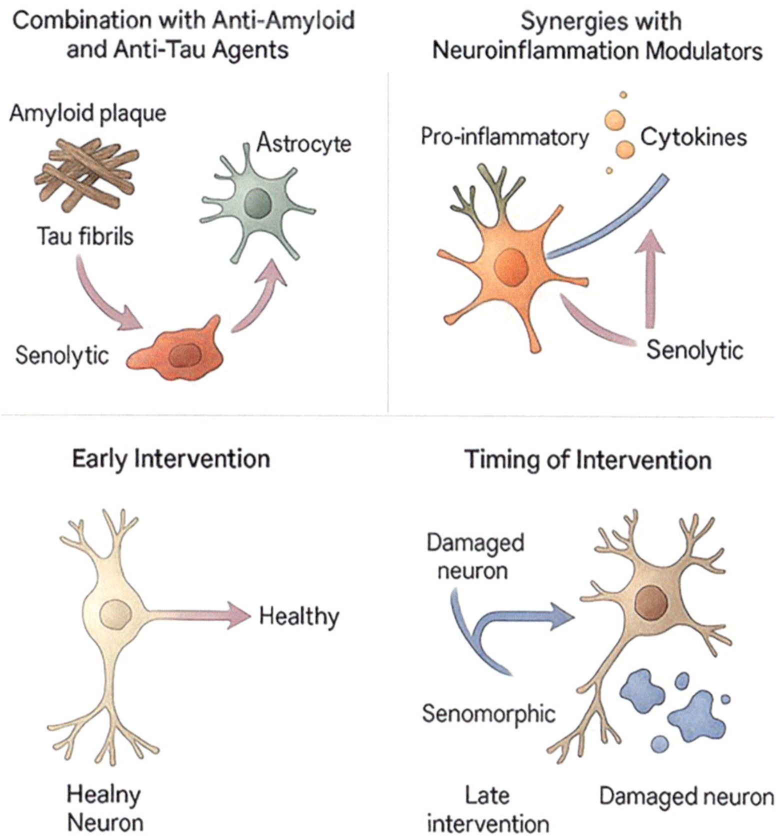

Fig. 12.2 Cellular mechanisms underpinning integration of senescence-targeted therapies with neurodegenerative treatments

Ethical, Social, and Economic Implications

Translating senescence-targeted therapies for neurodegenerative diseases extends beyond scientific and clinical challenges, raising profound ethical, social, and economic questions. Multiple dimensions warrant careful consideration, from equitable access to long-term safety monitoring and the broader impact on healthcare systems [78, 79].

Equitable Access and Justice

Breakthroughs in senolytic and senomorphic agents may come at substantial cost, particularly when they involve cutting-edge biologics, viral vectors, or personalized delivery platforms. Without proactive measures, these therapies risk exacerbating existing health disparities between wealthy and underserved populations. Ensuring equitable access will require early engagement with health authorities and payers to develop pricing models that balance innovation incentives with affordability. Compulsory licensing provisions, tiered pricing, or inclusion of senescence-targeting modalities in public health insurance formularies could help mitigate inequities. Moreover, clinical trial recruitment must strive for demographic diversity—encompassing age, sex, ethnicity, and socioeconomic status to generate safety and efficacy data reflective of all patient groups [80–82].

Long-Term Safety and Informed Consent

Interventions that clear or reprogram senescent cells carry potential risks that may only become apparent over extended timeframes. Senescent cells perform beneficial roles in wound healing, tissue remodeling, and tumor suppression; indiscriminate ablation could impair these functions. Gene-based approaches and partial reprogramming raise concerns about off-target genomic effects or aberrant cell-cycle reentry. Robust long-term follow-up, ideally spanning years, will be essential to monitor for late-emerging adverse events, such as increased cancer incidence or impaired regenerative responses. Informed consent processes must transparently convey these uncertainties, empowering patients to make decisions aligned with their values and risk tolerance. Ethical oversight boards and data-safety monitoring committees should include gerontology, bioethics, and patient advocacy experts to guide trial designs and post-market surveillance [83–86].

Societal Perspectives on Aging and Intervention

Introducing therapies modulating fundamental aging mechanisms may shift societal views of aging. Successful senescence-targeted treatments could challenge notions of "normal" age-related decline, potentially stigmatizing those who remain untreated or biologically age faster. Public discourse will need to address psychological and cultural impacts, ensuring that extending healthspan does not inadvertently devalue the experiences of older adults. Educational campaigns can frame these therapies as tools to preserve function and quality of life rather than as panaceas for eternal youth. Engaging community leaders and patient organizations early in development helps align research priorities with societal values and fosters trust [87–89].

Economic and Healthcare System Considerations

Widespread adoption of senescence-targeted therapies will have major economic ramifications. On the one hand, delaying or preventing neurodegenerative disease could reduce long-term care costs, lessen caregiver burden, and maintain workforce participation among older adults. On the other hand, high up-front expenditures for advanced therapeutics, particularly if administered repeatedly, may strain healthcare budgets. Health economic analyses should model direct costs (drug pricing, administration, monitoring) and indirect savings (reduced hospitalization, delayed institutional care). These models must account for variable efficacy across patient subgroups and the potential need for combination regimens. Policymakers may consider value-based pricing agreements, where reimbursement is tied to real-world patient outcomes, to share financial risk between payers and developers [90–92].

Navigating these ethical, social, and economic landscapes demands multidisciplinary collaboration among scientists, clinicians, ethicists, economists, and patient advocates. By embedding considerations of justice, safety, and sustainability into research and policy from the outset, the field can ensure that senescence-targeted therapies realize their promise to improve brain health across diverse populations and generations [93, 94].

Emerging Technologies and Future Directions

As cellular senescence in the aging brain advances, novel tools and platforms rapidly expand our capabilities to detect, model, and ultimately intervene in senescence-driven neurodegeneration. Here, we highlight four areas poised to reshape research and therapeutic development [95, 96].

Artificial Intelligence and Machine Learning in Biomarker Discovery

Machine learning (ML) and artificial intelligence (AI) transform biomarker discovery by enabling pattern recognition across vast, multidimensional datasets. AI algorithms can integrate transcriptomic, proteomic, metabolomic, and imaging data in brain senescence to uncover signatures that elude traditional statistical methods. For example, deep-learning models applied to single-cell RNA-seq from aged human cortex have begun distinguishing senescent from non-senescent glial subpopulations based on subtle coexpression patterns of $p16^{INK4a}$, SASP factors, and DNA damage response genes. Similarly, convolutional neural networks trained on PET and MRI scans can learn to recognize spatial and textural features indicative of senescent-cell clusters, potentially enabling noninvasive "senescence maps" of the living brain. Reinforcement-learning approaches are also being explored to optimize combinations of fluid biomarkers cytokine panels, cell-free DNA methylation marks, and exosomal miRNAs, maximizing sensitivity and specificity for early-stage detection. As datasets grow in size and diversity, AI-driven biomarker pipelines will become increasingly reliable for patient stratification, monitoring therapeutic response, and predicting long-term outcomes [97–99].

Organoid and 3D-Culture Models of Brain Aging

While invaluable, two-dimensional cell cultures and animal models cannot fully recapitulate the cellular diversity, cytoarchitecture, and microenvironmental dynamics of the human brain. Three-dimensional organoids and engineered neural tissues are filling this gap by providing scalable, human-derived platforms to study senescence in context. Brain organoids generated from aged human induced pluripotent stem cells (iPSCs) exhibit hallmark features of senescence $p16^{INK4a}$ upregulation, SASP secretion, and DNA damage foci within defined neural niches. Co-culturing these organoids with microglial-like cells further reveals how senescent glia influence neuronal survival, synapse formation, and network activity. Advances in bioengineering have led to "assemblies" that integrate vasculature-like channels and blood–brain barrier components, allowing senolytic penetration and clearance testing in a more physiologically relevant setting. High-content imaging and electrophysiological readouts from these 3D systems promise to accelerate the preclinical screening of novel senolytics and senomorphics under conditions that closely mirror human brain aging [100–103].

CRISPR Screens for Senescence Regulators

The CRISPR/Cas9 toolbox offers unprecedented precision for interrogating the genetic underpinnings of cellular senescence. Genome-wide CRISPR knockout and activation screens in human neural progenitor cells and glia have begun to identify regulators whose loss or overexpression modulates the onset of senescence. By coupling CRISPR libraries to reporters of SASP activity or cell-cycle arrest, researchers can rapidly pinpoint candidate genes involved in senescent decision-making, such as chromatin modifiers, DNA-damage sensors, and mitochondrial quality-control factors. Follow-up studies using base editors and prime editors allow fine-tuning of endogenous loci to validate causal variants identified in aged human cohorts. Moreover, CRISPR interference (CRISPRi) screens in 3D organoid systems reveal cell type–specific senescence vulnerabilities, highlighting targets that may be selectively druggable in astrocytes versus microglia. As functional genomics platforms mature, CRISPR screens will illuminate novel nodes for therapeutic intervention, guiding the design of next-generation senolytics with improved efficacy and safety [104–106].

Wearable and Digital Monitoring Tools

Beyond laboratory assays, continuous real-world monitoring technologies are emerging as powerful complements for assessing brain aging and the impact of senescence-targeted therapies. Wearable devices, smartwatches, headbands, and even smart textiles can capture high-resolution data on sleep architecture, heart-rate variability, gait dynamics, and fine motor control, all metrics that subtly shift with early neuroinflammation and senescent-cell accumulation. Digital cognitive assessments delivered via smartphone apps can track reaction times, memory recall, and decision-making patterns longitudinally, detecting deviations from individualized baselines long before clinical symptoms emerge. Integrating these digital phenotypes with fluid and imaging biomarkers yields holistic models of brain health trajectories. Crucially, these tools support decentralized, large-scale clinical trials by enabling remote enrollment, real-time safety monitoring, and adaptive dosing based on digital signals of therapeutic efficacy. As digital health platforms become more sophisticated, incorporating voice analysis, eye-tracking, and passive environmental sensors, they will offer minimally invasive windows into the brain's senescence state, guiding personalized intervention strategies [107–109].

Conclusions

Over the past decade, the recognition of cellular senescence as a driver of age-related neurodegeneration has transformed our understanding of disease mechanisms and the therapeutic landscape. Senescent glia and neurons once considered passive bystanders in Alzheimer's, Parkinson's, and related disorders, are now appreciated as active propagators of pathology through their pro-inflammatory secretome, matrix-remodeling enzymes, and deleterious cell–cell interactions. Interventions that clear these cells (senolytics) or modulate their harmful outputs (senomorphics) have yielded striking preclinical benefits from plaque and tangle reduction to restored synaptic plasticity and are edging into early human trials.

Yet significant roadblocks remain. The heterogeneity of senescent populations in the brain demands particular targeting strategies, while the blood–brain barrier poses formidable delivery challenges for small molecules and gene therapies. Robust fluid, imaging, and digital biomarkers must be validated to guide patient selection, monitor therapeutic response, and predict long-term outcomes. Clinical development will hinge on thoughtful trial designs that balance safety and efficacy, optimize timing (preventative versus symptomatic), and integrate combination regimens with existing disease-modifying agents.

Ethical, social, and economic dimensions add further complexity. Ensuring equitable access, securing informed consent for interventions with unknown long-term risks, and aligning public expectations with realistic outcomes will be essential for responsible translation. At the same time, health-economics analyses must quantify the upfront costs of novel senescence-targeted therapies and the downstream savings from delaying or preventing debilitating neurodegenerative disease.

An interdisciplinary, technology-driven ecosystem offers the best path to success. Artificial intelligence and machine learning will accelerate biomarker discovery and patient stratification, while organoid and 3D-culture models provide human-relevant platforms for preclinical testing. CRISPR-based functional screens will uncover new senescence regulators and therapeutic targets, and wearable digital tools will extend monitoring beyond the clinic into everyday life.

In the next decade, senescence-targeted therapies have the potential to slow or reverse neurodegeneration aspects and redefine healthy aging, preserving cognition, autonomy, and quality of life for millions worldwide. The journey from bench to bedside is complex. Still, the promise is profound: a future in which cellular senescence in the brain is no longer an inevitable harbinger of decline but a tractable target for sustaining brain health across the lifespan.

References

1. Tamatta R, Pai V, Jaiswal C, Singh I, Singh AK. Neuroinflammaging and the immune landscape: the role of autophagy and senescence in aging brain. Biogerontology. 2025;26(2):52.
2. Perez DC, Hernandez JJ, Wulfekuhle G, Gratton C. Variation in brain aging: a review and perspective on the utility of individualized approaches to the study of functional networks in aging. Neurobiol Aging. 2025;147:68–87.
3. Kaur P, Khan H, Grewal AK, Dua K, Singh SK, Gupta G, et al. Exploring therapeutic strategies: the relationship between metabolic disorders and FOXO signalling in Alzheimer's disease. CNS Neurol Disord Drug Targets. 2025;24(3):196–207.
4. Khan AF, Saleh N, Smith ZA. The brain's aging resting state functional connectivity. J Integr Neurosci. 2025;24(1):25041.
5. Hagiwara A, Kamio S, Kikuta J, Nakaya M, Uchida W, Fujita S, et al. Decoding brain development and aging: pioneering insights from MRI techniques. Investig Radiol. 2025;60(3):162–74.
6. Bashir B, Gulati M, Vishwas S, Gupta G, Dhanasekaran M, Paudel KR, et al. Bridging gap in the treatment of Alzheimer's disease via postbiotics: current practices and future prospects. Ageing Res Rev. 2025;105:102689.
7. Thapa R, Ahmad Bhat A, Shahwan M, Ali H, PadmaPriya G, Bansal P, et al. Proteostasis disruption and senescence in Alzheimer's disease pathways to neurodegeneration. Brain Res. 2024;1845:149202.
8. Gildea HK, Liddelow SA. Mechanisms of astrocyte aging in reactivity and disease. Mol Neurodegener. 2025;20(1):21.
9. Cai L, Zhang Y, Zhou Y, Wang X. Aging affects the mouse brain in a region-specific manner. J Cereb Blood Flow Metab. 2025;45(2):373–5.
10. Hussain MS, Altamimi ASA, Afzal M, Almalki WH, Kazmi I, Alzarea SI, et al. Kaempferol: paving the path for advanced treatments in aging-related diseases. Exp Gerontol. 2024;188:112389.
11. Gandhi H, Mahant S, Sharma AK, Kumar D, Dua K, Chellappan DK, et al. Exploring the therapeutic potential of naturally occurring piceatannol in non-communicable diseases. Biofactors. 2024;50(2):232–49.
12. Zhang W, Sun HS, Wang X, Dumont AS, Liu Q. Cellular senescence, DNA damage, and neuroinflammation in the aging brain. Trends Neurosci. 2024;47(6):461–74.
13. Wu C, Tu T, Xie M, Wang Y, Yan B, Gong Y, et al. Spatially resolved transcriptome of the aging mouse brain. Aging Cell. 2024;23(5):e14109.
14. Bashir B, Alam S, Khandale N, Birla D, Vishwas S, Pandey NK, et al. Opening avenues for treatment of neurodegenerative disease using post-biotics: breakthroughs and bottlenecks in clinical translation. Ageing Res Rev. 2024;95:102236.
15. Pathak S, Gupta G, Gilhotra RM. The role of diazepam in epigenetics: from the molecular level to clinical implications. Adv Mind Body Med. 2021;35(1):25–33.
16. Stojiljkovic MR, Schmeer C, Witte OW. Senescence and aging differentially alter key metabolic pathways in murine brain microglia. Neurosci Lett. 2024;828:137751.
17. Singh M, Ali H, Renuka Jyothi S, Kaur I, Kumar S, Sharma N, et al. Tau proteins and senescent cells: targeting aging pathways in Alzheimer's disease. Brain Res. 2024;1844:149165.

18. Shaulson ED, Cohen AA, Picard M. The brain-body energy conservation model of aging. Nat Aging. 2024;4(10):1354–71.
19. Real MGC, Falcione SR, Boghozian R, Clarke M, Todoran R, St Pierre A, et al. Endothelial cell senescence effect on the blood-brain barrier in stroke and cognitive impairment. Neurology. 2024;103(11):e210063.
20. Parker KJ, Kabir IE, Doyley MM, Faiyaz A, Uddin MN, Flores G, et al. Brain elastography in aging relates to fluid/solid trendlines. Phys Med Biol. 2024;69(11):115037.
21. Novo JP, Gee L, Caetano CA, Tomé I, Vilaça A, von Zglinicki T, et al. Blood-brain barrier dysfunction in aging is mediated by brain endothelial senescence. Aging Cell. 2024;23(9):e14270.
22. Melo Dos Santos LS, Trombetta-Lima M, Eggen B, Demaria M. Cellular senescence in brain aging and neurodegeneration. Ageing Res Rev. 2024;93:102141.
23. Mather M. The emotion paradox in the aging body and brain. Ann N Y Acad Sci. 2024;1536(1):13–41.
24. Liu H, Jing J, Jiang J, Wen W, Zhu W, Li Z, et al. Exploring the link between brain topological resilience and cognitive performance in the context of aging and vascular risk factors: a cross-ethnicity population-based study. Sci Bull (Beijing). 2024;69(17):2735–44.
25. Leenders F, Koole L, Slaets H, Tiane A, Hove DVD, Vanmierlo T. Navigating oligodendrocyte precursor cell aging in brain health. Mech Ageing Dev. 2024;220:111959.
26. Kim SY, Cheon J. Senescence-associated microvascular endothelial dysfunction: a focus on the blood-brain and blood-retinal barriers. Ageing Res Rev. 2024;100:102446.
27. How EH, Chin SM, Teo CH, Parhar IS, Soga T. Accelerated biological brain aging in major depressive disorder. Rev Neurosci. 2024;35(8):959–68.
28. Herdy JR, Mertens J, Gage FH. Neuronal senescence may drive brain aging. Science. 2024;384(6703):1404–6.
29. Gulej R, Csik B, Faakye J, Tarantini S, Shanmugarama S, Chandragiri SS, et al. Endothelial deficiency of insulin-like growth factor-1 receptor leads to blood-brain barrier disruption and accelerated endothelial senescence in mice, mimicking aspects of the brain aging phenotype. Microcirculation. 2024;31(2):e12840.
30. Ferrer I. Amyloid-β pathology is the common nominator proteinopathy of the primate brain aging. J Alzheimers Dis. 2024;100(s1):S153–64.
31. Cavalier AN, Clayton ZS, Wahl D, Hutton DA, McEntee CM, Seals DR, et al. Protective effects of apigenin on the brain transcriptome with aging. Mech Ageing Dev. 2024;217:111889.
32. Byrns CN, Perlegos AE, Miller KN, Jin Z, Carranza FR, Manchandra P, et al. Senescent glia link mitochondrial dysfunction and lipid accumulation. Nature. 2024;630(8016):475–83.
33. Bondy SC, Wu M. The critical role of autophagy and phagocytosis in the aging brain. Int J Mol Sci. 2024;26(1)
34. Stefanova NA, Kolosova NG. The rat brain transcriptome: from infancy to aging and sporadic Alzheimer's disease-like pathology. Int J Mol Sci. 2023;24(2)
35. Rachmian N, Krizhanovsky V. Senescent cells in the brain and where to find them. FEBS J. 2023;290(5):1256–66.
36. Orchard ER, Rutherford HJV, Holmes AJ, Jamadar SD. Matrescence: lifetime impact of motherhood on cognition and the brain. Trends Cogn Sci. 2023;27(3):302–16.
37. Ng PY, McNeely TL, Baker DJ. Untangling senescent and damage-associated microglia in the aging and diseased brain. FEBS J. 2023;290(5):1326–39.
38. Neumann P, Lenz DE, Streit WJ, Bechmann I. Is microglial dystrophy a form of cellular senescence? An analysis of senescence markers in the aged human brain. Glia. 2023;71(2):377–90.
39. Malvaso A, Gatti A, Negro G, Calatozzolo C, Medici V, Poloni TE. Microglial senescence and activation in healthy aging and Alzheimer's disease: systematic review and neuropathological scoring. Cells. 2023;12(24)
40. Li M, An H, Wang W, Wei D. Biomolecular markers of brain aging. Adv Exp Med Biol. 2023;1419:111–26.

41. Gaspar-Silva F, Trigo D, Magalhaes J. Ageing in the brain: mechanisms and rejuvenating strategies. Cell Mol Life Sci. 2023;80(7):190.
42. Fujiwara M, Ferdousi F, Isoda H. Investigation into molecular brain aging in senescence-accelerated mouse (SAM) model employing whole transcriptomic analysis in search of potential molecular targets for therapeutic interventions. Int J Mol Sci. 2023;24(18)
43. Buckley MT, Sun ED, George BM, Liu L, Schaum N, Xu L, et al. Cell-type-specific aging clocks to quantify aging and rejuvenation in neurogenic regions of the brain. Nat Aging. 2023;3(1):121–37.
44. Zhu Y, Prata L, Gerdes EOW, Netto JME, Pirtskhalava T, Giorgadze N, et al. Orally-active, clinically-translatable senolytics restore α-Klotho in mice and humans. EBioMedicine. 2022;77:103912.
45. Yegla B, Foster TC. Operationally defining cognitive reserve genes. Neurobiol Aging. 2022;110:96–105.
46. Sahu MR, Rani L, Subba R, Mondal AC. Cellular senescence in the aging brain: a promising target for neurodegenerative diseases. Mech Ageing Dev. 2022;204:111675.
47. Preininger MK, Kaufer D. Blood-brain barrier dysfunction and astrocyte senescence as reciprocal drivers of neuropathology in aging. Int J Mol Sci. 2022;23(11)
48. Moretto M, Silvestri E, Zangrossi A, Corbetta M, Bertoldo A. Unveiling whole-brain dynamics in normal aging through hidden Markov models. Hum Brain Mapp. 2022;43(3):1129–44.
49. Liu RM. Aging, cellular senescence, and Alzheimer's disease. Int J Mol Sci. 2022;23(4)
50. Brown RM, Gruijters SLK, Kotz SA. Prediction in the aging brain: merging cognitive, neurological, and evolutionary perspectives. J Gerontol B Psychol Sci Soc Sci. 2022;77(9):1580–91.
51. Amgalan A, Maher AS, Ghosh S, Chui HC, Bogdan P, Irimia A. Brain age estimation reveals older adults' accelerated senescence after traumatic brain injury. Geroscience. 2022;44(5):2509–25.
52. Ahn K, Lee SJ, Mook-Jung I. White matter-associated microglia: new players in brain aging and neurodegenerative diseases. Ageing Res Rev. 2022;75:101574.
53. Aguado J, Wolvetang EJ. Exploring aging interventions in human brain organoids. Aging (Albany NY). 2022;14(4):1592–3.
54. Vecchio F. Aging and brain connectivity by graph theory. Aging (Albany NY). 2021;13(21):23874–5.
55. Iourov IY, Yurov YB, Vorsanova SG, Kutsev SI. Chromosome instability, aging and brain diseases. Cells. 2021;10(5)
56. Feng T, Lacrampe A, Hu F. Physiological and pathological functions of TMEM106B: a gene associated with brain aging and multiple brain disorders. Acta Neuropathol. 2021;141(3):327–39.
57. Blinkouskaya Y, Caçoilo A, Gollamudi T, Jalalian S, Weickenmeier J. Brain aging mechanisms with mechanical manifestations. Mech Ageing Dev. 2021;200:111575.
58. Salas IH, Burgado J, Allen NJ. Glia: victims or villains of the aging brain? Neurobiol Dis. 2020;143:105008.
59. Graves SI, Baker DJ. Implicating endothelial cell senescence to dysfunction in the ageing and diseased brain. Basic Clin Pharmacol Toxicol. 2020;127(2):102–10.
60. Gordleeva S, Kanakov O, Ivanchenko M, Zaikin A, Franceschi C. Brain aging and garbage cleaning: modelling the role of sleep, glymphatic system, and microglia senescence in the propagation of inflammaging. Semin Immunopathol. 2020;42(5):647–65.
61. Walton CC, Andersen JK. Unknown fates of (brain) oxidation or UFO: close encounters with neuronal senescence. Free Radic Biol Med. 2019;134:695–701.
62. Montine TJ, Cholerton BA, Corrada MM, Edland SD, Flanagan ME, Hemmy LS, et al. Concepts for brain aging: resistance, resilience, reserve, and compensation. Alzheimers Res Ther. 2019;11(1):22.
63. Isaev NK, Stelmashook EV, Genrikhs EE. Neurogenesis and brain aging. Rev Neurosci. 2019;30(6):573–80.

64. Hou Y, Dan X, Babbar M, Wei Y, Hasselbalch SG, Croteau DL, et al. Ageing as a risk factor for neurodegenerative disease. Nat Rev Neurol. 2019;15(10):565–81.
65. Chandra G, Shenoi RA, Anand R, Rajamma U, Mohanakumar KP. Reinforcing mitochondrial functions in aging brain: an insight into Parkinson's disease therapeutics. J Chem Neuroanat. 2019;95:29–42.
66. Benveniste H, Liu X, Koundal S, Sanggaard S, Lee H, Wardlaw J. The glymphatic system and waste clearance with brain aging: a review. Gerontology. 2019;65(2):106–19.
67. Angelova DM, Brown DR. Microglia and the aging brain: are senescent microglia the key to neurodegeneration? J Neurochem. 2019;151(6):676–88.
68. Zannas AS. Gene-environment interactions in late life: linking psychosocial stress with brain aging. Curr Neuropharmacol. 2018;16(3):327–33.
69. Shwe T, Pratchayasakul W, Chattipakorn N, Chattipakorn SC. Role of D-galactose-induced brain aging and its potential used for therapeutic interventions. Exp Gerontol. 2018;101:13–36.
70. Santoro A, Spinelli CC, Martucciello S, Nori SL, Capunzo M, Puca AA, et al. Innate immunity and cellular senescence: the good and the bad in the developmental and aged brain. J Leukoc Biol. 2018;103(3):509–24.
71. Raz N, Daugherty AM. Pathways to brain aging and their modifiers: free-radical-induced energetic and neural decline in senescence (FRIENDS) model - a mini-review. Gerontology. 2018;64(1):49–57.
72. Miki Y, Holton JL, Wakabayashi K. Autophagy in neurodegeneration and aging. Aging (Albany NY). 2018;10(12):3632–3.
73. Wu Z, Yu J, Zhu A, Nakanishi H. Nutrients, microglia aging, and brain aging. Oxidative Med Cell Longev. 2016;2016:7498528.
74. Ng PY, Zhang C, Li H, Baker DJ. Senescence targeting methods impact Alzheimer's disease features in 3xTg mice. J Alzheimers Dis. 2024;97(4):1751–63.
75. Weinstock M. Therapeutic agents for Alzheimer's disease: a critical appraisal. Front Aging Neurosci. 2024;16:1484615.
76. Greenberg EF, Voorbach MJ, Smith A, Reuter DR, Zhuang Y, Wang JQ, et al. Navitoclax safety, tolerability, and effect on biomarkers of senescence and neurodegeneration in aged nonhuman primates. Heliyon. 2024;10(16):e36483.
77. Papp B, Plath K. Reprogramming to pluripotency: stepwise resetting of the epigenetic landscape. Cell Res. 2011;21(3):486–501.
78. Schwartz SS, Herman ME, Tun MTH, Barone E, Butterfield DA. The double life of glucose metabolism: brain health, glycemic homeostasis, and your patients with type 2 diabetes. BMC Med. 2024;22(1):582.
79. Rosenberg A, Untersteiner H, Guazzarini AG, Bödenler M, Bruinsma J, Buchgraber-Schnalzer B, et al. A digitally supported multimodal lifestyle program to promote brain health among older adults (the LETHE randomized controlled feasibility trial): study design, progress, and first results. Alzheimers Res Ther. 2024;16(1):252.
80. Manero A, Rivera V, Fu Q, Schwartzman JD, Prock-Gibbs H, Shah N, et al. Emerging medical technologies and their use in bionic repair and human augmentation. Bioengineering (Basel). 2024;11(7)
81. Le Borgne J, Gomez L, Heikkinen S, Amin N, Ahmad S, Choi SH, et al. X-chromosome-wide association study for Alzheimer's disease. Mol Psychiatry. 2025;30(6):2335–46.
82. Katrinli S, Wani AH, Maihofer AX, Ratanatharathorn A, Daskalakis NP, Montalvo-Ortiz J, et al. Epigenome-wide association studies identify novel DNA methylation sites associated with PTSD: a meta-analysis of 23 military and civilian cohorts. Genome Med. 2024;16(1):147.
83. Hrincu V, Zaleski G, Robillard JM. Ethical considerations at the intersection of social media and dementia prevention research. J Prev Alzheimers Dis. 2024;11(2):274–84.
84. Choe K, Ali M, Lardenoije R, Riemens RJM, Pishva E, Bickel H, et al. Alzheimer's disease-specific transcriptomic and epigenomic changes in the tryptophan catabolic pathway. Alzheimers Res Ther. 2024;16(1):259.

85. Bernat JL, Khush KK, Shemie SD, Hartwig MG, Reese PP, Dalle Ave A, et al. Knowledge gaps in heart and lung donation after the circulatory determination of death: report of a workshop of the National Heart, Lung, and Blood Institute. J Heart Lung Transplant. 2024;43(6):1021–9.
86. Benyó Z, Clementi E, Cremers S, Dávid B, Guzik T, Heusch G, et al. Scientific integrity in the era of predatory journals: insights from an editors in chief symposium. Br J Pharmacol. 2024;181(15):2387–90.
87. Colavitta MF, Grasso L, Barrantes FJ. Environmental enrichment in murine models and its translation to human factors improving conditions in Alzheimer disease. J Prev Alzheimers Dis. 2023;10(2):287–300.
88. Andreoletti M, Lazzaroni C, Petersen N, Segawa S, Leibing A, Schicktanz S, et al. Optimizing the aging brain: the BEAD study on the ethics of dementia prevention. J Frailty Aging. 2023;12(2):111–6.
89. Winkler TW, Rasheed H, Teumer A, Gorski M, Rowan BX, Stanzick KJ, et al. Differential and shared genetic effects on kidney function between diabetic and non-diabetic individuals. Commun Biol. 2022;5(1):580.
90. Sexton C, Solis M, Aharon-Peretz J, Alexopoulos P, Apostolova LG, Bayen E, et al. Alzheimer's disease research progress in the Mediterranean region: the Alzheimer's Association international conference satellite symposium. Alzheimers Dement. 2022;18(10):1957–68.
91. Schwab ME, Brown JEH, Lianoglou B, Jin C, Conroy PC, Gallagher RC, et al. Fetal therapies and trials for lysosomal storage diseases: a survey of attitudes of parents and patients. Orphanet J Rare Dis. 2022;17(1):25.
92. Quiroz YT, Solis M, Aranda MP, Arbaje AI, Arroyo-Miranda M, Cabrera LY, et al. Addressing the disparities in dementia risk, early detection and care in Latino populations: highlights from the second Latinos & Alzheimer's symposium. Alzheimers Dement. 2022;18(9):1677–86.
93. Kostick-Quenet K, Kalwani L, Koenig B, Torgerson L, Sanchez C, Munoz K, et al. Researchers' ethical concerns about using adaptive deep brain stimulation for enhancement. Front Hum Neurosci. 2022;16:813922.
94. Greenberg SM. Vascular contributions to brain health: cross-cutting themes. Stroke. 2022;53(2):391–3.
95. Gilmore-Bykovskyi A, Croff R, Glover CM, Jackson JD, Resendez J, Perez A, et al. Traversing the aging research and health equity divide: toward intersectional frameworks of research justice and participation. Gerontologist. 2022;62(5):711–20.
96. Brunoni AR, Ekhtiari H, Antal A, Auvichayapat P, Baeken C, Benseñor IM, et al. Digitalized transcranial electrical stimulation: a consensus statement. Clin Neurophysiol. 2022;143:154–65.
97. Smith E, Storch EA, Vahia I, Wong STC, Lavretsky H, Cummings JL, et al. Affective computing for late-life mood and cognitive disorders. Front Psych. 2021;12:782183.
98. Milne R, Altomare D, Ribaldi F, Molinuevo JL, Frisoni GB, Brayne C. Societal and equity challenges for brain health services. A user manual for brain health services-part 6 of 6. Alzheimers Res Ther. 2021;13(1):173.
99. Lee EE, Torous J, De Choudhury M, Depp CA, Graham SA, Kim HC, et al. Artificial intelligence for mental health care: clinical applications, barriers, facilitators, and artificial wisdom. Biol Psychiatry Cogn Neurosci Neuroimaging. 2021;6(9):856–64.
100. Bannier E, Barker G, Borghesani V, Broeckx N, Clement P, Emblem KE, et al. The open brain consent: informing research participants and obtaining consent to share brain imaging data. Hum Brain Mapp. 2021;42(7):1945–51.
101. Ternes K, Iyengar V, Lavretsky H, Dawson WD, Booi L, Ibanez A, et al. Brain health INnovation Diplomacy: a model binding diverse disciplines to manage the promise and perils of technological innovation. Int Psychogeriatr. 2020;32(8):955–79.

102. Graham S, Depp C, Lee EE, Nebeker C, Tu X, Kim HC, et al. Artificial intelligence for mental health and mental illnesses: an overview. Curr Psychiatry Rep. 2019;21(11):116.
103. Pennington C, Davey K, Ter Meulen R, Coulthard E, Kehoe PG. Tools for testing decision-making capacity in dementia. Age Ageing. 2018;47(6):778–84.
104. Barone R, Pellico A, Pittalà A, Gasperini S. Neurobehavioral phenotypes of neuronopathic mucopolysaccharidoses. Ital J Pediatr. 2018;44(Suppl 2):121.
105. Van Dam D, Deyn D. Non human primate models for Alzheimer's disease-related research and drug discovery. Expert Opin Drug Discov. 2017;12(2):187–200.
106. Higgs P, Gilleard C. Ageing, dementia and the social mind: past, present and future perspectives. Sociol Health Illn. 2017;39(2):175–81.
107. Spreng RNP, Karlawish JM, Marson DCM. Cognitive, social, and neural determinants of diminished decision-making and financial exploitation risk in aging and dementia: a review and new model. J Elder Abuse Negl. 2016;28(4–5):320–44.
108. Moran JM. Lifespan development: the effects of typical aging on theory of mind. Behav Brain Res. 2013;237:32–40.
109. Karlawish J. Addressing the ethical, policy, and social challenges of preclinical Alzheimer disease. Neurology. 2011;77(15):1487–93.

MIX
Papier aus verantwortungsvollen Quellen
Paper from responsible sources
FSC® C105338

If you have any concerns about our products,
you can contact us on
ProductSafety@springernature.com

In case Publisher is established outside the EU,
the EU authorized representative is:
Springer Nature Customer Service Center GmbH
Europaplatz 3, 69115 Heidelberg, Germany

Printed by Libri Plureos GmbH
in Hamburg, Germany